SAE International's Dictionary of ADAS and Connected Vehicles

SAE International's Dictionary of ADAS and Connected Vehicles

JON M. QUIGLEY, AMOL GULVE, JAYALEKSHMI KRISHNAMOORTHY

Warrendale, Pennsylvania, USA

400 Commonwealth Drive
Warrendale, PA 15096-0001 USA
E-mail: CustomerService@sae.org
Phone: 877-606-7323 (inside USA and Canada)
724-776-4970 (outside USA)
FAX: 724-776-0790

Copyright © 2025 SAE International. All rights reserved.

No part of this publication may be reproduced, stored in a retrieval system, transmitted, in any form or by any means, electronic, mechanical, photocopying, recording, or otherwise, or used for text and data mining, AI training, or similar technologies, without the prior written permission of SAE. For permission and licensing requests, contact SAE Permissions, 400 Commonwealth Drive, Warrendale, PA 15096-0001 USA; e-mail: copyright@sae.org; phone: 724-772-4028.

Library of Congress Catalog Number 2025946101
http://dx.doi.org/10.4271/9781468607864

Information contained in this work has been obtained by SAE International from sources believed to be reliable. However, neither SAE International nor its authors guarantee the accuracy or completeness of any information published herein and neither SAE International nor its authors shall be responsible for any errors, omissions, or damages arising out of use of this information. This work is published with the understanding that SAE International and its authors are supplying information but are not attempting to render engineering or other professional services. If such services are required, the assistance of an appropriate professional should be sought.

ISBN-Print 978-1-4686-0785-7
ISBN-PDF   978-1-4686-0786-4
ISBN-epub   978-1-4686-0787-1

To purchase bulk quantities, please contact: SAE Customer Service

E-mail: CustomerService@sae.org
Phone: 877-606-7323 (inside USA and Canada)
724-776-4970 (outside USA)
Fax: 724-776-0790

Visit the SAE International Bookstore at books.sae.org

Publisher
Sherry Dickinson Nigam

Product Manager
Amanda Zeidan

Production and Manufacturing Associate
Michelle Silberman

Contents

Preface vii
Acknowledgments viii
List of Acronyms x
A 1
B 19
C 25
D 63
E 79
F 101
G 113
H 117
I 125
L 157
M 169
N 175
O 181
P 191
R 209

S	233
T	267
U	291
V	297
W	325
Y	329

Index	331
About the Authors	341

Preface

The convergence of advanced driver assistance systems (ADAS) and connected vehicle technologies marks a transformative era in automotive innovation, fundamentally enhancing vehicle safety, efficiency, and overall driving experience. This book, *SAE International's Dictionary of ADAS and Connected Vehicles,* has been meticulously compiled to serve as a comprehensive and authoritative resource for engineers, researchers, automotive professionals, and enthusiasts operating in these rapidly evolving domains.

Within these pages, you will find clear and precise definitions of terms, technologies, and systems pertinent not only to ADAS but also to the expanding landscape of connected vehicles. This includes vehicle-to-vehicle (V2V), vehicle-to-infrastructure (V2I), and vehicle-to-everything (V2X) communication technologies. The dictionary reflects the latest industry standards and innovations, bridging the gap between complex technical jargon and practical understanding. By fostering greater collaboration and knowledge sharing across disciplines, it supports those involved in cutting-edge system design, regulatory compliance, and anyone seeking to deepen their knowledge of modern automotive technologies.

The landscape of automotive technology is advancing at an unprecedented pace, and with it, the language describing these innovations must be equally dynamic and accessible. SAE International remains committed to supporting the global community by providing reliable technical resources that promote innovation, safety, and progress in the development of mobility solutions. We hope that this dictionary empowers you to navigate and contribute to the future of transportation with confidence and clarity.

Acknowledgments

We have been affiliated with SAE International for many years; our first book was published in 2016. Even before that, we were on many standards committees. Jon personally still owns his 1997 three-book volume of the Ground Standards, believing that if he was going to be in the industry, he should have some knowledge about it.

An organization like SAE International plays a pivotal role in advancing mobility technology by bringing together engineers, technical experts, and industry stakeholders from around the world. It fosters engineering collaboration, standard development, and knowledge sharing in fields such as automotive, aerospace, and commercial vehicles. By providing a global platform for the exchange of ideas, SAE International helps drive innovation, ensure safety and quality, and support professional growth. Its technical standards and educational programs set benchmarks for industry performance and reliability, enabling organizations and individuals to stay at the forefront of technological progress and regulatory compliance.

Over the past several years, Jon has had the privilege of collaborating closely with SAE International on the publication of five additional books. The partnership with Monica Nogueira, Sherry Nigam, and Amanda Zeidan has been especially rewarding—their expertise, dedication, and unwavering support have significantly elevated the quality and impact of each project. It has truly been a pleasure working alongside such talented professionals.

Amol Gulve's contributions to the latest book have been particularly valuable. His deep expertise in ADAS and autonomous vehicles, combined with a broad industry perspective, played a critical role in shaping the content and ensuring its relevance to today's mobility challenges.

The list of people Jon owes is quite long, dating back to his childhood. Harold, Chuck, Dave, and Eddie all showed me a better way at a time when I could have easily not ended up where I am now without them. They have all four saved my metaphoric bacon in my late adolescence and early adulthood. I could never thank them enough and am very glad to have known them for these many decades. Each of their families effectively adopted me as their own, and I am grateful forever.

Amol would like to extend his sincere gratitude to Jon Quigley, whose leadership and steadfast support were instrumental in driving this initiative. I am also deeply appreciative of my industry colleagues for their insights and collaboration, which enhanced the depth and relevance of this work. My heartfelt thanks go to the SAE editorial team for their commitment to advancing engineering knowledge and for their exceptional stewardship in bringing this publication to fruition. Lastly, I am profoundly grateful to my wife, Amruta, and my daughters, Aisha and Navya, whose encouragement and belief in this journey inspired me to make this meaningful contribution to the automotive industry.

Jaya owes her gratitude to her husband Renjith, your unwavering belief in me motivated me to contribute to this book, when I nearly gave up. Spandana, your light and laughter reminded me why I write and whom I write for. I am deeply grateful to my parents and brother for their constant encouragement, and to Ramya, Nikhitha, Rajasree, and Gaynel for standing by me during one of the hardest years of my life.

List of Acronyms

2G	Second-Generation
3D	Three Dimensional
3DCNNs	3D Convolutional Neural Networks
2G	Second-Generation
3G	Third-Generation
4G	Fourth-Generation
5G	Fifth-Generation
AACN	Advanced Automatic Collision Notification
ABS	Anti-Lock Braking System
ACC	Adaptive Cruise Control
ACN	Automatic Crash Notification
ACV	Autonomous and Connected Vehicle
ADA	Americans with Disabilities Act
ADAS	Advanced Driver Assistance Systems
ADB	Adaptive Driving Beam
ADR	Australia Design Rules
AEB	Automatic Emergency Braking
AEBS	Automated Emergency Braking System
AFS	Adaptive Front Lighting System
AHB	Automatic High Beam
AI	Artificial Intelligence
AMR	Autonomous Mobile Robot
ANN	Artificial Neural Network
ANSI	American National Standards Institute
AOI	Area of Interest
API	Application Programming Interface
AR	Augmented Reality
ASEAN	Association of Southeast Asian Nations
ASF	Advanced Software Framework
ATISs	Advanced Traveler Information Systems
AUTOSAR	Automotive Open System Architecture
AV	Autonomous Vehicle

AVP	Automated Valet Parking
BAC	Blood Alcohol Concentration
BAS	Brake Assist Systems
BbW	Brake-by-Wire
BEVs	Battery Electric Vehicles
BIM	Back-End Integration Manager
BLE	Bluetooth Low Energy
BMS	Battery Management System
BrAC	Breath Alcohol Concentration
BSM	Basic Safety Message
C2X	Car-to-X
CACC	Cooperative Adaptive Cruise Control
CAM	Cooperative Awareness Message
CAN	Controller Area Network
CAV	Connected Autonomous Vehicles
CCA	Cooperative Collision Avoidance
CCPA	California Consumer Privacy Act
CD	Compact Disc
CDA	Cooperative Driving Automation
CGW	Central Gateway
CICAS	Cooperative Intersection Collision Avoidance System
CICAS-V	Cooperative Intersection Collision Avoidance System for Violation
CIM	Cooperative Intersection Management
CIS	Cooperative Intersection Safety
C-ITSs	Cooperative Intelligent Transport Systems
CMVSS	Canadian Motor Vehicle Safety Standards
CNNs	Convolutional Neural Networks
CPD	Child Presence Detection
CPS	Cyber-Physical System
CPU	Central Processing Unit
CRM	Customer Relationship Management
CRS	Current Road Segment
CSA	Curve Speed Assistance
CSW	Curve Speed Warning
CTA	Cross-Traffic Alert
C-V2X	Cellular Vehicle to Everything
CVC	Cooperative Vehicle Communication
CVI	Connected Vehicle Infrastructure
CVO	Commercial Vehicle Operation
CV	Connected Vehicle
CVS	Cooperative Vehicle Security

CVSS	Cooperative Vehicle Safety Systems
CVT	Cooperative Vehicle Testing
DbW	Drive-by-Wire
DCFC	DC Fast Charging
DDS	Data Distribution Services
DDT	Dynamic Driving Task
DENM	Decentralized Environmental Notification Message
DL	Deep Learning
DMS	Driver Monitoring Systems
DOTA	Download Over-the-Air
DRO	Dynamic Route Optimization
DSRC	Dedicated Short-Range Communications
DTC	Diagnostic Trouble Codes
DVD	Digital Video Disc
EARFCN	E-UTRA Absolute Radio Frequency Channel Number
EBA	Emergency Brake Assistance
EBC	Electronic Brake control
EBD	Electronic Brake force Distribution
EBS	Emergency Braking System
ECG	Electrocardiogram
ECN	Electric Charging Network
ECU	Electronic Control Unit
EDA	Event-Driven Architecture
EDRs	Event Data Recorders
EEBL	Emergency Electronic Brake Light
E-Horizon	Electronic Horizon
ELK	Emergency Lane Keeping
ERP	Enterprise Resource Planning
ESA	Evasive Steering Assist
ESC	Electronic Stability Control
e-Sim	e-Subscriber Identity Module
ESS	Emergency Stop Signal
ETC	Electronic Toll Collection
EVC	Emergency Vehicle Communication
EVP	Emergency Vehicle Priority
EV	Electric Vehicle
EVSE	Electric Vehicle Supply Equipment
EVTM	Emergency Vehicle Traffic Management
EVTSP	Emergency Vehicle Traffic Signal Priority
FCW	Forward Collision Warning
FMCSA	Federal Motor Carrier Safety Administration
FMVSS	Federal Motor Vehicle Safety Standards

FOD	Feature-on-Demand
FOTA	Firmware Over-the-Air
FoV	Field of View
FOW	Forward Obstruction Warning
FYL	Functional Years Lost
GDPR	General Data Protection Regulation
GHG	Greenhouse Gas
GIS	Geographic Information System
GNSS	Global Navigation Satellite System
GPS	Global Positioning System
HDMI	High-Definition Multimedia Interface
HEVs	Hybrid Electric Vehicles
HMI	Human Machine Interface
HMD	Head-Mounted Display
HOV	High Occupancy Vehicle
HRMS	Human Resources Management Systems
HUD	Head-Up Display
HVAC	Heating, Ventilation, and Air Conditioning
IC	Instrument Cluster
ICE	Internal Combustion Engine
IDS	Intrusion Detection Systems
IID	Ignition Interlock Device
IIHS	Insurance Institute for Highway Safety
IMA	Intersection Movement Assist
IMU	Inertial Measurement Unit
IoT	Internet of Things
IPGS	Intelligent Parking Guidance System
IR	Infrared
ISA	Intelligent Speed Adaptation
ISS	Injury Severity Score
IT	Information Technology
ITS	Intelligent Transport System
IVI	In-Vehicle Information
IVIS	Vehicle-in-Vehicle Information System
JIS	Japanese Industrial Standards
LCC	Lane Centering Control
LDW	Lane Departure Warning
LED	Light-Emitting Diode
LiDAR	Light Detection and Ranging
LIN	Local Interconnect Network
LKA	Lane-Keeping Assistance
LLVIP	Low-Light Visible Infrared and Paired

LoRaWAN	Long-Range Wide Area Network
LPR	License Plate Recognition
LSTMs	Long Short-Term Memory Networks
LTA	Left Turn Assist
LTE	Long-Term Evolution
LVDS	Low-Voltage Differential Signaling
MaaS	Mobility as a Service
MAC	Medium Access Control
MAP	Map Data
MD	Message Dispatcher
MEC	Multi-Access Edge Computing
MFM	Mobile Fleet Management
ML	Machine Learning
MOT	Multi-Object Tracker
MQTT	Message Queueing Telemetry Transport
MTT	Multiple Target Tracker
NCAP	New Car Assessment Program
NHTSA	National Highway Traffic Safety Administration
NLP	Natural Language Processing
NLU	Natural Language Understanding
NRS	Next Road Segment
NVS	Night Vision System
OBD	Onboard Diagnostics
OBFCM	Onboard Fuel/Energy Consumption Monitoring
OBM	Onboard Emission Monitoring
OBU	Onboard Unit
ODD	Operational Design Domain
OEDR	Object and Event Detection and Response
OEM	Original Equipment Manufacturer
OTA	Over-the-Air
PAS	Parking Assistance System
PCA	Principal Component Analysis
PERCLOS	Percentage of Eye Closure
PHEVs	Plug-in Hybrid Electric Vehicles
PIN	Personal Identification Number
PIPL	Personal Information Protection Law
PMT	Passenger Miles Traveled
POI	Point of Interest
PSM	Personal Safety Message
RADAR	Radio Detection and Ranging
RCTA	Rear Cross Traffic Alert
RESS	Rechargeable Energy Storage System

RL	Reinforcement Learning
RNN	Recurrent Neural Network
ROA	Rear Occupant Alert
ROM	Read-Only Memory
RSA	Road Sign Assist
RSU	Roadside Unit
RTCM-104	Radio Technical Commission for Maritime
RTK	Real-Time Kinematic
RTOS	Real-Time Operating System
RWIS	Road Weather Information System
SbW	Steer-by-Wire
SCMS	Security Credential Management System
SDC	Software-Defined Chassis
SDC	Software-Defined Cockpit
SDC	Software-Defined Connectivity
SDCAV	Software-Defined Connected and Autonomous Vehicle
SDN	Software-Defined Networking
SDS	Software-Defined Suspension
SDT	Software-Defined Telemetry
SDTM	Software-Defined Traffic Management
SDVD	Software-Defined Vehicle Diagnostics
SDVS	Software-Defined Vehicle
SDX	Software-Defined Anything
SIM	Subscriber Identity Module
SLA	Speed Limit Assist
SLAM	Simultaneous Localization and Mapping
SOA	Service-Oriented Architecture
SOTIF	Safety of the Intended Functionality
SPaT	Signal Phase and Timing
SRM	Service Request Message
SVS	Surround View System
TBP	Traffic Behavior Prediction
TCA	Traffic Congestion Assistance
TCS	Traction Control System
TCU	Telematics Control Unit
TJA	Traffic Jam Assist
TJP	Traffic Jam Pilot
TMC	Traffic Management Centers
TMS	Traffic Management System
TPMS	Tire Pressure Monitoring System
TSC	Traffic Signal Control
TSP	Traffic Signal Preemption

TSP	Traffic Signal Priority
TSR	Traffic Sign Recognition
TSS	Traffic Signal Synchronization
TTC	Time to Collision
TTI	Time to Impact
TTS	Text-to-Speech
UBI	Usage-Based Insurance
UI	User Interface
UL	Underwriters Laboratories
UNECE	United Nations Economic Commission for Europe
URLLC	Ultra-Reliable Low-Latency Communication
USB	Universal Serial Bus
UTC	Coordinated Universal Time
UX	User Experience
V2B	Vehicle-to-Building
V2B	Vehicle-to-Business
V2C	Vehicle-to-Cloud
V2C	Vehicle-to-Cyclist
V2D	Vehicle-to-Device
V2E	Vehicle-to-Edge
V2G	Vehicle-to-Grid
V2H	Vehicle-to-Home
V2I	Vehicle-to-Infrastructure
V2N	Vehicle-to-Network
V2P	Vehicle-to-Pedestrian
V2R	Vehicle-to-Roadside
V2V	Vehicle-to-Vehicle
V2X	Vehicle-to-Everything
VCU	Vehicle Control Unit
VIN	Vehicle Identification Number
VIP	Vehicle Integration Platform
VPA	Virtual Personal Assistant
VRU	Vulnerable Road User
WAVE	Wireless Access in Vehicular Environments
WLAN	Wireless Local Area Network
WMI	World Manufacturer Identifier
WSMP	WAVE Short Message Protocol
YOLO-PD	You Only Look Once—Pedestrian Detection

ABS (Anti-lock Braking System)

An ABS is a safety and control technology used in automobiles, including autonomous and connected vehicles (ACVs), to prevent the wheels from locking up during emergency braking situations. This enables the driver to maintain steering control and helps reduce the stopping distances on both dry and slippery surfaces.

ABS operates by continuously monitoring the speed of each wheel through sensors. If the system detects a wheel is rotating significantly slower than the others—a condition indicative of impending lock-up—it momentarily reduces the brake force to that wheel. This adjustment happens in rapid succession (many times per second), which prevents the wheels from locking and skidding, thereby allowing the vehicle to maintain traction with the road.

ACC (Adaptive Cruise Control)

ACC is an advanced driver assistance system (ADAS) integrated into connected vehicles (CVs), designed to enhance driving comfort and safety. Unlike traditional cruise control, ACC utilizes sensors, such as radar or LiDAR, to automatically adjust the vehicle's speed in response to the surrounding traffic conditions. The system maintains a preset cruising speed set by the driver and, through real-time data from sensors, can autonomously slow down or accelerate to keep a safe following distance from the vehicle ahead. ACC contributes to a smoother and more adaptive driving experience, promoting efficient traffic flow and reducing driver workload, particularly in congested or varying traffic environments (Figure A.1).

For more information on this, see **SAE J2945/6_202310** and **SAE J2399_202110** [1, 2].

FIGURE A.1 Clearance between vehicles [1].

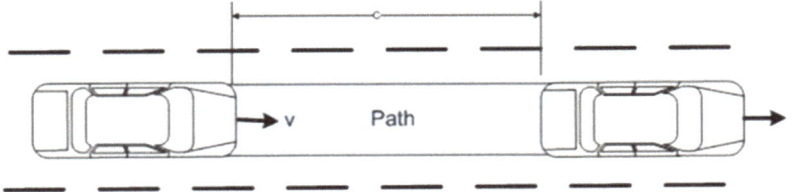

Adaptive Headlights

Adaptive headlights [adaptive driving beam (ADB)] are intelligent lighting systems that enhance driving visibility and safety. These headlights adjust their direction and range dynamically based on various factors, such as vehicle speed, steering input, and environmental conditions. Connected to sensors and cameras, adaptive headlights can anticipate curves, intersections, and changes in elevation, directing the light beam to follow the path of the road and illuminate potential obstacles (**Figure A.2**).

FIGURE A.2 Opposing and preceding vehicle test drive maximum illuminance levels [3].

Distance (m) ADB Vehicle to Opposing Vehicle Driver/Rider's Eye	Maximum Illuminance Opposing Vehicle Driver/Rider's Eye (lux)	Maximum Illuminance Preceding Vehicle Rear View Mirror, Driver Side View Mirror, and Passenger Side View Mirror (lux)
155	0.3	4.0
120	0.3	4.0
60	0.7	8.9
30	1.8	18.9

Adaptive headlights improve the driver's visibility without causing glare to oncoming traffic by adapting to the driving situation in real time. This technology contributes to a safer driving experience by providing optimal illumination in different driving scenarios, ultimately enhancing road safety, reducing driver fatigue, and improving overall visibility during nighttime and challenging driving conditions.

For more information on this, see **J3069** [3].

ADAS (Advanced Driver Assistance Systems)

ADAS refers to an automotive system that uses vehicle-mounted sensors and communication technologies to enhance its safety and driving capabilities through the integration of intelligent systems. CVs equipped with ADAS leverage various sensors, cameras, radar, and communication interfaces to gather and share real-time data with other vehicles, infrastructure, and central systems. The goal is to improve situational awareness, provide driver assistance, and enable cooperative functionalities to enhance overall road safety and efficiency.

For more information, see *Effectiveness of Advanced Driver Assistance Systems in Preventing System-Relevant Crashes* [4].

ADS (Automated Driving System)

ADS comprises hardware and software that sustainably perform the complete dynamic driving task (DDT). This includes systems that perform lateral vehicle control via steering, longitudinal control via acceleration and deceleration, object detection, tracking, environment monitoring, trajectory planning, and conspicuity strategies (**Figure A.3**).

For more information, see **SAE J3016** *Taxonomy and Definitions for Terms Related to Driving Automation Systems for On-Road Motor Vehicles* [6].

FIGURE A.3 ADS [5].

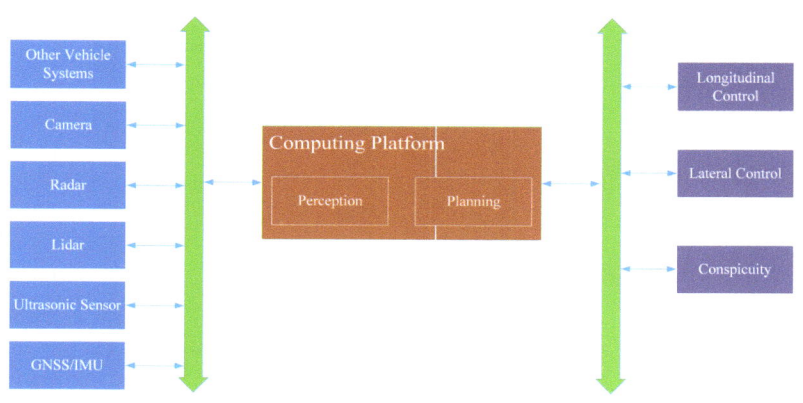

Advanced Navigation Systems

Advanced navigation systems refer to sophisticated and intelligent technologies designed to provide comprehensive and enhanced guidance for drivers. These systems leverage real-time data, often obtained through global positioning system (GPS), vehicle-to-infrastructure (V2I) communication, and other sensors, to offer dynamic and adaptive route planning. For more information, see *Data Dictionary for Advanced Traveler Information Systems (ATISs)* [7].

Critical features of advanced navigation systems include the following:

- Real-Time Traffic Updates: Integration with traffic data sources enables the system to provide up-to-the-minute information on traffic conditions, allowing drivers to navigate congested areas and choose the most efficient routes.
- Predictive Analysis: These systems can use historical and real-time data to predict future traffic patterns and recommend optimal routes, contributing to time and fuel efficiency.
- Integration with Connectivity Platforms: Connected to broader connectivity platforms, advanced navigation systems can incorporate information about points of interest (POIs), weather conditions, and live updates, offering a comprehensive and personalized navigation experience.
- Augmented Reality (AR) Navigation: Some advanced systems utilize AR technology to overlay navigation instructions onto the driver's view of the road, enhancing situational awareness and simplifying complex maneuvers.
- Voice Recognition and Natural Language Processing (NLP): Interaction with the navigation system is facilitated through voice commands, allowing drivers to focus on the road while receiving guidance or making adjustments.

Advanced Parking Assistance

Advanced parking assistance in CVs refers to sophisticated technologies and systems designed to assist drivers in parking maneuvers. These systems utilize various sensors, cameras, and connectivity features to provide enhanced assistance during parallel, perpendicular, and angular parking scenarios (**Figure A.4**). For additional information, see *Design of Automatic Parallel Parking System Based on Multi-Point Preview Theory* [8].

FIGURE A.4 Multi-point preview model of automatic parking [8].

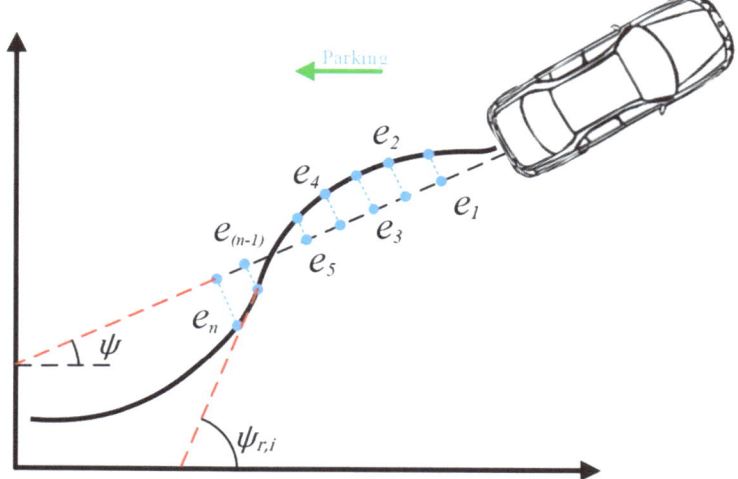

In certain implementations, the system can take control of steering, acceleration, and braking to autonomously navigate the vehicle into a parking space, thereby reducing the driver's burden.

Vehicles are equipped with ultrasonic sensors and cameras around the vehicle. The system can detect obstacles and provide visual or audible alerts to help drivers maneuver safely into parking spaces. Sensors and connectivity allow the system to identify available parking spaces and provide real-time information to the driver, facilitating quicker and more efficient parking. In some advanced systems, drivers can control the parking process remotely using a smartphone app, allowing for precise parking in tight spaces. With cameras positioned around the vehicle, the system can provide a virtual 360° view, aiding the driver in avoiding obstacles and navigating parking spaces with increased confidence. These advanced vehicles can function as a valet, with the support of sensors mounted in the parking infrastructure in addition to the in-vehicle sensors, and guide the vehicle to a designated parking spot.

Advanced Traffic Management

Advanced traffic management is an integrated system that utilizes technology and data to optimize and streamline traffic flow, enhance safety, and improve overall transportation efficiency (**Figure A.5**). This interconnected system leverages real-time information, including V2I communication, sensors, and traffic monitoring devices.

FIGURE A.5 Advanced traffic management systems (TMSs) monitor the state of traffic and take alternative actions [3].

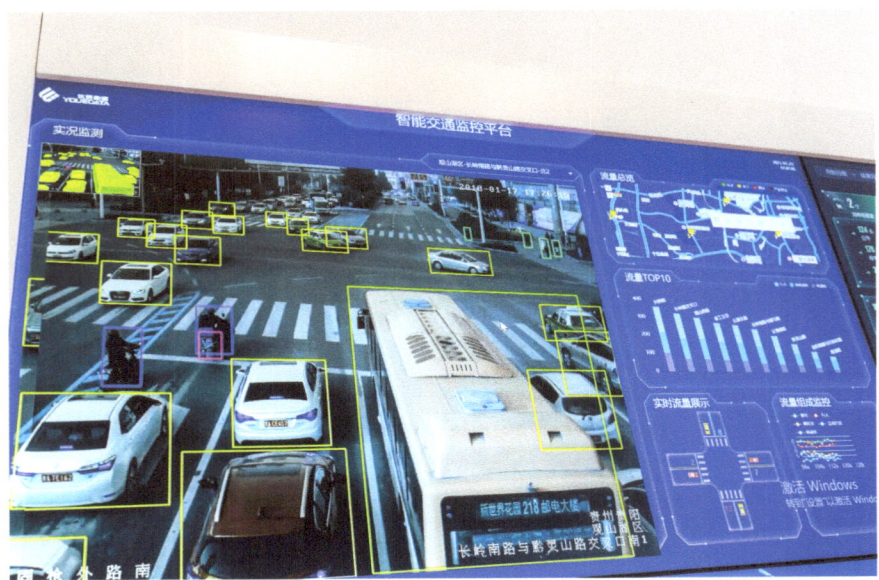

helloabc/Shutterstock.com.

The system continuously monitors traffic conditions using sensors and data analytics, identifying real-time congestion, incidents, and anomalies. Adaptive traffic signal systems adjust signal timings based on real-time traffic conditions, optimizing signal phasing to reduce congestion and improve the overall flow of vehicles. By analyzing historical and real-time data, advanced TMSs can predict traffic patterns, allowing for proactive measures to alleviate potential congestion and enhance traffic efficiency.

Communication between the TMS and CVs enables personalized routing information, real-time updates, and coordination to improve the overall traffic experience for drivers. Rapid detection and response to incidents, accidents, or road hazards contribute to quicker resolution and minimization of traffic disruptions. The system helps in providing an unobstructed corridor to emergency vehicles in reaching their destination. Coordination with public transportation systems facilitates efficient integration, providing seamless transitions between private and public transportation modes.

AEB (Automatic Emergency Braking)

AEB is an advanced safety technology designed to enhance vehicle safety by autonomously applying the brakes in situations where a collision with an obstacle or another vehicle is imminent. AEB systems utilize various sensors and cameras to monitor the vehicle's surroundings continuously, and when a potential collision risk is detected, the system intervenes to mitigate or prevent the collision.

- Sensor Fusion: AEB systems integrate data from multiple sensors, including radar, cameras, and sometimes LiDAR, to provide a comprehensive view of the vehicle's environment.
- Collision Warning: AEB includes a collision warning system that alerts the driver with visual or audible signals when an imminent collision is detected.
- Autonomous Braking: In emergencies where the driver does not respond to the collision warning, AEB autonomously applies the vehicle's brakes to either prevent the collision or reduce its severity.
- Pedestrian and Cyclist Detection: Many AEB systems are equipped to detect pedestrians and cyclists in addition to other vehicles, extending collision avoidance capabilities to vulnerable road users (VRUs).
- Integration with Forward Collision Warning (FCW): AEB often works in conjunction with FCW systems, enhancing the overall safety net by providing both warnings and autonomous braking.

AHB (Automatic High Beam)

AHB is an automotive lighting technology designed to enhance nighttime visibility and safety by automatically controlling the activation and deactivation of a vehicle's high-beam headlights. AHB systems use sensors to detect the headlights of oncoming or preceding vehicles and adjust the headlight illumination accordingly, optimizing visibility without causing glare to other drivers.

AHB systems continuously monitor the road ahead using sensors, assessing the lighting conditions and the presence of other vehicles. When the road is clear, and no oncoming or preceding vehicles are detected, AHB systems automatically activate the high beams to maximize visibility. As soon as oncoming headlights or preceding taillights are detected, AHB systems swiftly cut the high beam in the driving field of the oncoming drivers, to prevent glare, ensuring the safety and comfort of other drivers. Once the road is clear again, AHB systems restore the high beams to provide optimal illumination.

For additional information, see **J3069** [3]. See also **Adaptive Headlights**.

AI (Artificial Intelligence)

AI is a scientific and technical discipline based on computer science that develops methods to enable computational systems to generate outputs for a set of tasks defined by humans. AI integrates advanced computational algorithms and machine learning (ML) techniques to enable vehicles to perceive, analyze, and respond to their environment. AI plays a transformative role in enhancing the intelligence and autonomy of connected cars, contributing to various aspects of driving, safety, and overall user experience (UX).

Key applications of AI in CVs include the following:
- Autonomous Driving: AI algorithms enable vehicles to perceive their surroundings through sensors such as cameras, LiDAR, and radar, allowing for real-time decision-making and navigation without human intervention.
- Predictive Maintenance: AI analyzes vehicle data to identify potential issues and schedule proactive maintenance, reducing the risk of breakdowns and optimizing vehicle performance.
- NLP: Integration of NLP allows for seamless interaction between occupants and the vehicle's systems through voice commands, enhancing the user interface (UI) and overall in-car experience.
- ADAS: AI powers ADAS features such as lane-keeping assistance (LKA), collision avoidance, and ACC, contributing to enhanced safety and driver assistance.
- Traffic Prediction and Routing: AI algorithms analyze real-time and historical traffic data to predict congestion, optimize routes, and provide accurate travel time estimates for improved navigation.
- Behavioral Analysis: AI can analyze driver behavior and preferences, personalizing the in-car experience and contributing to a more adaptive and user-centric environment.

ANNs (Artificial Neural Networks)

ANNs represent a subset of AI technologies employed to emulate the learning and decision-making processes inspired by the human brain. ANNs are utilized to process complex datasets, make predictions, and optimize various functionalities within CVs.

Key applications of ANN in CVs include the following:
- Sensor Data Processing: ANNs are employed to process and interpret data from various sensors, such as cameras, LiDAR, and radar, enhancing the vehicle's ability to perceive and understand its surroundings.
- Autonomous Driving: ANNs play a crucial role in enabling autonomous driving by processing input data, recognizing patterns, and making real-time decisions to navigate the vehicle safely.

- Predictive Analytics: ANNs analyze historical and real-time data to predict future events, contributing to applications such as predictive maintenance, traffic prediction, and energy optimization.
- Gesture and Voice Recognition: Neural networks recognize and interpret gestures and voice commands, facilitating natural and intuitive interaction between occupants and vehicle systems.
- Driver Monitoring Systems (DMSs): ANNs can analyze driver behavior, attention, and fatigue through data from in-vehicle sensors, often cameras, contributing to safety features such as driver alertness monitoring.
- Customized UX: By learning from user preferences and behavior, ANNs contribute to personalizing the in-car experience, adjusting settings, providing route preference based on learned navigation patterns and selections, and providing tailored recommendations to occupants.

Antenna

A device on an automobile or any other type of system that is equipped with communication technologies allows it to exchange data with other vehicles, infrastructure, and external networks. In a CV, the antenna is a crucial component that facilitates wireless communication. It sends and receives signals to connect with other vehicles, roadside infrastructure, and communication networks. There are multiple antennas in vehicles to receive signals of different frequencies to enable communication over a broad range of wireless technologies.

Antennas enable communication between nearby vehicles, allowing them to share information about their speed, location, and other relevant data. This communication enhances safety and allows cooperative driving scenarios. Antennas also facilitate communication between vehicles and roadside infrastructure, such as traffic lights, road signs, and other smart city components. This communication helps optimize traffic flow and provides drivers with real-time information.

Vehicle-to-X (V2X) communication encompasses communication between vehicles and any entity that may affect the vehicle, including vehicle-to-vehicle (V2V), vehicle-to-infrastructure (V2I), and vehicle-to-pedestrian (V2P). Antennas play a crucial role in establishing and maintaining these diverse connections. Antennas are also used for in-vehicle wireless connectivity, allowing passengers to access the Internet, stream media, and use various infotainment services. This includes antennas for technologies like Wi-Fi, Bluetooth, and cellular networks [9]. GPS consists of antennas that allow the vehicle to receive signals from GPS satellites, providing accurate location information for navigation and location-based services.

AR (Augmented Reality)

AR in the automotive context refers to the integration of digital information, graphics, or sensory enhancements into the real-world driving environment. This technology overlays computer-generated content onto the driver's physical surroundings, providing an enhanced and interactive driving experience.

- Head-Up Display (HUD): AR is often implemented through HUDs, projecting vital information such as speed, navigation directions, and safety warnings onto the vehicle's windshield. This allows drivers to access crucial data without diverting their attention from the road.
- Navigation and Wayfinding: AR enhances navigation systems by superimposing arrows, street names, and other guidance information directly onto the driver's field of view (FoV), facilitating more intuitive route guidance.
- Object Recognition: AR can recognize and highlight objects in the external environment, providing contextual information about traffic signs, landmarks, and potential hazards.
- Driver Assistance Systems: AR is integrated into ADAS to display real-time information about the vehicle's surroundings, including distance to obstacles, lane-keeping guidance, and collision warnings.
- Virtual Displays and Controls: AR can create virtual displays within the vehicle, enabling gesture-based or touch-sensitive controls for multimedia, climate, and other vehicle settings.

Assisted Driving

Assisted driving refers to a level of vehicle automation where advanced technologies and systems support the driver in various aspects of the driving process. Also known as Level 2 (partial driving automation) according to **SAE J3016** [6] classification, assisted driving systems provide a combination of automated features, including ACC, LKA, lane centering control (LCC), and automated braking.

For more information, see *V2X Sensor-Sharing for Cooperative and Automated Driving* [10].

- ACC: Assisted driving systems maintain a set speed but also automatically adjust the vehicle's speed to maintain a safe following distance from the vehicle ahead.
- LKA: These systems utilize sensors and cameras to help drivers stay within their lane by providing gentle steering inputs or alerts when an unintentional lane departure is detected.

- LCC: This system utilizes onboard cameras, Global Navigation Satellite System (GNSS) sensors, and onboard high-definition maps to maintain the vehicle in the center of the lane.
- Automated Braking: Assisted driving incorporates automated braking systems that can intervene in emergencies, reducing the severity of a collision or preventing it altogether.
- Traffic Jam Assist (TJA): Some assisted driving systems offer functionality in slow-moving traffic, automatically controlling acceleration, braking, and steering within predefined conditions.
- Collision Avoidance Systems: These systems use sensors to detect potential collisions and can autonomously apply the brakes or take evasive actions to avoid or mitigate the impact.

Automated Driving

Automated driving, often referred to as autonomous driving or self-driving, represents a revolutionary advancement in automotive technology. In this technology, vehicles can navigate and operate without direct human input. ADSs leverage a combination of sensors, cameras, radar, and advanced algorithms to perceive the environment, make decisions, and control the vehicle. SAE International defined levels of automated driving in **SAE J3016** [6].

- Level 0—No Driving Automation: The human driver is solely responsible for maintaining control of the vehicle.
- Level 1—Driver Assistance: Basic driver assistance features, such as ACC or LKA, are automated, but the driver remains engaged and responsible for overall vehicle control.
- Level 2—Partial Driving Automation: The vehicle can control both steering and acceleration/deceleration simultaneously under certain conditions, requiring the driver to remain attentive and ready to take control at any time.
- Level 3—Conditional Driving Automation: The vehicle can perform most driving tasks under specific conditions, allowing the driver to disengage but must be ready to take over when prompted by the system.
- Level 4—High Driving Automation: The vehicle can operate autonomously in predefined scenarios or environments without driver intervention. Outside of these conditions, human control may be necessary.
- Level 5—Full Driving Automation: The vehicle is fully autonomous and capable of handling all driving tasks under all conditions. No human input is required.

Automotive Middleware

Automotive middleware is a specialized software layer that facilitates communication and integration between electronic systems and components within a vehicle's architecture. This middleware acts as a mediator, allowing various automotive applications, sensors, control units, and electronic modules to exchange information seamlessly. It plays a crucial role in ensuring interoperability and efficient communication in modern vehicles' complex network of electronic systems.

Automotive middleware supports standardized communication protocols, such as controller area network (CAN), FlexRay, Ethernet, and others, ensuring compatibility and efficient data exchange among electronic components. Middleware enables diverse automotive software applications and systems to communicate, even if they are developed by different manufacturers or adhere to different standards, fostering interoperability. In many automotive applications, especially those related to safety and control systems, middleware provides real-time communication capabilities, ensuring timely and accurate data exchange between components. Automotive middleware often incorporates security features to protect communication channels and prevent unauthorized access, ensuring the integrity and safety of the vehicle's electronic systems. As vehicles become more complex and incorporate advanced features, automotive middleware is designed to be scalable, accommodating the growing number of electronic components and applications.

Automotive Operating Systems

Modern vehicles have many electronic control units (ECUs) that control vehicle functions such as propulsion and steering and auxiliary functions such as infotainment. They run complex software that utilizes a standardized architecture such as Automotive Open System Architecture (AUTOSAR) or proprietary software (SW) architecture. One of the key SW architecture elements is an operating system. Automotive operating systems have real-time capabilities. They allocate hardware resources, schedule tasks, manage applications, and ensure the safe operation of SW applications. In automotive ECUs, both safety-critical and non-safety applications coexist in the same architecture. Automotive operating systems play a critical role in ensuring the required resources, such as processor time and memory, are available for executing safety-critical applications.

Automotive Vehicle Legislation

Automotive legislation is the law and regulation that govern aspects of vehicles, such as safety, emission, crashworthiness, and theft protection. Vehicles must satisfy the rules of the country, for which it is to be approved for sale. Certain geographies have testing agencies that approve the vehicles based on the regional regulations. This process is referred to as type approval. In other countries, the manufacturers self-certify that the vehicles meet the regulations.

Autonomous Driving

Autonomous driving refers to the capability of vehicles to navigate themselves from a source to a destination and perform all driving functions independently and self-sufficiently (Figure A.6). The term autonomy refers to the capability of being self-sufficient. The term is not recommended for use in the context of vehicles as there is always a human element, even in the case of Level 4 and Level 5 ADS [6].

FIGURE A.6 Levels of driving automation [5].

Autonomous Parking

Autonomous parking is a function in which the vehicle navigates itself from a driving lane into a final parking position through longitudinal and lateral control actions. An onboard processing unit generates the longitudinal and lateral control commands when selected to park in parallel, perpendicular, or angular parking slots.

Autonomous Software

Autonomous software is the collection of software modules that enable an autonomous vehicle (AV) to perform perception of the driving environment, trajectory planning, and actuator commands. This is also referred to as the autonomy stack. The autonomy stack is distributed. Sensing elements have software that processes images, radar detection, and LiDAR point clouds. Some sensors are capable of generating objects and lane information, which is then broadcast to the in-vehicle network. In specific architectures, the raw sensor data are processed by the autonomy compute platform itself. The autonomy stack also includes the software layers that perform data transfer across these various elements of the autonomy system. They consist of deterministic software and software based on AI.

AV (Autonomous Vehicle)

AVs, also known as self-driving or driverless vehicles, are vehicles equipped with advanced sensor systems and computational technologies that enable them to navigate and operate without human input. These vehicles use a combination of hardware and software, sensors, ML models, and complex algorithms to perceive the environment and make decisions.

AV Simulation

AV simulation refers to mimicking the complete ADS or parts of it in a virtual environment. Simulation is a powerful method for verifying ADS at various levels. It is used in ADS verification for the categories of model-in-the-loop, software-in-the-loop, hardware-in-the-loop, and vehicle-in-the-loop. The real-world driving data are analyzed using big data techniques, and relevant scenarios are generated to be used in a virtual environment.

AV Technology

AV technology refers to the integrated systems and methodologies employed in designing, developing, and deploying vehicles that can operate without human intervention. This technology encompasses various components such as sensors, cameras, light detection and ranging (LiDAR, Lidar, or LIDAR), radar, and AI to perceive the environment, make decisions, and navigate safely.

Key Components:

1. **Sensors and Cameras:** These devices gather real-time data about the vehicle's surroundings, including obstacles, traffic conditions, road signs, and lane markings.
2. **LiDAR:** This remote sensing method uses light in the form of a pulsed laser to measure variable distances to the objects in the environment, providing a precise three-dimensional (3D) reconstruction of the objects in the surrounding environment. These reconstructions are termed as point clouds.
3. **Radar:** Radars detect the distance and speed of objects around the vehicle. Radars are crucial for ACC, blind spot monitoring, and collision avoidance systems.
4. **AI:** AI algorithms process the data collected from sensors and cameras to interpret the environment, make decisions, and learn from new situations through ML techniques.

AV technology is applied in various sectors, including personal transportation, public transit, delivery robots, and logistics. These vehicles aim to enhance safety by reducing human error, increasing efficiency in transport systems, and improving mobility for the non-driving population.

AVP (Automated Valet Parking)

AVP is an advanced automotive technology that enables vehicles to autonomously navigate and park in designated parking areas without direct human intervention (Figure A.7). This innovative system leverages a combination of vehicle-mounted sensors, such as ultrasonic sensors, radars, and cameras, and infrastructure-mounted sensors, such as lidars, and communication technologies, to facilitate a seamless and efficient parking experience. The vehicle is guided to a designated parking spot in the parking structure by connected technologies, and the parking maneuver is executed based on the vehicle-mounted sensors.

For additional information, see *Towards High Accuracy Parking Slot Detection for Automated Valet Parking System* [11].

FIGURE A.7 Two stages in AVP. A typical fully automated parking system includes modules of slot detection, path planning, path following, and ego vehicle's pose estimation [11].

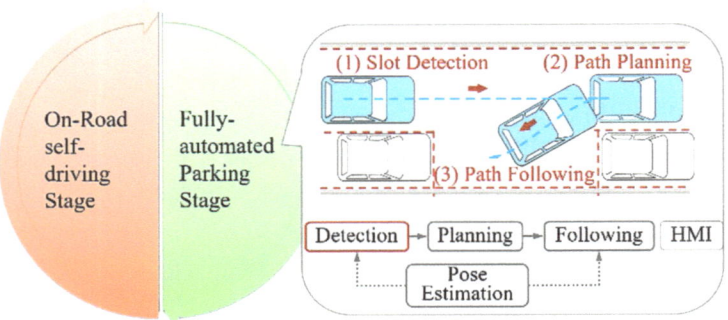

Axes of Motion Vehicles

The axes of motion refer to the 3D coordinate system used to describe a vehicle's orientation and movement within its environment (**Figure A.8**). These axes are fundamental in understanding and controlling vehicle dynamics, especially in ADSs.

1. **Longitudinal Axis (X-axis):**
   - **Orientation:** Runs from the front to the back of the vehicle (bumper to bumper).
   - **Motion:** Controls forward and backward movement, such as acceleration and braking.

2. **Lateral Axis (Y-axis):**
   - **Orientation:** Extends from one side of the vehicle to the other (side to side).
   - **Motion:** Governs movements like lane changes and sidestepping obstacles.

3. **Vertical Axis (Z-axis):**
   - **Orientation:** Runs from the bottom to the top of the vehicle (ground to roof).
   - **Motion:** Influences actions such as suspension response to road irregularities and vehicle roll during turns.

FIGURE A.8 Diagram showing vehicle axes of motion [12].

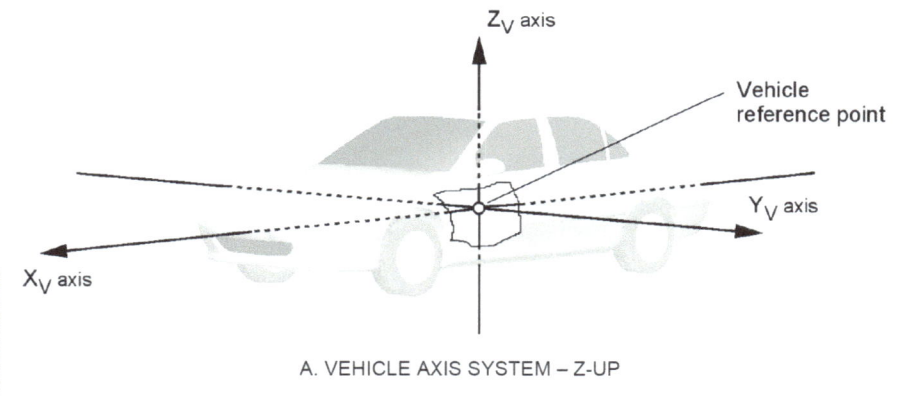

A. VEHICLE AXIS SYSTEM – Z-UP

REFERENCES

1. Society of Automotive Engineers, "J2399 Adaptive Cruise Control (ACC) Operating Characteristics and User Interface," SAE Publishing, Warrendale, PA, 2012.

2. Society of Automotive Engineers, "J2945 Performance Requirements for Cooperative Adaptive Cruise Control (CACC) and Platooning," SAE Publishing, Warrendale, PA, 2023.

3. Society of Automotive Engineers, "J3069 Adaptive Driving Beam System," SAE Publishing, Warrendale, PA, 2021.

4. Spicer, R., Vahabaghaie, A., Murakhovsky, D., Bahouth, G. et al., "Effectiveness of Advanced Driver Assistance Systems in Preventing System-Relevant Crashes," *SAE Int. J. Adv. & Curr. Prac. in Mobility* 3, no. 4 (2021): 1697-1701, doi:https://doi.org/10.4271/2021-01-0869.

5. Shuttleworth, J., "SAE News," January 7, 2019, accessed July 15, 2024, https://www.sae.org/news/2019/01/sae-updates-j3016-automated-driving-graphic.

6. Society of Automotive Engineers, "J3016 Taxonomy and Definitions for Terms Related to Driving Automation Systems for On-Road Motor Vehicles," SAE Publishing, Warrendale, PA, 2021.

7. Society of Automotive Engineers, "J2353 Data Dictionary for Advanced Traveler Information Systems (ATIS)," SAE Publishing, Warrendale, PA, 2019.

8. Shi, J., Wu, J., Zhu, B., Li, J. et al., "Design of Automatic Parallel Parking System Based on Multi-Point Preview Theory," SAE Technical Paper 2018-01-0604 (2018), doi:https://doi.org/10.4271/2018-01-0604.

9. Dai, D., "On-Glass Antenna for Connected Vehicle Communications," *SAE Int. J. Adv. & Curr. Prac. in Mobility* 2, no. 5 (2020): 2956-2962, doi:https://doi.org/10.4271/2020-01-1370.

10. Society of Automotive Engineers, "J3224 V2X Sensor-Sharing for Cooperative and Automated Driving," SAE Publishing, Warrendale, PA, 2022.

11. Yang, Q., Chen, H., Su, J., and Li, J., "Towards High Accuracy Parking Slot Detection for Automated Valet Parking System," SAE Technical Paper 2019-01-5061 (2019), doi:https://doi.org/10.4271/2019-01-5061.

12. Society of Automotive Engineers, "J670 Vehicle Dynamics Terminology," SAE Publishing, Warrendale, PA, 2022.

B

Big Data

Big data refers to extensive, diverse, complex data sets growing over time. Since it combines structured and unstructured data, specialized data mining software is required to perform data analytics. Big data are processed with the help of big data technologies. Big data technologies include methods and tools for data storage, data mining, data analysis, and data visualization. An example of big data is the data collected by sensors fitted in CAVs. Fleets with multiple vehicles generate petabytes of data, which are processed using big data technologies. Automotive organizations perform data analysis and define feature enhancements based on collected data.

Biometric Authentication

Biological features are unique for every individual. Examples of such features are fingerprints, iris, and voice [1]. When these features are used to identify an individual by a security system, it is called biometric authentication. The system learns the features of the person it needs to authenticate. This is stored in the memory. Every time a person tries to access the system, the person's feature is compared to the stored feature. Only if it matches, access is permitted (Figure B.1).

FIGURE B.1 Fishbone diagram for the voice biometric success [1].

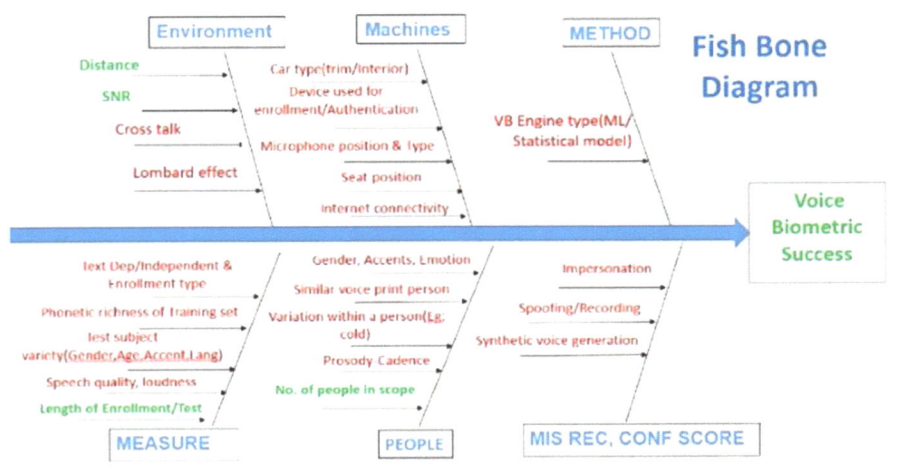

Biometric Vehicle Access

When a vehicle uses biometric authentication, the biological characteristics of permitted users are stored in the vehicle. When a driver needs to access the vehicle, she/he presents the biological feature, which is received by a biometric sensor such as a fingerprint reader or camera for facial features. If it matches the stored feature, then access is permitted.

For more information, see *Evaluation of Voice Biometrics for Identification and Authentication* [1].

Biometric Vehicle Start

When biometric authentication is used to authorize a person to start the vehicle, the biometric characteristics are compared with the prestored traits of the person. Upon matching traits, the vehicle start is permitted. Typical sensors for such an authentication are fingerprint and facial feature scanners.

For more information, see *Evaluation of Voice Biometrics for Identification and Authentication* [1].

Blind Spot Detection

The blind spot is the region around the vehicle that is not directly visible to the driver, in the side view mirrors or rearview mirror [2]. Even though it is typically referred to in the rear-side region, blockage of the driver's view due to other parts of the vehicle, such as the A-pillar, also results in a blind spot in the front. The structural design, the width of the A-pillar, and the placement of seats within the vehicle are factors that affect blind spots. A blind spot due to the A-pillar obstructs the driver's peripheral vision and blocks pedestrians, bicyclists, and motorcyclists. Larger vehicles have larger blind spots.

Radars or cameras fitted on the sides of the vehicles detect the presence of other vehicles in the blind spot. The ADAS/ADS software processes this and generates warnings in the side mirrors [3].

Blind Spot Intervention

Blind spot intervention is a function of ADAS/ADS. When a vehicle is detected in the blind spot and the driver intends to move into the lane, the function intervenes by providing counter-steering or opposite-wheel braking. Driver intention is determined by steering angle, turn indicator, and inertial measurement unit (IMU) inputs.

Blockchain

Blockchain in the automotive industry refers to a decentralized digital ledger technology used to securely and transparently record transactions and data exchanges within a network. It enhances data integrity, security, and trust by ensuring that information once entered cannot be altered. Blockchain ledger is a distributed database of transactions and is controlled and distributed by participating units in the blockchain.

A system that integrates blockchain technology with the architecture of vehicles enhances security, transparency, and efficiency in various operations, including data management, communication, and transaction processing (**Figure B.2**) [4].

FIGURE B.2 Architecture for full-architecture ECUs [4].

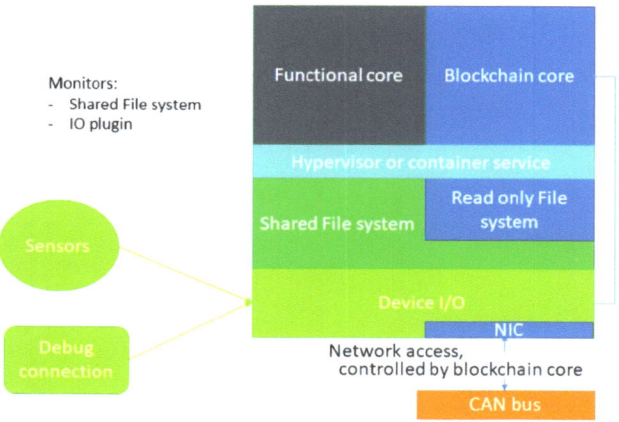

Blockchain, when used in the automotive supply chain, provides traceability and transparency, ensuring the authenticity of parts and compliance with regulations. Vehicle identity and history provide a tamper-proof record of vehicle data, including ownership, service history, and mileage, which can be crucial for used car sales, warranty management, and insurance purposes. Telematics and usage-based insurance (UBI) facilitate the secure sharing of vehicle data for calculating insurance premiums based on vehicle usage while ensuring data privacy.

BMS (Battery Management System)

In the automotive context, a BMS maintains the rechargeable energy storage system (RESS) of the vehicle in a safe operating region. It monitors the cells and battery packs during charging and discharging (vehicle operation) for potential thermal runaway conditions [5]. It controls the opening and closing of contactors that connect the high-voltage battery packs to the high-voltage network (**Figure B.3**).

FIGURE B.3 Key features of BMS [1].

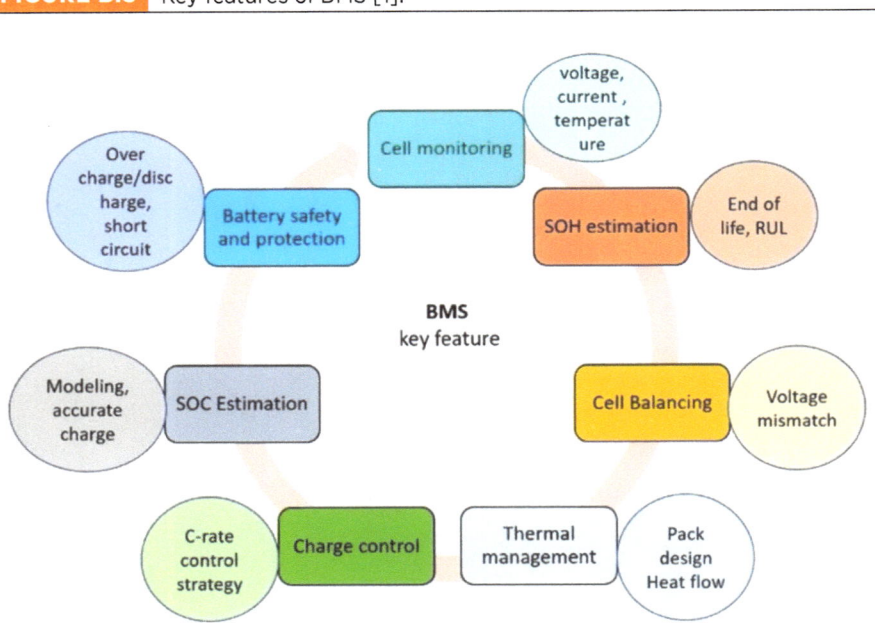

Brake Assist

Brake assist (also known as brake assist systems [BAS]) is an ADAS feature in CVs that enhances the vehicle's braking capability during emergencies. It detects circumstances when a driver initiates a panic stop and automatically applies maximum braking force to reduce the stopping distance. This system is particularly effective in preventing or mitigating the severity of a collision [6].

The system uses sensors to monitor the speed at which the brake pedal is pressed. If the action is quicker than a preset threshold, indicative of an emergency, the system intervenes by applying additional brake force to assist in braking. This maximizes the braking efficiency without causing the brakes to lock, working in conjunction with ABS.

BSM (Basic Safety Message)

BSM is used to exchange vehicle-state-related safety data among V2X applications in the data format specified in the **SAE J2735** message set. Part 1 data are sent every time a BSM message is broadcast, and Part 2 data are optional. Part 1 data contain safety-critical data such as speed, heading, acceleration, and steering wheel angle.

For more information, see **SAE J2735** *V2X Communications Message Set Dictionary* [7].

REFERENCES

1. Bekkanti, N., Busch, L., and Amman, S., "Evaluation of Voice Biometrics for Identification and Authentication," SAE Technical Paper 2021-01-0262 (2021), doi:https://doi.org/10.4271/2021-01-0262.

2. Yoshihira, M., Watanabe, S., Nishira, H., and Kishi, N., "Developing an Autonomous Vehicle Control System for Intersections Using Obstacle/Blind Spot Detection Frames," SAE Technical Paper 2016-01-0143 (2016), doi:https://doi.org/10.4271/2016-01-0143.

3. Bhosale, T., Attar, A., Warang, P., and Patil, R., "Object Detection for Autonomous Guided Vehicle," *ASEAN Journal of Science and Engineering* 2, no. 3 (2021): 209-216.

4. Chan, K., Pasco, M., and Cheng, B., "Towards a Blockchain Framework for Autonomous Vehicle System Integrity," *SAE Int. J. Transp. Cyber. & Privacy* 4, no. 1 (2021): 19-38, doi:https://doi.org/10.4271/11-04-01-0002.

5. Society of Automotive Engineers, "J3073 Battery Thermal Management," SAE Publishing, Warrendale, PA, 2016.

6. Hirose, T., Taniguchi, T., Hatano, T., Takahashi, K. et al., "A Study on the Effect of Brake Assist Systems (BAS)," *SAE Int. J. Passeng. Cars - Mech. Syst.* 1, no. 1 (2009): 729-735, doi:https://doi.org/10.4271/2008-01-0824.

7. Society of Automotive Engineers, "J2735 V2X Communications Message Set Dictionary," SAE Publishing, Warrendale, PA, 2023.

C2X (Car-to-X)

C2X is a subset of V2X communication, which means vehicles-to-everything. V2X is an umbrella term indicating communication between a vehicle and another vehicle, infrastructure elements, pedestrians, or other VRUs [1]. A vehicle that is capable of V2X is equipped with onboard units (OBUs) that transmit V2X messages over dedicated short-range communication (DSRC) or cellular-V2X (C-V2X) technologies. They receive messages from other vehicles or infrastructure elements. Some applications of V2X are cooperative ACC (CACC), red light violation warning, electronic toll collection (ETC), etc.

CA (Connected and Autonomous) Data Analytics

CA data analytics, also known as connected and autonomous data analytics, refers to the specialized field of data analytics focused on processing, interpreting, and deriving actionable insights from the vast amounts of data generated by CAVs. This field encompasses various data sources, including vehicle sensors, communication systems, user interactions, and external environmental inputs.

Key components for data collection are as follows:

1. Sensors: Includes data from LiDAR, radar, cameras, ultrasonic sensors, and other onboard sensors.
2. Telematics: Information from V2X communication, such as V2V and V2I data.
3. User Data: Data generated from user interactions and infotainment systems.

Data Processing:

1. Real-Time Analysis: Immediate data processing for real-time decision-making and autonomous driving functions.
2. Batch Processing: Analysis of large datasets over time to identify patterns, trends, and insights.

Issues with Data Processing [2]:

1. Quality of Data—Accuracy and completeness
2. Volume of Data—Scalability and storage
3. Velocity of Data—Real-time processing and streaming data
4. Variety of Data—Diverse formats and unstructured data
5. Veracity of Data—Data quality assurance and data bias
6. Security and Privacy—Data protection and cybersecurity

Data Storage:

1. Edge Computing: Localized processing and storage on the vehicle to reduce latency.
2. Cloud Computing: Centralized storage and processing in cloud platforms for extensive analysis and long-term data retention.

Data Analytics Techniques (**Figure C.1**):

1. ML: Algorithms to predict and enhance vehicle behavior and performance.
2. Big Data Analytics: Techniques to handle and analyze massive volumes of data efficiently.
3. Predictive Analytics: Methods to foresee potential issues and optimize maintenance schedules.

FIGURE C.1 Data analytics flow [2].

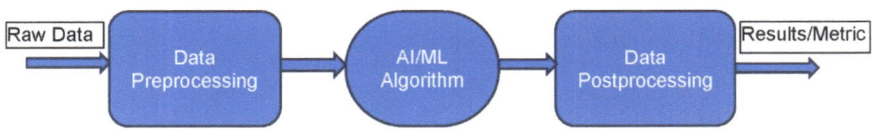

© SAE International.

CACC (Cooperative ACC)

CACC is an ADAS that extends the functionality of traditional ACC by incorporating V2V communication (**Figure C.2**). This technology enables vehicles to adjust speed automatically and maintain a safe following distance from the vehicle ahead based on onboard sensor data and information shared between CVs. CACC is pivotal in developing ACV technologies, aiming to enhance road safety, improve traffic flow, and reduce fuel consumption [3].

FIGURE C.2 CACC overview [3].

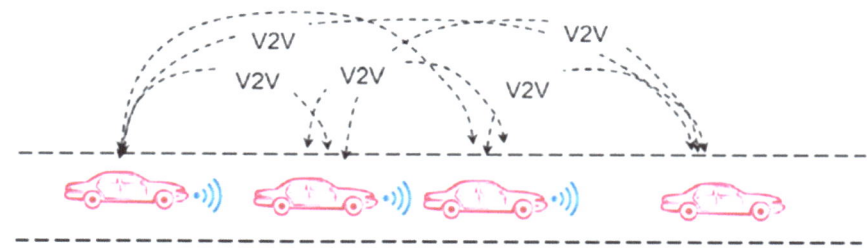

CACC systems use a combination of radar, cameras, and V2V communication to maintain a safe following distance accurately. By receiving data from the vehicle ahead and, potentially, from several vehicles in front, CACC can anticipate speed changes more effectively than traditional systems that rely solely on direct sensor input. This cooperative element enables tighter vehicle spacing and coordinated braking and acceleration, resulting in more efficient traffic flow and a reduced risk of accidents.

Cameras

Cameras are one of the primary sensors for ADSs (**Figure C.3**). They use optical principles to perceive the surroundings. They employ image processing algorithms or neural networks to classify objects and lanes in the surroundings. Cameras are used in vehicles in a single-lens (mono) or dual-lens (stereo) configuration.

FIGURE C.3 ADAS hardware configuration using multi-core microcontroller (MCU) [4].

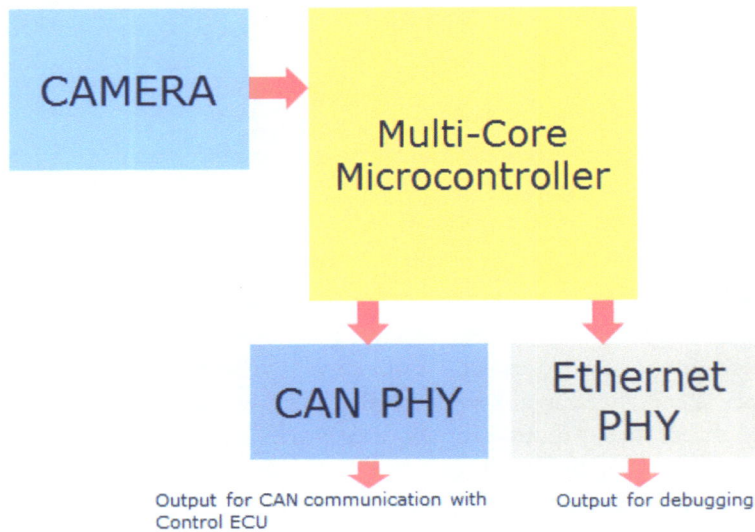

Carbon Footprint

The carbon footprint of CVs refers to the total amount of greenhouse gases (GHGs), particularly carbon dioxide (CO_2), emitted directly or indirectly through the life cycle of CVs (**Figure C.4**). This encompasses emissions from the production of the vehicle, the generation of electricity or fuel used to power the vehicle, and the end-of-life vehicle disposal process. In the context of CVs, which are equipped with advanced technologies for communication, data exchange, and sometimes autonomous operation, the carbon footprint also includes the energy consumed by these technologies throughout the vehicle's operational life (**Figure C.5**).

FIGURE C.4 Steps of the design process and the possibility for carbon footprint calculation [5].

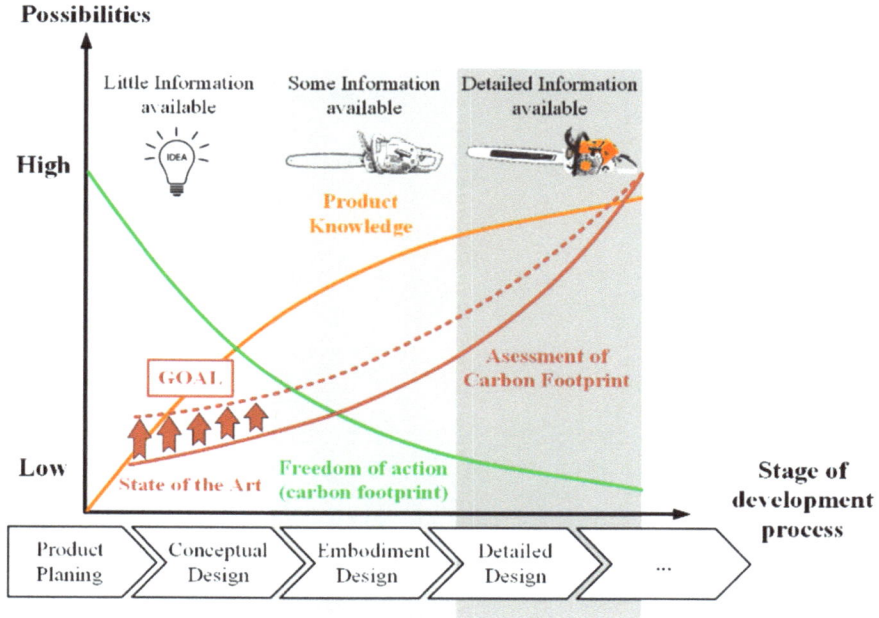

FIGURE C.5 Product life cycle emissions [5].

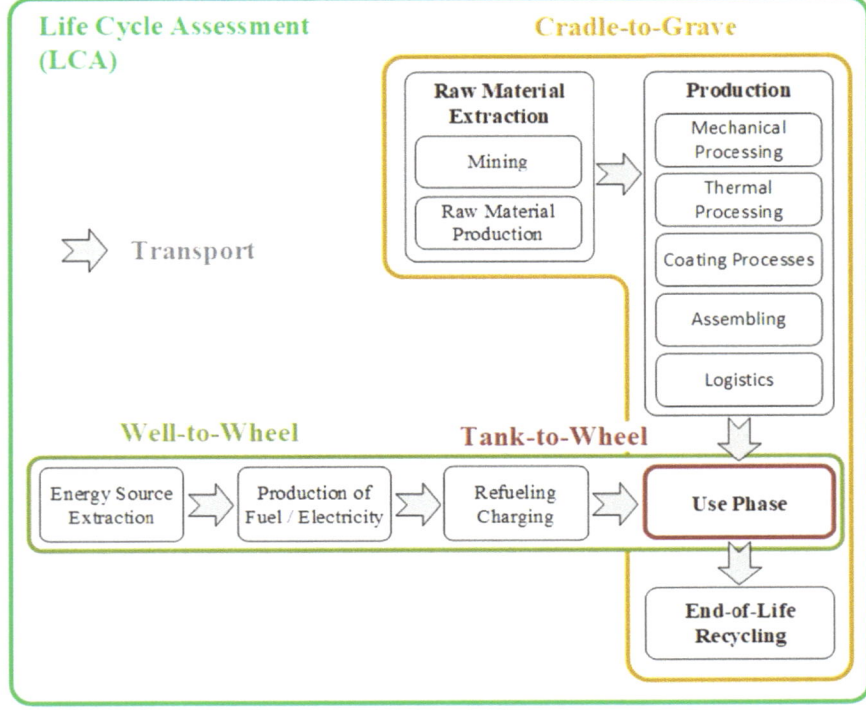

Carpooling

Carpooling refers to the shared use of a car by the driver and one or more passengers, typically to commute to work, school, or similar destinations. This reduces each person's travel costs, such as fuel costs, tolls, and the stress of driving. Carpooling is a significant component of urban mobility strategies aimed at reducing traffic congestion, environmental impact, and transportation-related expenses and transportation agencies incentivize this globally. Many regions have high-occupancy vehicle (HOV) lanes which are designated for more than two occupants.

The integration of carpooling with AV technology and improved telematics systems may further revolutionize how carpooling is implemented, making it more efficient and appealing to a broader audience. This could lead to more structured and reliable carpooling systems that could integrate seamlessly with public transport and other mobility services.

CCA (Cooperative Collision Avoidance)

CCA refers to an advanced safety mechanism within the domain of CVs, designed to prevent or mitigate the severity of vehicular collisions. This technology leverages the communication capabilities of CVs—vehicles equipped with Internet access and usually a wireless local area network (WLAN)—that allow them to communicate with each other (V2V) and with infrastructure (V2I) (**Figure C.6**). By sharing information about their speed, direction, and position in real time, CCA systems can identify potential collision scenarios early and take proactive measures to avoid them.

FIGURE C.6 Occluded pedestrian crossing a roadway (V2V) use case [6].

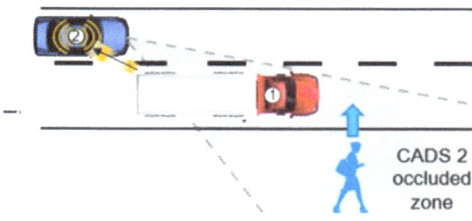

CCA systems continuously monitor the vehicle's environment and communicate with nearby vehicles and traffic infrastructure. When a potential collision threat is detected, the system can alert the driver through visual, auditory, or haptic signals or, in more advanced systems, automatically initiate actions such as braking or steering adjustments to avoid the collision [6].

As the automotive industry progresses toward fully AVs, CCA is poised to play a pivotal role in ensuring the safety of these vehicles and their passengers. Ongoing advancements in communication technologies and AI are expected to enhance the effectiveness and reliability of CCA systems, making roads safer for everyone.

Charging Infrastructure

Charging infrastructure refers to the network of charging stations and associated equipment required to supply electric power for recharging electric vehicles (EVs), including battery EVs (BEVs) and plug-in hybrid EVs (HEVs) (PHEVs) [7]. This infrastructure is a critical component in the ecosystem of CVs, enabling the widespread adoption and use of electric mobility solutions by providing accessible, efficient, and reliable charging options (**Figure C.7**).

FIGURE C.7 Example of charging infrastructure.

Charging Station

A charging station, also known as EV supply equipment (EVSE), is a device that supplies electrical power to recharge plug-in EVs, including electric cars, neighborhood EVs, and plug-in hybrids [7].

- Home Charging: Lower power charging stations designed for residential use, typically overnight.
- Public Charging: Higher power stations located in public places provide faster charging options for EV users on the go.
- Commercial and Workplace Charging: Stations installed in commercial properties such as parking lots of supermarkets, department stores, and workplaces facilitate easy charging for employees and customers.

The development of faster, more efficient charging technologies and expanding charging networks are critical for supporting the widespread adoption of EVs. Continued advancements and standardization efforts by organizations like SAE International will play a pivotal role in this evolution (**Figure C.8**).

FIGURE C.8 There are a variety of charging station examples.

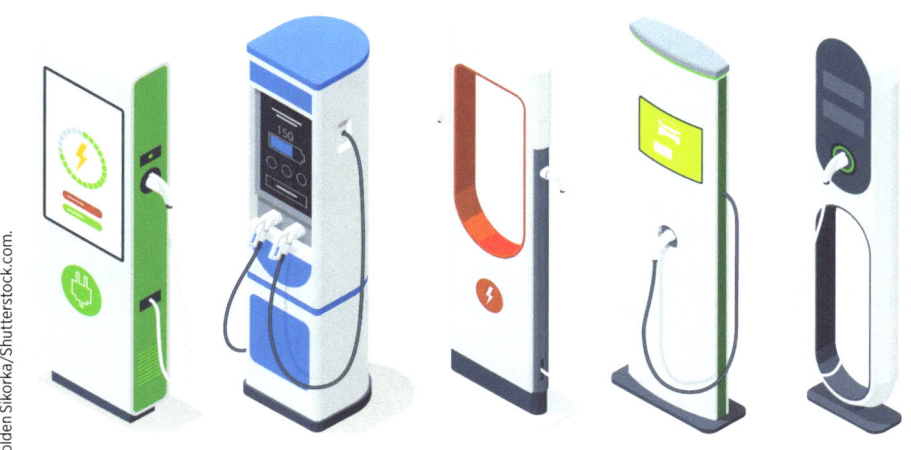

CICAS (Cooperative Intersection Collision Avoidance System)

A CICAS is an ADAS designed to prevent collisions at intersections by enabling communication between vehicles and infrastructure (**Figure C.9**). It leverages V2V and V2I technologies to provide real-time information and warnings to drivers about potential collision risks.

FIGURE C.9 Illustration of permissive left turns with opposing traffic [8].

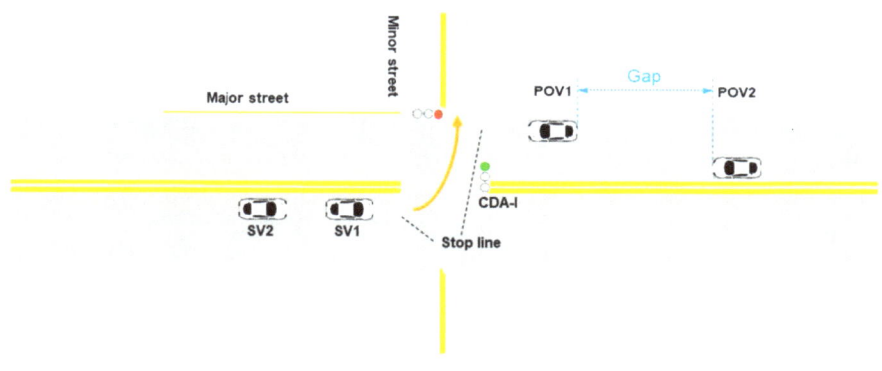

CICAS uses real-time data to predict potential collisions based on the movement and trajectory of vehicles approaching an intersection. The system alerts drivers to potential collision risks through visual, auditory, or haptic feedback, enabling them to take corrective action. In some cases, CICAS can autonomously control vehicle functions (e.g., braking or steering) to prevent a collision if the driver does not respond to warnings in time.

CIM (Cooperative Intersection Management)

CIM is an advanced system designed to optimize traffic flow and enhance safety at intersections by enabling real-time communication and coordination between AVs, CVs, and traffic infrastructure. CIM utilizes V2V and V2I communication technologies to manage the movement of vehicles through intersections efficiently, reducing delays and preventing accidents.

CIM can be particularly effective in busy urban areas where traffic density is high, and intersections are frequent. CIM can prioritize the passage of emergency vehicles through intersections, ensuring rapid response times without disrupting overall traffic flow. CIM manages the interaction of autonomous, connected, and traditional human-driven vehicles at intersections, supporting their coexistence.

CIS (Cooperative Intersection Safety)

CIS refers to advanced technologies and strategies used in ACVs to enhance safety at intersections. CIS leverages V2X communication, which includes V2V, V2I, and V2P interactions, to share critical information in real time.

Intersection movement assist (IMA) system alerts drivers or autonomous systems of potential collisions at intersections by predicting the paths of vehicles and pedestrians. Left turn assist (LTA) helps drivers or autonomous systems make safer left turns by providing alerts about oncoming traffic, including vehicles and bicycles that may be in blind spots. Traffic signal coordination synchronizes vehicle movements with traffic signal timings, reducing the likelihood of accidents caused by signal violations or misjudged signal changes. Pedestrian detection and warning systems use sensors and communication technology to detect pedestrians near or in the intersection and warn drivers and pedestrians. Emergency vehicle preemption allows emergency vehicles to communicate with traffic signals to ensure they have a clear path through intersections, enhancing response times and safety. Collision avoidance systems use data from multiple vehicles and infrastructure to predict and prevent potential collisions at intersections.

C-ITS (Cooperative Intelligent Transport System)

C-ITSs refer to a subset of intelligent transport systems (ITSs) where vehicles, infrastructure, and other road users share information to improve road safety, efficiency, and sustainability. This information exchange is through V2X communication technologies, which include V2V, V2I, V2P, and vehicle-to-network (V2N) communications (**Figure C.10**).

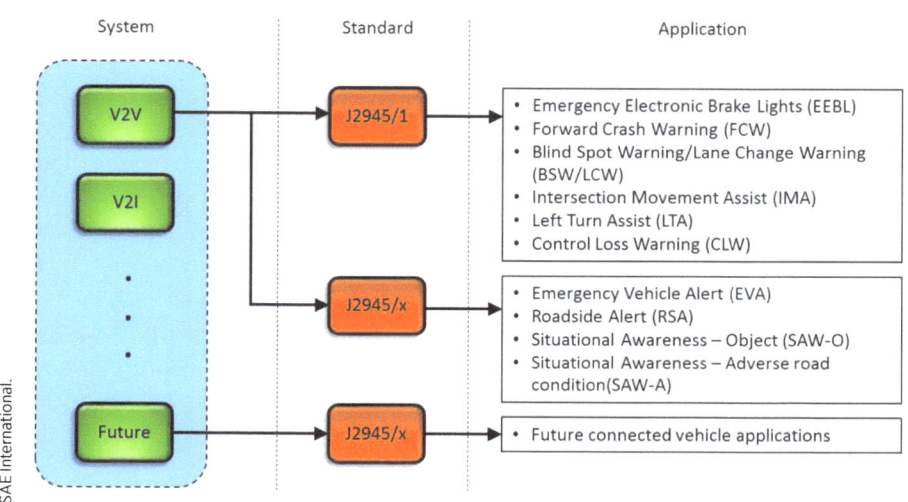

FIGURE C.10 Relationship of the system and J2945/x standard [9].

V2V communication enables direct information exchange between vehicles. This helps share critical information such as speed, heading, and location to prevent collisions and improve traffic flow. V2I communication facilitates communication between vehicles and road infrastructure, such as traffic signals, road signs, and toll gates. This can optimize traffic light timings, provide real-time traffic updates, and enhance road safety. V2P communication focuses on the interaction between vehicles and VRUs, such as pedestrians and cyclists [9]. It helps alert drivers to pedestrians' presence and movement to avoid accidents. V2N communication connects vehicles to broader networks, including cloud services, for accessing real-time data and services such as navigation, traffic management, and infotainment.

Collision avoidance systems alert drivers to potential collisions with other vehicles or obstacles. Traffic signal priority prioritizes emergency vehicles or public transport at traffic signals to improve response times and service efficiency. Dynamic traffic management adjusts traffic signal timings and lane usage based on real-time traffic conditions. Eco-driving assistance provides feedback to drivers on how to drive more efficiently to reduce fuel consumption and emissions.

Cloud Computing

Cloud computing is a model that enables ubiquitous, convenient, on-demand network access to a shared pool of configurable computing resources, which can be rapidly provisioned and released with minimal management effort or interaction with the service provider. Storage, processing, memory, network bandwidth, and software are examples of cloud resources.

Cloud Platform

A cloud platform in the context of connected and AVs refers to the use of cloud computing infrastructure to support vehicle functionalities, including data storage, processing, and software updates. These platforms facilitate the integration of vehicle data with other services, enhance vehicle operations through real-time data analytics, and enable over-the-air (OTA) updates (**Figure C.11**).

An example of the application of a cloud platform is the normalization of GHG emissions per passenger miles traveled (PMT). This is a common metric for comparing the environmental efficiency of two alternative trip modes, which is different from vehicle miles traveled [10].

FIGURE C.11 Relevant metrics should be highlighted for various parties including citizens, cities and enterprises, and mobility service providers.

Applications:
- Data Management: Cloud platforms manage vast amounts of data from CVs, from operational data to user behavior analytics.
- Software Management: Supports OTA software updates, bug fixes, and feature upgrades without requiring physical access to the vehicle.
- Enhanced Connectivity: Facilitates better integration of vehicles with smart city infrastructure, other vehicles, and various consumer devices.

The role of cloud platforms in automotive technologies is expected to grow, driven by advancements in cloud computing, fifth-generation (5G) technology, and the increasing incorporation of AI and ML algorithms. These technologies will enhance cloud platforms' capabilities to support more complex and AV functions.

Cloud-Based Fleet Management

Cloud-based fleet management refers to using cloud computing technologies to remotely oversee, coordinate, and maintain a fleet of vehicles [11]. This advanced system facilitates real-time data collection, analysis, and management of vehicle and driver information, enhancing operational efficiency and reducing costs [12].

Components [12]:
- Vehicle Telematics: Devices installed in vehicles that collect and transmit data such as location, speed, and diagnostic codes.
- Cloud Storage: A central repository where all collected data are stored and processed.
- UI: Web-based dashboards and mobile apps that allow fleet managers to visualize and interact with the data.
- Real-Time Tracking: Enables continuous monitoring of vehicle locations and routes using GPS technology.
- Maintenance Alerts: Automated notifications about vehicle maintenance needs based on mileage, engine status, or diagnostic trouble codes (DTCs).
- Driver Performance Monitoring: Analysis of driving patterns to assess driver behavior and ensure compliance with safety standards.
- Fuel Management: Tracking and analyzing fuel usage to identify cost-saving opportunities.
- Compliance Management: Assisting with regulatory compliance related to vehicle inspections, emissions, and driver hours.

Cloud-Based Vehicle Management

Cloud-based vehicle management is a comprehensive approach to managing individual vehicles or fleets that leverages cloud computing technology. This system facilitates centralized control and monitoring of vehicle operations, performance, and maintenance through Internet-based platforms, providing access to real-time vehicle data and analytics [10].

- Vehicle Diagnostics: Remote vehicle health monitoring, including engine performance, battery life, and other critical systems.
- Performance Analysis: Tracking and analyzing vehicle performance metrics to optimize fuel efficiency and operational effectiveness.
- Maintenance Scheduling: Automated scheduling of maintenance based on real-time data and predictive analytics.
- Security Features: Enhanced vehicle security through remote monitoring, including real-time alerts for unauthorized use or geofence breaches.

Collaborative Driving

Collaborative driving refers to the concept and implementation of systems in which the driver and automated vehicle technologies actively cooperate [13]. This approach, aligned with standards, integrates human inputs and ADAS to enhance driving safety, efficiency, and comfort [14]. The system dynamically shifts control between the driver and the vehicle based on real-time conditions, driver intent, and environmental factors (**Figure C.12**).

FIGURE C.12 Millstones of the intelligent connected passenger vehicles [13].

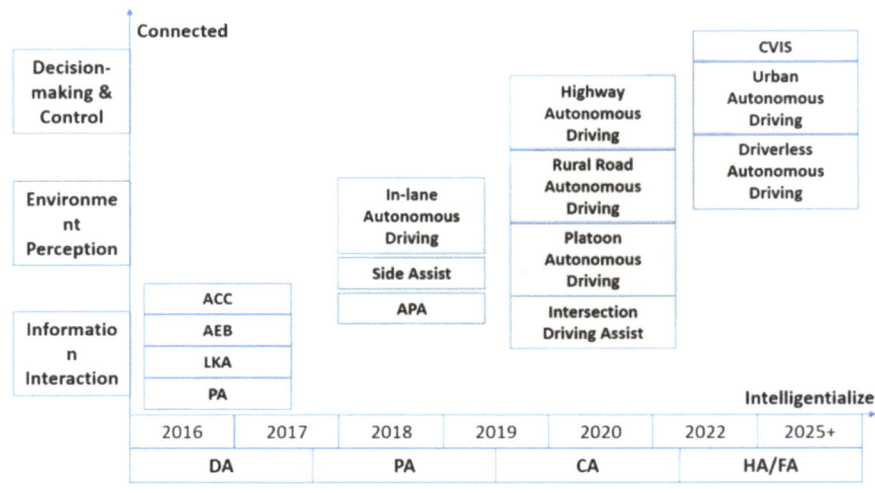

Collision Avoidance Systems

Collision avoidance systems are ADAS designed to prevent or reduce the severity of a collision. These systems use sensors, cameras, and radar technology to detect potential hazards on the road, such as vehicles, pedestrians, or obstacles, and automatically take action to avoid them. Actions may include alerts to the driver, automatic braking, or steering adjustments.

V2X technologies enable collision avoidance in CVs. Each vehicle broadcasts the current road segment (CRS), the next road segment (NRS), the current lane number, and the arrival time and the exit time, to all the other vehicles in its communication range (**Figure C.13**). Vehicles are also assumed to have access to a GPS with locally generated Radio Technical Commission for Maritime (RTCM-104) corrections to achieve real-time kinematic (RTK) solution [15].

FIGURE C.13 Vehicles share information to ensure collisions are avoided [15].

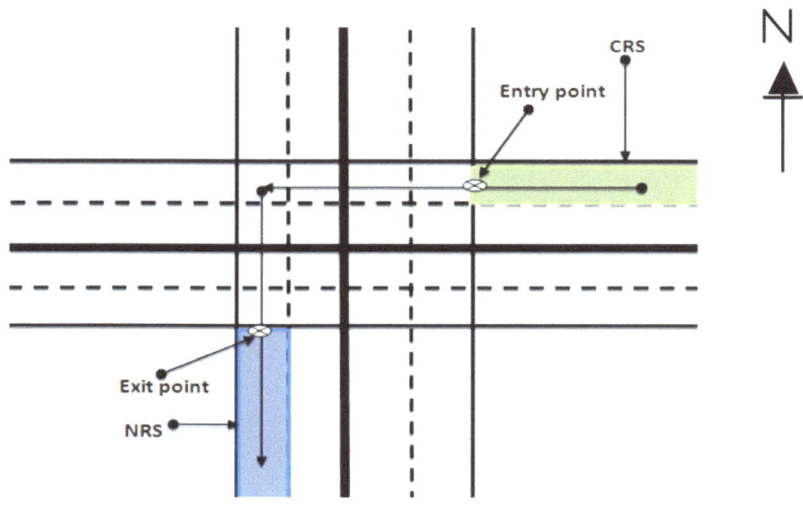

© SAE International.

Collision Energy Management

Collision energy management refers to the systems and strategies employed in vehicle design and safety to manage and mitigate the energy generated during a vehicle collision. This concept primarily focuses on reducing the severity of injuries to occupants and minimizing damage to the vehicle through controlled deformation and energy absorption techniques.

Structural design involves designing vehicle structures such as crumple zones, which are engineered to absorb and dissipate collision energy away from the passenger compartment. Material selection uses advanced materials and technologies that enhance energy absorption capabilities, such as high-strength steels and composites. Safety features include the integration of passive safety systems like airbags and seat belts that help manage the energy experienced by occupants. Vehicle crashworthiness enhances a vehicle's ability to protect its occupants by effectively managing the energy during collisions. Pedestrian safety implementing exterior vehicle designs minimizes injury to pedestrians in the event of a collision by managing impact energy [16].

Balancing strength and weight engineering challenges include designing vehicle structures that are strong enough to manage collision energy while maintaining a lightweight design for overall vehicle efficiency.

Collision Impact Reduction

Collision impact reduction refers to systems and technologies designed to minimize the severity of an impact during a vehicular collision (**Figure C.14**). These include active safety features that prevent collisions and passive systems that reduce damage and injury when a collision is unavoidable based on time to collision (TTC) [17].

FIGURE C.14 Severe injury/fatality risk for drivers in frontal crashes (11-1) O'clock [17].

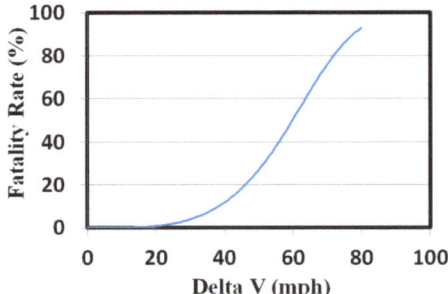

Active safety systems employ technologies such as AEB, electronic stability control (ESC), and ADAS that actively intervene to prevent collisions or reduce their severity. Passive safety features include airbags, crumple zones, seatbelts, and energy-absorbing steering systems designed to protect occupants by reducing the forces experienced during a crash.

Collision Mitigation System

Collision mitigation systems are active safety systems that detect an impending collision, issue warnings, apply brakes, augment steering, and activate measures to decrease collision severity, such as seat belt pretensioning and seat movement. They utilize cameras, LiDARs, and radar sensors to predict the TTC in front, rear, or cross traffic based on sensor placement [18]:

$$\text{Time to Collision}(\text{TTC}) = \left(\frac{\text{Relative distance}(\text{ft})}{\text{Relative speed}(\text{ft}/\text{s})}\right)$$

Collision Reconstruction

Collision reconstruction involves investigating, analyzing, and drawing conclusions about the causes and events of a vehicle collision. This includes using data from vehicle systems, such as data loggers and event data recorders (EDRs), to reconstruct the events leading up to, during, and following a crash [19].

The system is designed to ensure the accuracy and integrity of the data collected from various system sensors gathered via the EDRs. They store time-stamped data for a certain time duration before and after a collision. To reconstruct a collision, the data stored in multiple ECUs are combined together.

With advancements in vehicle technology, particularly the rise of connected and AVs, the field of collision reconstruction is undergoing significant evolution. More sophisticated data, tools, and methodologies are being developed to handle the increased complexity of modern vehicle systems and to utilize the extensive data these systems provide. CAVs record video data in addition to various signal information, which helps in accident reconstruction, continuous monitoring, and identifying near-miss collisions.

Collision Sensing System

A collision sensing system refers to an integrated set of sensors and software used in vehicles to detect the potential of a collision with another object, vehicle, pedestrian, or obstacle (**Figure C.15**). These systems are crucial components of ADAS and are designed to alert drivers or autonomously take action to avoid or mitigate collisions [20].

FIGURE C.15 Block diagram of the camera-based forward collision alert (FCA) [20].

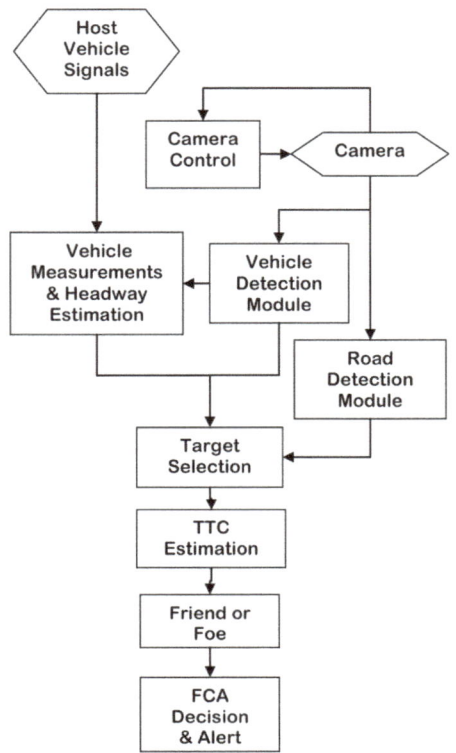

Collision sensing systems typically involve a combination of radar, LiDAR, cameras, and ultrasonic sensors, each contributing to a comprehensive view of the vehicle's surroundings. These systems analyze sensor data to predict potential collisions, provide warnings to the driver, or automatically initiate actions such as braking or steering adjustments to prevent accidents.

These systems command the braking systems to apply brakes when an imminent collision is detected automatically, and the driver has not yet reacted. Additionally, an audible and visual FCW alerts the driver to potential collisions with objects in the vehicle's forward path. Blind spot detection monitors areas of the vehicle that are difficult for a driver to see.

Collision Severity Prediction

Collision severity prediction refers to the methodologies and technologies used to forecast the potential severity of vehicle collisions. This prediction is based on vehicle speed, impact angle, vehicle mass, and safety features and results in an injury severity score (ISS) [21]. Technology is crucial for enhancing precrash safety measures and mitigating the effects of collisions.

Automated emergency braking systems (EBSs) (AEBSs) use collision severity prediction to adjust the braking force and timing to minimize impact. Advanced automatic collision notification (AACN) enhances the effectiveness of emergency response by providing first responders with predicted injury severity information, potentially improving patient outcomes.

Collision Severity Reduction

Collision severity reduction refers to the strategies and technologies employed to minimize the impact and consequences of traffic collisions. This includes reducing the likelihood of injury and the extent of damage in the event of a crash [17].

Pre-collision systems (PCSs) use AEB and collision avoidance systems to detect potential collisions and take preemptive actions to mitigate the impact. Vehicle structural design and material use enhance the absorption and dissipation of impact forces more effectively, protecting occupants during a crash. Advanced seatbelt designs and airbag systems adapt based on the severity of the crash to reduce occupant injuries.

With advancements in ACV technologies, there is potential for significant improvements in collision severity reduction. Technologies like V2X communication could enable more sophisticated and proactive systems to prevent accidents before they occur.

Collision Warning System

Collision warning system is an ADAS installed in vehicles to warn the driver of an impending collision [22]. The warning is provided in visual, audible, or haptic modalities. This system uses sensors such as radar, LiDAR, or cameras and a processing unit to calculate the time to impact (TTI) with another vehicle. The warning is provided to the driver if the TTI is less than a parameterized value (**Figure C.16**). These warnings are provided for forward collisions and reverse collisions based on the direction of motion of the system vehicle [23].

FIGURE C.16 This is an image of a simulated scenario including cars in the opposite and the same moving direction of AV. The camera is mounted on the roof of AV [23].

© SAE International.

Computer Vision

Computer vision is the capability of computers or computing platforms to recognize, analyze, and derive conclusions from visual data. Computer vision is an essential technology in connected automated vehicles. Visual data could be 2D or 3D images from multiple cameras, point clouds from LiDARs, or video data. Computer vision tries to reconstruct the world model around the vehicle as the human eye and brain do.

Connected Intersection

Connected intersections communicate with vehicles, pedestrians, bicyclists, and back end traffic infrastructure. These intersections enable operators to change traffic strategies based on density and other localized conditions [24].

Connected intersections are often signalized and have a roadside unit (RSU) communicating with the TMS. The intersection broadcasts messages to advertise the services available in the intersection. It also broadcasts the Signal Phase and Timing (SPaT), MapData (MAP), and position correction messages [25].

A CV has an OBU that receives messages about the intersection's services. The vehicle will use the onboard sensor data and messages received from the intersection to perform connected and AV functions (**Figure C.17**).

FIGURE C.17 Zoning control [24].

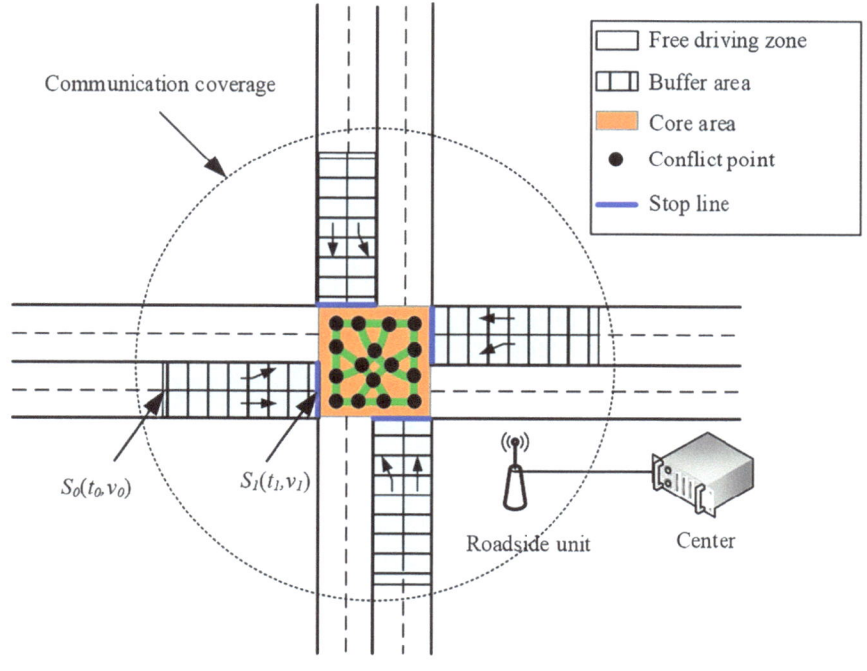

Connected Vehicle Ecosystem

The CV ecosystem refers to the integrated network of vehicles, infrastructure, communication technologies, and data services that enable vehicles to communicate with each other (V2V), with infrastructure (V2I), and with other entities (V2X). This ecosystem enhances road safety and traffic efficiency, providing services and applications to drivers and passengers.

Components:

1. **Vehicles:** Equipped with sensors, communication modules, and data processing units that enable connectivity.
2. **Infrastructure:** Includes smart traffic signals, road sensors, and communication hubs that interact with vehicles.
3. **Communication Technologies:** Protocols and networks such as 5G, DSRC, and long-term evolution (LTE) enable real-time data exchange.
4. **Data Services:** Cloud platforms and data analytics services that process and analyze data collected from vehicles and infrastructure to provide actionable insights.
5. **Applications and Services:** Safety applications (collision warnings and emergency braking), mobility services (real-time traffic information and route optimization), and infotainment (streaming services and Internet access).

Key Features:
- **Interconnectivity:** Vehicles communicate with each other and the infrastructure to share real-time information about traffic, road conditions, and potential hazards.
- **Data Sharing:** Continuous data exchange between vehicles and centralized data platforms to enhance decision-making and service provision.
- **Automation Support:** Provides a robust data and communication framework to facilitate developing and deploying autonomous driving technologies.
- **Enhanced Safety:** Improves road safety by enabling predictive and preventive measures through real-time data and communication.
- **Traffic Efficiency:** Optimizes traffic flow and reduces congestion by coordinating vehicle movements and traffic signals.

Cooperative V2C (Vehicle-to-Cloud)

Cooperative V2C is a communication system in ACVs where data are exchanged between vehicles and cloud-based services. This connectivity enhances vehicle operations, safety, and UX by leveraging cloud computing capabilities for data processing, storage, and real-time information dissemination.

Real-time traffic data from the cloud help vehicles avoid congestion and choose optimal routes. Remote diagnostics, which includes vehicle health and performance data, can be uploaded to the cloud, where it is analyzed to predict maintenance needs and prevent breakdowns. V2C supports autonomous driving by providing real-time updates and high-definition maps from the cloud, enabling safer and more efficient self-driving operations.

Cooperative V2G (Vehicle-to-Grid)

Cooperative V2G refers to a system in which EVs interact collaboratively with the power grid. This interaction involves bidirectional power flow, enabling EVs to draw energy from the grid for charging and supply stored energy back to the grid when needed. The cooperative aspect implies coordinated efforts among multiple vehicles, charging stations, and grid operators to optimize energy distribution, enhance grid stability, and improve the overall efficiency of energy use.

EVs can receive electricity from the grid and return excess stored energy back to the grid. By supplying energy during peak demand periods, EVs can help balance load and prevent grid overloads. Coordinated V2G systems optimize multiple vehicles' charging and discharging cycles to ensure efficient energy use and reduce costs. V2G facilitates the integration of renewable energy sources, such as solar and wind, by storing excess energy and supplying it when production is low. Owners of EVs can earn incentives or payments for participating in V2G programs, offsetting the cost of vehicle ownership and contributing to the economic viability of renewable energy systems.

Cooperative V2X (Vehicle-to-Everything)
Cooperative V2X refers to a comprehensive communication system that enables vehicles to interact with each other (V2V), with infrastructure (V2I), with pedestrians (V2P), and with networks (V2N). This communication paradigm aims to enhance traffic efficiency, safety, and the driving experience by sharing real-time information among all entities within the transportation ecosystem.

- V2V: This technology allows direct communication between vehicles to share data such as speed, position, and direction, which can help in collision avoidance and traffic flow optimization.
- V2I: This technology facilitates communication between vehicles and road infrastructure, such as traffic lights, signs, and toll booths, to improve traffic management and road safety.
- V2P: This technology enhances safety for pedestrians and cyclists by enabling vehicles to communicate with personal devices carried by pedestrians, alerting them to potential dangers.
- V2N: This technology connects vehicles to cellular networks and the Internet, providing access to cloud services, real-time traffic information, and other data that can enhance navigation and infotainment systems.

Cooperative Vehicle Safety
Cooperative vehicle safety refers to using advanced communication technologies and protocols that enable vehicles to share critical safety information with roadside infrastructure. This information exchange aims to enhance the overall safety of the transportation system by preventing accidents, minimizing traffic congestion, and improving emergency response times.

V2V communication allows vehicles to communicate with each other to share information such as speed, position, and direction. This allows for real-time updates and can help prevent collisions by alerting drivers or autonomous systems to potential hazards.

V2I communication allows vehicles to interact with roadside infrastructure, such as traffic lights, road signs, and sensors. This communication can help optimize traffic flow, provide real-time traffic updates, and enhance road safety by adjusting signal timings and warnings about road conditions.

V2X communication is an all-encompassing term that includes V2V and V2I, as well as communication with other entities such as pedestrians (V2P) and networks (V2N). V2X aims to create a comprehensive safety network by integrating various communication channels and traffic participants.

These systems use the data exchanged through V2V and V2I to predict and prevent potential collisions. For instance, if a vehicle ahead suddenly brakes, the following vehicle can receive an alert and take preemptive action.

Cooperative vehicle safety systems (CVSSs) can facilitate the movement of emergency vehicles by communicating with other vehicles and traffic signals to clear the path, thereby reducing response times and improving outcomes in critical situations.

Traffic management shares real-time data, and CVSSs can help manage traffic flow, reduce congestion, and improve the overall efficiency of the transportation network. This can lead to a smoother and safer driving experience for all road users.

CPS (Cyber-Physical System)

CPSs are integrated systems that combine computational algorithms and physical components, enabling them to interact with the physical world and each other through a network. These systems seamlessly integrate physical processes with computational resources and communication capabilities, forming the backbone of advanced technologies such as ACVs.

CPS tightly integrates physical components like sensors, actuators, and mechanical parts with software algorithms and digital networks. These systems operate in real time, responding to physical processes and external stimuli instantaneously to maintain system performance and safety. CPS relies on robust communication networks to facilitate data exchange between system components and external entities, ensuring coordinated operation. They often include embedded systems that control physical processes, such as vehicle engine control units (ECUs). CPS typically features closed-loop feedback systems, where physical actions influence computational processes and vice versa, allowing for adaptive and autonomous control.

For ADAS and AVs, CPS plays a crucial role by enabling the following functionalities. CPS integrates various sensors (e.g., LiDAR, radar, and cameras) to perceive the vehicle's environment, providing real-time navigation and obstacle detection data. Computational algorithms process sensor data to make decisions, such as lane keeping, ACC, and collision avoidance, ensuring safe and efficient vehicle operation. CPS facilitates communication between vehicles (V2V), infrastructure (V2I), and other entities (V2X), enhancing situational awareness and enabling coordinated actions. By combining real-time data processing and ML, CPS allows for the autonomous control of vehicles, reducing the need for human intervention and improving safety and efficiency. CPS monitors vehicle health and performance, enabling predictive maintenance and reducing the likelihood of system failures.

CPU (Central Processing Unit)

A CPU is a set of very complex electronic circuits that executes the software stack in a computer. It consists of elements such as an arithmetic and logic unit, a control unit, registers, memory, a memory management unit, and a clock. The software is connected to instructions set corresponding to the processor in the CPU and is executed.

Connected and AVs have multiple processing units that receive data from sensors through in-vehicle networks and compute actuator commands. They can also process data, images, video, and point clouds and run ML models.

Cross-Platform Integration

Cross-platform integration refers to the process of enabling different ACV (ADAS) systems and platforms to work together seamlessly (**Figure C.18**). This involves integrating various software applications, hardware components, and communication protocols to ensure that data and functionalities are shared effectively across different systems and devices through integrated vehicle health management (IVHM) [26].

FIGURE C.18 Overview of IVHM deployment process with JA6268 [26].

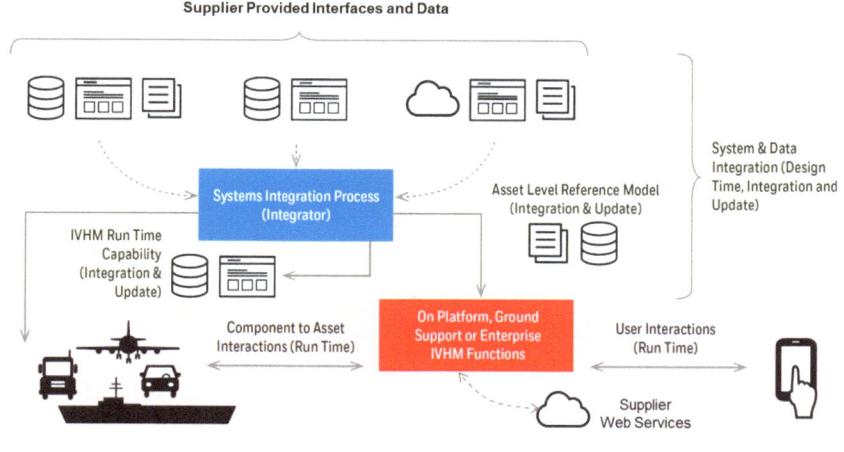

Cross-platform integration ensures that different software applications used in ADAS are compatible and can communicate with each other seamlessly. This includes operating systems, middleware, and application software.

Like software interoperability, hardware interoperability involves integrating various hardware components, such as sensors, processors, and communication modules, so they can function together efficiently. This involves standardizing interfaces.

Implementing common communication protocols allows different systems and devices to share information. Protocols like CAN, Ethernet, and wireless standards (e.g., 5G and Wi-Fi) are crucial for effective integration.

It involves facilitating data exchange between different systems and managing these data to ensure accuracy, security, and real-time availability. This includes data from V2V and V2I communications.

CSA (Curve Speed Assistance)

CSA is an ADAS feature designed to enhance vehicle safety and comfort by automatically adjusting the vehicle's speed when navigating curves (**Figure C.19**). This system leverages data from various sensors, including GPS, cameras, and radar, to assess the curvature of the road ahead and determine the optimal speed for safely and comfortably negotiating the curve.

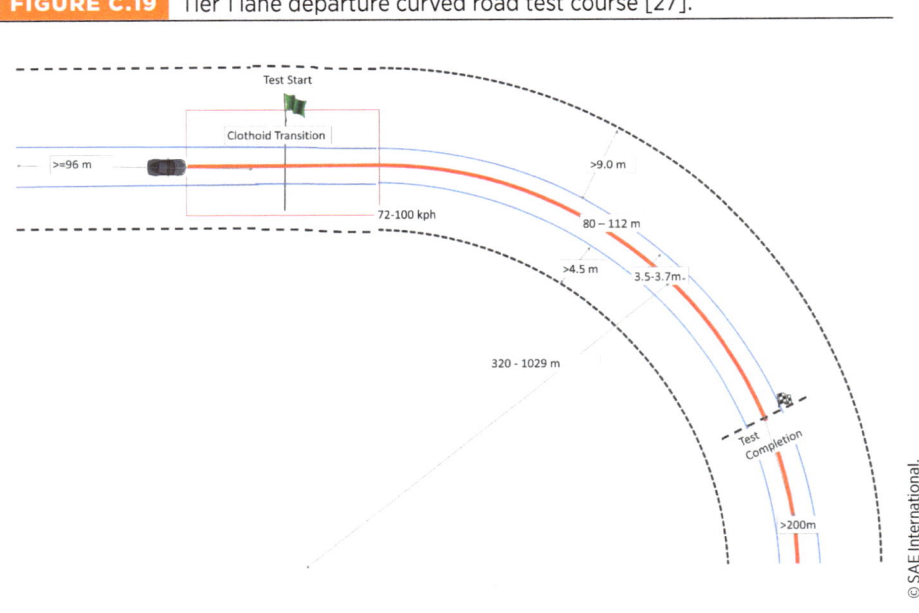

FIGURE C.19 Tier 1 lane departure curved road test course [27].

CSA is beneficial in regions with many winding roads or variable driving conditions. It is often integrated with other ADAS features like ACC and LKA to provide a comprehensive safety and driving assistance package.

CSA gathers real-time information about the vehicle's current speed, position, and trajectory using onboard sensors and GPS data through data collection. The system identifies upcoming curves by analyzing map data and road conditions through cameras and other sensors. Based on the curvature and other relevant factors like road surface conditions and traffic, CSA calculates the appropriate speed for the curve and adjusts the vehicle's speed accordingly. This is typically done through automatic throttle control and, if necessary, gentle braking.

CSW (Curve Speed Warning)

CSW is an ADAS feature designed to enhance vehicle safety by alerting drivers when their vehicle's speed is too high for an approaching curve. Utilizing a combination of GPS data, digital maps, and vehicle sensors, CSW evaluates the vehicle's current speed and the curvature of the upcoming road segment to determine if the driver needs to reduce speed to safely navigate the curve.

When the system detects that the vehicle is approaching a curve at a speed that may compromise safety, it warns the driver. The warning prompts the driver to decelerate to a safer speed. Some advanced CSW systems may integrate with the vehicle's ACC to adjust the vehicle's speed automatically if the driver does not respond in time.

CSW is commonly used in both passenger vehicles and commercial fleets to improve road safety and driver performance. It is particularly beneficial in regions with winding roads or frequent changes in road curvature.

GPS and digital maps provide accurate positioning and detailed road geometry information, including the radius and angle of upcoming curves. Vehicle sensors monitor the vehicle's speed and dynamic behavior, such as acceleration and deceleration rates. Human–machine interface (HMI) communicates warnings to the driver through visual, auditory, or haptic signals.

CTA (Cross-Traffic Alert)-Rear

CTA-rear is an ADAS feature designed to enhance vehicle safety by detecting and alerting the driver to traffic approaching from the sides, while the vehicle is in reverse or at low speeds. This system is beneficial when backing out of parking spaces or driveways, where visibility may be limited.

CTA utilizes radar sensors or cameras typically mounted on the rear corners of the vehicle to monitor the areas perpendicular to the vehicle's path. When the system detects an approaching vehicle from sides, it provides the driver with visual and/or audible alerts. Some systems may include haptic feedback, such as seat vibrations or steering wheel pulses. CTA can be integrated with other ADAS features, such as rear cross-traffic brake assist, which can automatically apply the brakes if the driver does not respond to the alerts in time.

CTAs have various use cases. One is navigating parking lots when backing out of a parking space with limited visibility due to parked vehicles or other obstructions. This includes residential driveways while reversing onto a street with oncoming traffic from both sides, in urban areas, or in environments where pedestrian and cyclist traffic can be unpredictable and not easily seen.

C-V2X (Cellular Vehicle-to-Everything)

C-V2X refers to a communication technology that enables vehicles to connect with each other, roadside infrastructure, pedestrians, and the network [28]. Utilizing cellular networks, including the newer 5G technology, C-V2X facilitates real-time, high-speed data exchange to improve road safety, traffic efficiency, and support for autonomous driving features.

CVC (Cooperative Vehicle Communication)

CVC refers to the technology that enables vehicles to communicate with each other (cooperative driving automation [CDA]) and with infrastructure elements to improve road safety, traffic efficiency, and the overall driving experience (Figure C.20). This communication is facilitated through various wireless technologies and protocols, allowing vehicles to share critical information in real time [29].

FIGURE C.20 Relation of CDA and transportation system agents' function [29].

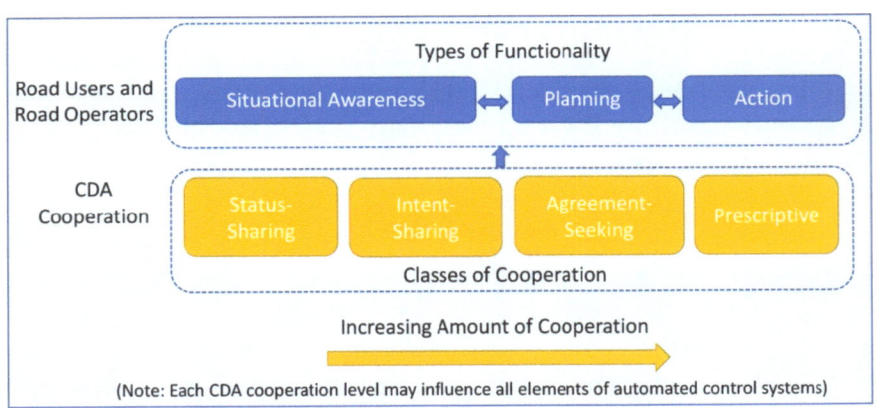

V2V communication enables vehicles to share data, such as speed, location, and direction, to prevent collisions and improve traffic flow. V2I communication enables interaction between vehicles and road infrastructure, such as traffic lights and road signs, thereby optimizing traffic management and providing drivers with real-time information about road conditions. V2X communication is an encompassing term that encompasses V2V, V2I, and communication with other entities, such as pedestrians and cyclists, thereby enhancing overall safety and mobility.

DSRCs are a wireless communication protocol designed for automotive use. It provides low-latency data exchange [9]. C-V2X uses cellular networks to enable communication between vehicles and infrastructure, offering a wider range and higher capacity. 5G networks enhance the capabilities of V2X communication by providing faster speeds, lower latency, and greater connectivity.

CV (Connected Vehicle) Platform

CV platforms are frameworks in which vehicle manufacturers integrate various applications to enhance the UX of CVs. They include an onboard computing system, software solutions, feature-related apps, cloud support, and data analytics.

CVI (Connected Vehicle Infrastructure)

CVI refers to the physical and digital systems that enable communication between vehicles and external entities such as other vehicles, roadways, traffic signals, and cloud services (**Figure C.21**). This infrastructure is crucial for implementing CV technologies and facilitating the exchange of information to enhance safety, mobility, and efficiency in transportation networks.

FIGURE C.21 Geographical illustration of probe data configuration reporting [30].

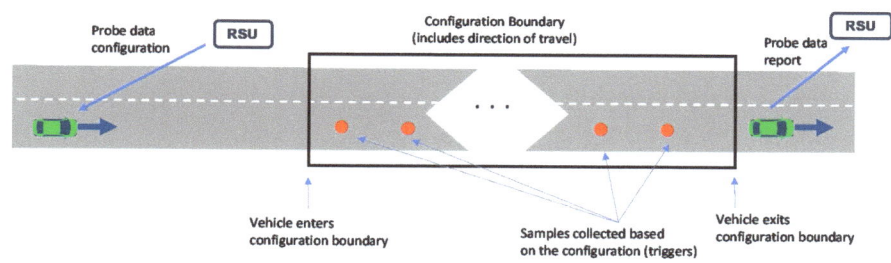

Components [30]:

- **RSUs:** Devices installed along roadways that communicate with vehicles and TMS.
- **OBUs:** Devices within vehicles enable communication with RSUs and other vehicles.

- **Traffic Management Centers (TMCs):** These facilities monitor and manage traffic flow, leveraging data from CVs and infrastructure.
- **Communication Networks:** Systems such as DSRC, cellular networks (e.g., fourth generation (4G) and 5G), and the IoT that facilitate data transmission.

Functions and Applications:
- **V2I:** Communication between vehicles and road infrastructure, enabling real-time traffic signal updates, toll payments, and road hazard warnings.
- **V2V:** Direct communication between vehicles to share information about speed, position, and intent, reducing collision risks and enhancing cooperative driving.
- **V2X:** Broader communication encompassing V2I, V2V, and communication with pedestrians, cyclists, and cloud services, enabling a comprehensive networked transportation system.

CV (Connected Vehicle)

CVs are equipped with Internet access and various sensors that allow them to communicate with each other (V2V), with infrastructure (V2I), and with other external networks and systems. This connectivity enhances the functionality and safety of ADAS and paves the way for fully autonomous driving.

1. Traffic Management: Vehicles can share information about traffic conditions, road hazards, and optimal routes, helping to reduce congestion and travel time.
2. Collision Avoidance: V2V communication allows vehicles to share their speed and position, enabling systems to predict and prevent potential collisions thereby increasing road transportation safety.
3. Smart Infrastructure Interaction: V2I communication supports interactions with traffic lights, road signs, and other infrastructure, optimizing traffic signal timing and reducing stops.
4. Remote Diagnostics and Updates: CVs can receive OTA updates and send diagnostic information to manufacturers, ensuring the latest software improvements and maintenance alerts.

Technologies Involved:
- **DSRC:** A technology enabling direct V2V and V2I communication within a short range.
- **Cellular Networks (C-V2X):** Using existing cellular networks to facilitate wider-range communication between vehicles and infrastructure.
- **Sensors and Internet of Things (IoT) Devices:** Includes GPS, cameras, LiDAR, and radar to gather and transmit data.

CVS (Cooperative Vehicle Security)

CVS refers to the integrated system of technologies and protocols designed to protect ACVs from cyberthreats while ensuring secure communication and coordination between vehicles and infrastructure. CVS encompasses a range of security measures, including encryption, authentication, data integrity, and anomaly detection, to prevent unauthorized access, data breaches, and malicious attacks on vehicular networks.

Key components of the system include encryption that ensures that data transmitted between vehicles and infrastructure are secure and inaccessible to unauthorized parties. Encryption algorithms protect the confidentiality and integrity of information such as vehicle positions, speed, and other critical data. Authentication mechanisms verify the identity of vehicles and infrastructure components to ensure that only trusted entities can participate in the communication network. Authentication mechanisms prevent unauthorized devices from gaining access to the system. Data integrity mechanisms maintain the accuracy and consistency of data as they are transmitted and received. Data integrity checks detect and mitigate any tampering or corruption of information during transmission. This includes anomaly detection, which identifies unusual patterns or behaviors in the network that could indicate potential security threats. Anomaly detection systems monitor the communication network in real time to detect and respond to potential cyber-attacks.

CVSS (Cooperative Vehicle Safety System)

CVSSs are advanced technologies designed to enhance road safety by enabling vehicles to communicate with each other and the infrastructure around them. These systems leverage V2X communication, which includes V2V, V2I, V2P, and V2N interactions.

V2V communication enables direct communication between vehicles to share information about their speed, position, and heading. This helps in collision avoidance by providing real-time alerts about potential hazards, such as sudden braking or lane changes.

V2I communication facilitates interaction between vehicles and road infrastructure, such as traffic signals, road signs, and highway management systems. This optimizes and enhances traffic flow and safety by providing data on road conditions, traffic light status, and upcoming construction zones. Examples for this include dynamic speed limit displays that update speed limits based on current road conditions and in-vehicle red light violation warning systems.

V2P communication connects vehicles with VRUs, including pedestrians and cyclists, through mobile devices or wearable technology. This improves safety by alerting drivers to the presence of pedestrians, especially in low-visibility conditions or at busy intersections.

V2N communication integrates vehicles with broader communication networks to access real-time information from cloud services and other data sources. It supports various safety and efficiency applications, such as dynamic route guidance, emergency response coordination, and OTA updates.

The applications of V2X include intersection collision warning, which alerts drivers or autonomous systems of potential collisions at intersections by sharing vehicle trajectories and speeds. It also includes emergency vehicle alert, which notifies surrounding vehicles of approaching emergency vehicles, allowing them to yield and clear the way. Blind spot/lane change warning provides alerts about vehicles in blind spots or approaching rapidly from behind, assisting in safer lane changes. Lastly, a road hazard warning communicates real-time information about road hazards, such as debris, ice, or accidents, to approaching vehicles.

CVT (Cooperative Vehicle Testing)

CVT is the systematic evaluation and validation process where multiple ACVs are tested simultaneously under coordinated scenarios. This testing method ensures vehicles' seamless interaction, safety, and efficiency in a shared environment.

Fundamental to the system's effectiveness is coordination and communication, starting with vehicle communication with each other and with the infrastructure using V2V and V2I communication technologies. This ensures synchronized actions and decisions. For example, CVT involves replicating real-world driving conditions, including traffic, weather, and road variations, to test the vehicles' responses in a controlled yet realistic environment.

The primary goal is to assess and enhance ACVs' safety protocols and operational efficiency. This includes evaluating collision avoidance systems, ACC, and coordinated maneuvers like lane changes and merging. Data sharing and analysis of extensive data collected from all participating vehicles are then analyzed to identify performance issues, optimize algorithms, and improve vehicle cooperation strategies. CVT helps ensure that ACVs meet the required safety standards and regulatory guidelines before they are deployed on public roads.

Cybersecurity Auditing

Cybersecurity auditing in the context of ADAS and CVs involves systematically examining and evaluating the security measures implemented within the vehicle's electronic systems and communication networks. This process aims to identify vulnerabilities, ensure compliance with security standards, and safeguard against cyberthreats that could compromise the vehicle's functionality and data integrity.

With modern vehicles' increasing complexity and connectivity, cybersecurity auditing is crucial to protect against cyber-attacks that could compromise safety, privacy, and operational integrity. Effective cybersecurity audits help maintain the trust of consumers and stakeholders, ensuring that vehicles operate securely in a connected environment (**Figure C.22**).

FIGURE C.22 Relation of CDA and transportation system agents' function [31].

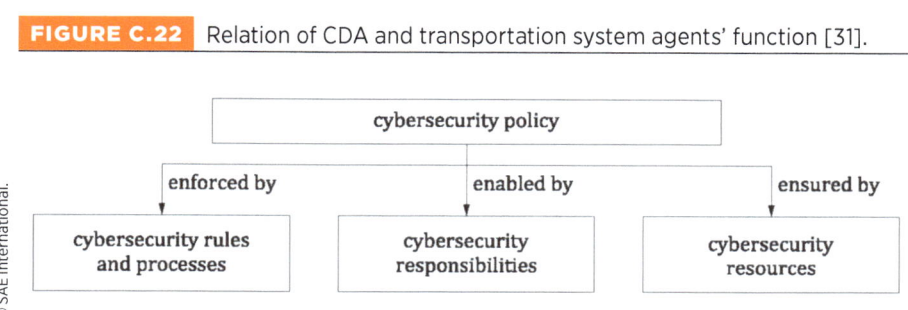

Compliance verification ensures that the vehicle's cybersecurity measures adhere to industry standards and regulations, such as the International Organization for Standardization (ISO)/SAE 21434 [31] and the United Nations Economic Commission for Europe (UNECE) WP.29. It also verifies that security policies and procedures are properly implemented and maintained. Threat analysis evaluates potential cyberthreats and their impact on the vehicle's systems. Additionally, threat models are developed to understand how different attack vectors could exploit vulnerabilities. Incident response planning prepares for potential cyber-incidents by developing and testing response plans. This ensures that procedures are in place to detect, respond to, and recover from cybersecurity breaches. Lastly, continuous monitoring implements ongoing monitoring of the vehicle's systems to detect and respond to emerging threats in real time. It uses intrusion detection systems (IDSs) and other monitoring tools to maintain security vigilance [32].

Cybersecurity for Vehicles

Cybersecurity for vehicles refers to the measures, protocols, and practices employed to protect the electronic systems, communication networks, and data of connected and AVs (CAVs) from malicious attacks, unauthorized access, and potential threats [31].

As vehicles become increasingly connected and autonomous, they are more susceptible to cyberthreats (**Figure C.23**). Ensuring robust cybersecurity for vehicles is crucial to protect the safety, privacy, and reliability of the vehicle systems, passengers, and broader transportation infrastructure [31]. Effective cybersecurity measures help prevent potential risks such as remote hijacking, data theft, and disruption of vehicle operations, thereby maintaining public trust and the integrity of the automotive ecosystem [33].

FIGURE C.23 Example of an attack path derived by attack tree analysis [31].

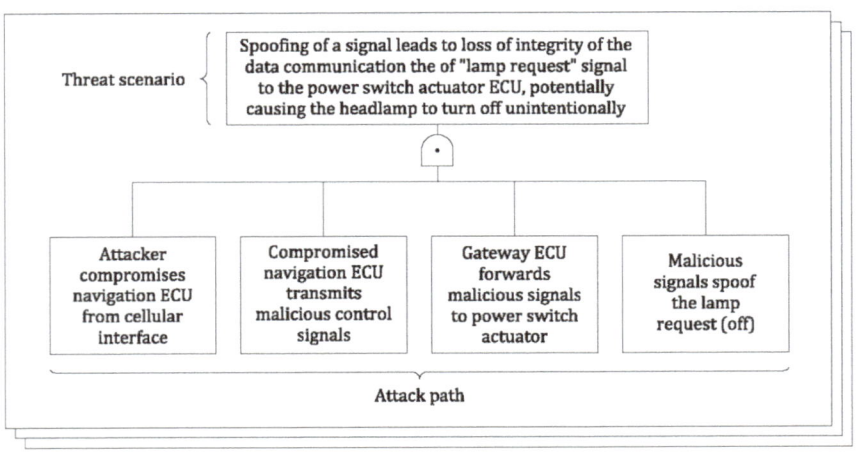

Secure communication protocols ensure that data exchanged between vehicles, infrastructure, and other devices are encrypted and authenticated to prevent interception and tampering. IDS monitors vehicle systems for unusual activities or anomalies that could indicate a security breach or malicious activity. Firewall and gateway protection use firewalls and secure gateways to segregate critical vehicle functions from noncritical ones, limiting the attack surface.

Software and firmware updates implement secure OTA update mechanisms to ensure that vehicle software and firmware can be updated regularly and securely to patch vulnerabilities. Access control enforces strict access control policies, ensuring that only authorized individuals or systems can access sensitive vehicle functions and data. Encryption and data protection ensure sensitive vehicle data are protected during transmission against data breaches and unauthorized access. Threat intelligence and response continuously gather and analyze threat intelligence, to avoid potential cyberthreats and establish a robust incident response plan to mitigate and recover from attacks.

Cyclist Detection

Cyclist detection is an ADAS technology designed to identify and track cyclists on the road. This system utilizes a combination of sensors, including cameras, radar, and LiDAR, to detect and predict cyclists' movements. The goal is to enhance cyclists' safety by providing timely warnings to the driver or autonomously taking corrective actions to avoid collisions [34].

High-resolution cameras capture real-time images and videos of the surrounding environment. Computer vision algorithms process these images to identify cyclists based on their shape, movement, and distinguishing features (Figure C.24).

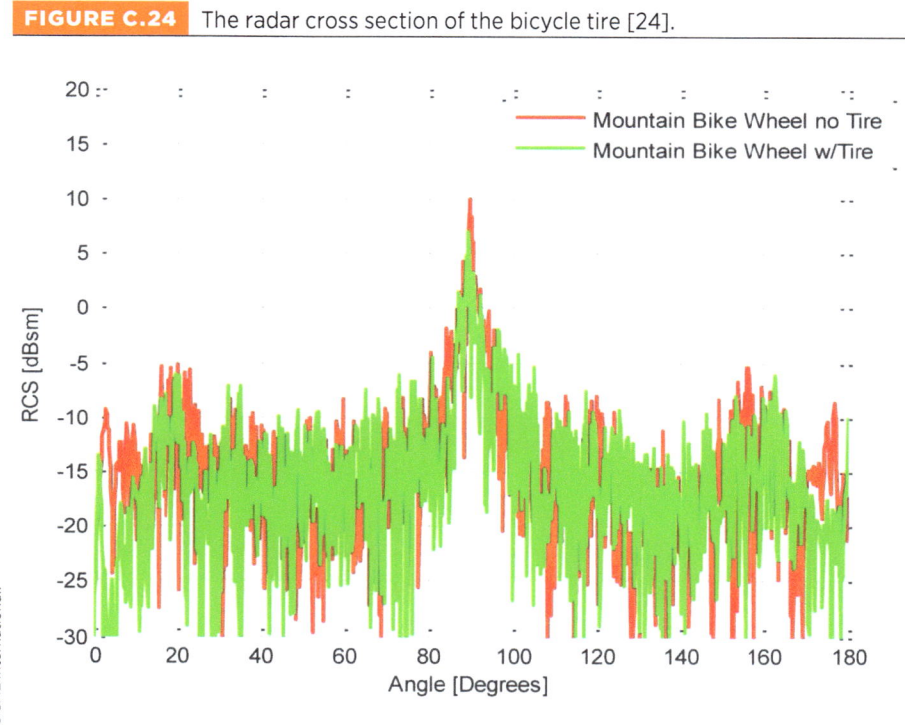

FIGURE C.24 The radar cross section of the bicycle tire [24].

Radar sensors emit radio waves that bounce off objects and return to the sensor. The system measures the time it takes for the waves to return, allowing it to detect the distance, speed, and direction of moving cyclists. LiDAR uses laser pulses to create detailed 3D maps of the environment. This helps accurately identify cyclists' size, shape, and position in relation to the vehicle.

Advanced ML and AI algorithms process the sensor data to accurately detect cyclists, predict their movements, and assess potential collision risks. The system continuously monitors the vehicle's surroundings to detect cyclists and track their movements, predicts potential collision scenarios, and warns the driver or takes autonomous actions, such as brakes or steering adjustments, to avoid accidents. The system culminates features like pedestrian detection, lane-keeping assist, and AEB to enhance overall road safety.

REFERENCES

1. Hakim, M. and Kreuer, D.D., "Car-to-X Communication Flexible Integration of Antenna Systems," *ATZelektronik Worldwide* 9, no. 1 (2014): 22-27.

2. Society of Automotive Engineers, "J3298 Artificial Intelligence Data for Ground Vehicle Applications," SAE Publishing, Warrendale, PA, 2024.

3. Society of Automotive Engineers, "J2945/6 Performance Requirements for Cooperative (CACC) and Platooning," SAE Publishing, Warrendale, PA, 2023.

4. Kim, S., Kim, J., Nugraha, D., Wan, V. et al., "Automotive ADAS Camera System Configuration Using Multi-Core Microcontroller," SAE Technical Paper 2015-01-0023 (2015), doi:https://doi.org/10.4271/2015-01-0023.

5. Merschak, S., Hehenberger, P., Schmidt, S., and Kirchberger, R., "Considerations of Life Cycle Assessment and the Estimate of Carbon Footprint of Powertrains," SAE Technical Paper 2020-32-2314 (2020), doi:https://doi.org/10.4271/2020-32-2314.

6. Society of Automotive Engineers, "J3251 Cooperative Driving Automation (CDA) Feature: Perception Status Sharing for Occluded Pedestrian Collision Avoidance," SAE Publishing, Warrendale, PA, 2023.

7. Quigley, J.M. and Starkey, F., *SAE International's Dictionary of Electric Vehicles* (Warrendale, PA: SAE International, 2024).

8. Society of Automotive Engineers, "J3282 Cooperative Infrastructure CDA Feature: Cooperative Permissive Left Turn across Opposing Traffic with Infrastructure Guidance," SAE Publishing, Warrendale, PA, 2024.

9. Society of Automotive Engineers, "J2945 Dedicated Short Range Communication (DSRC) Systems Engineering Process Guidance for SAE J2945/X Documents and Common Design Concepts," SAE Publishing, Warrendale, PA, 2017.

10. Meroux, D., Telenko, C., Jiang, Z., and Fu, Y., "Towards Design of Sustainable Smart Mobility Services through a Cloud Platform," *SAE Int. J. Adv. & Curr. Prac. in Mobility* 1, no. 3 (2020): 1717-1727, doi:https://doi.org/10.4271/2020-01-1048.

11. Chominsky, W. and Quigley, J.M., *SAE International's Dictionary of Commercial Vehicles* (Warrendale, PA: SAE International Publishing, 2024).

12. Thomas, S., Dubey, A., Viassolo, D., and Zanette, M., "Digital Fleet Management: A Scalable Cloud Framework Based on Data-Driven Prediction Models," in *Annual Conference of the PHM Society*, Houston, TX, 2020.

13. Bie, W., Li, K., Zhang, R., Huang, Y. et al., "Research on the Development Trend of Brain Controlled Cars," SAE Technical Paper 2018-01-1587 (2018), doi:https://doi.org/10.4271/2018-01-1587.

14. Ran, B., Cheng, Y., Li, S., Li, H. et al., "Classification of Roadway Infrastructure and Collaborative Automated Driving System," *SAE Int. J. CAV* 6, no. 4 (2023): 387-395, doi:https://doi.org/10.4271/12-06-04-0026.

15. Azimi, S., Bhatia, G., Rajkumar, R., and Mudalige, P., "Vehicular Networks for Collision Avoidance at Intersections," *SAE Int. J. Passeng. Cars - Mech. Syst.* 4, no. 1 (2011): 406-416, doi:https://doi.org/10.4271/2011-01-0573.

16. Society of Automotive Engineers, "J128 Occupant Restraint System Evaluation - Passenger Cars and Light-Duty Trucks," SAE Publishing, Warrendale, PA, 2011.

17. Nusholtz, G., Hsu, T., Gracián, M., Prado, J. et al., "Forward Collision Warning Timing in Near Term Applications," *SAE Int. J. Trans. Safety* 1, no. 2 (2013): 467-479, doi:https://doi.org/10.4271/2013-01-0727.

18. Harrington, S. and Handzic, D., "An Evaluation of the Forward Collision Warning System and Evasive Steering Maneuvers Using a 2016 Volvo XC90," SAE Technical Paper 2023-01-0627 (2023), doi:https://doi.org/10.4271/2023-01-0627.

19. Scanlon, J., Page, K., Sherony, R., and Gabler, H., "Using Event Data Recorders from Real-World Crashes to Investigate the Earliest Detection Opportunity for an Intersection Advanced Driver Assistance System," SAE Technical Paper 2016-01-1457 (2016), doi:https://doi.org/10.4271/2016-01-1457.

20. Raphael, E., Kiefer, R., Reisman, P., and Hayon, G., "Development of a Camera-Based Forward Collision Alert System," *SAE Int. J. Passeng. Cars - Mech. Syst.* 4, no. 1 (2011): 467-478, doi:https://doi.org/10.4271/2011-01-0579.

21. Ejima, S., Goto, T., Zhang, P., Cunningham, K. et al., "Injury Severity Prediction Algorithm Based on Select Vehicle Category for Advanced Automatic Collision Notification," SAE Technical Paper 2022-01-0834 (2022), doi:https://doi.org/10.4271/2022-01-0834.

22. Society of Automotive Engineers, "J3029 Forward Collision Warning and Automatic Emergency Braking Test Procedure and Minimum Performance Requirements - Truck and Bus," SAE Publishing, Warrendale, PA, 2023.

23. Fekr, P., Abedi, V., Dargahi, J., and Zadeh, M., "A Forward Collision Warning System Using Deep Reinforcement Learning," SAE Technical Paper 2020-01-0138 (2020), doi:https://doi.org/10.4271/2020-01-0138.

24. Chen, Z., Lin, P., He, H., and Zhuo, F., "Intersection Coordination Control Strategy for Intelligent Connected Vehicle," SAE Technical Paper 2020-01-5226 (2020), doi:https://doi.org/10.4271/2020-01-5226.

25. Society of Automotive Engineers, "J2735 V2X Communications Message Set Dictionary," SAE Publishing, Warrendale, PA, 2023.

26. Society of Automotive Engineers, "JA6268 Design and Run-Time Information Exchange for Health-Ready Components," SAE Publishing, Warrendale, PA, 2023.

27. Society of Automotive Engineers, "J3240 Passenger Vehicle Lane Departure Warning, Lane Keeping Assistance, and Lane Centering Assistance Systems Test Procedure," SAE Publishing, Warrendale, PA, 2023.

28. Elangovan, V., Xiang, W., and Liu, S., "A Multiagency Long Short-Term Model Beamforming Prediction Model for Cellular Vehicle to Everything," *SAE Int. J. CAV* 6, no. 4 (2023): 459-472, doi:https://doi.org/10.4271/12-06-04-0030.

29. Society of Automotive Engineers, "J3216 Taxonomy and Definitions for Terms Related to Cooperative Driving Automation for On-Road Motor Vehicles," SAE Publishing, Warrendale, PA, 2021.

30. Society of Automotive Engineers, "J2945/C Requirements for Probe Data Collection Applications," SAE Publishing, Warrendale, PA, 2022.

31. Society of Automotive Engineers, "ISO/SAE 21434 Road Vehicles - Cybersecurity Engineering," SAE Publishing, Warrendale, PA, 2021.

32. Code Intelligence, "How Fuzzing Complements Static Code Analysis (SAST) When Testing Automotive Software," White Paper, accessed December 13, 2024, https://7466322.fs1.hubspotusercontent-na1.net/hubfs/7466322/White%20Paper%20Fuzz%20Testing%20%26%20ISO%2021434.pdf.

33. Society of Automotive Engineers, "J3061 Cybersecurity Guidebook for Cyber-Physical Vehicle Systems," SAE Publishing, Warrendale, PA, 2021.

34. Yi, Q., Chien, S., Brink, J., Niu, W. et al., "Development of Bicycle Surrogate for Bicyclist Pre-Collision System Evaluation," SAE Technical Paper 2016-01-1447 (2016), doi:https://doi.org/10.4271/2016-01-1447.

Data Analytics

Data analytics in ADAS and CVs refers to the process of collecting, processing, and analyzing large volumes of data generated by vehicle sensors, communication systems, and external sources [1]. These data enable real-time decision-making, improve vehicle performance, enhance safety features, and support predictive maintenance [2].

ADAS and CVs are equipped with LiDAR, radar, cameras, and ultrasonic sensors, which continuously collect data about the vehicle's environment and operations (**Figure D.1**). These data are crucial for obstacle detection, lane keeping, ACC, and automated parking. Additionally, CVs receive and transmit data through V2V and V2I communications.

Data analytics helps detect and analyze potential hazards, enabling preventive measures or automatic corrections to avoid accidents. It can identify patterns that precede accidents and adjust ADAS responses accordingly. Analyzing data from various sensors and systems enables the optimization of fuel efficiency, battery life in EVs, and overall vehicle performance. By analyzing operational data, potential failures can be identified and predicted before they occur, thereby reducing downtime and maintenance costs.

Data analytics can enhance the driving experience by learning driver preferences and automatically adjusting vehicle settings, such as climate control, seat adjustments, and multimedia settings. By integrating V2I communications, data analytics can improve traffic flow and reduce congestion by adjusting traffic signals based on real-time traffic data.

Data analytics in ADAS and CVs face several challenges, including data privacy concerns, the need for robust cybersecurity measures to protect against hacking, and the requirement for high data integrity and reliability. Additionally, processing vast amounts of data requires significant computational power, data transmission speeds for safety-critical applications, and sophisticated algorithms.

FIGURE D.1 The learning agent ADAS is auxiliary in delivering better data allocation and proposing information on it [3].

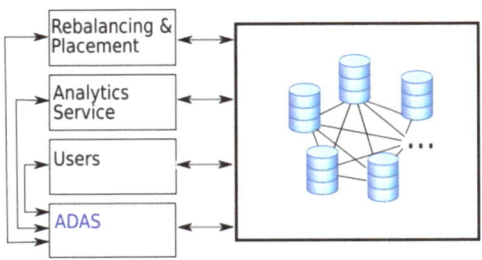

https://doi.org/10.1051/epjconf/201921406012.

Data Fusion

Data fusion in the context of ADAS and CVs refers to integrating and analyzing data from multiple sources to improve vehicle decision-making, performance, and safety [4]. This includes sensors such as radar, LiDAR, cameras, ultrasonic, GPS, IMU, and V2X communications [5].

The primary goal of data fusion is to create a comprehensive and accurate representation of the vehicle's surroundings and its status within those surroundings. By synthesizing data from diverse sources, data fusion enhances the reliability, redundancy, and precision of the information available to the vehicle's control systems.

Multiple sensors and communication systems gather data about the vehicle's external and internal environment. Raw data are filtered and normalized to ensure compatibility across different data types and formats. Data from various sources are combined using algorithms that manage discrepancies, often employing techniques from AI and ML. The integrated data are correlated to detect patterns, identify objects, and assess situations, supporting the vehicle's ADAS's real-time decisions. The insights gained from data fusion are used to execute vehicle controls, such as steering, braking, and acceleration, and provide drivers with warnings or alerts.

Data Labeling

Data labeling involves tagging raw data with meaningful and informative labels to make it usable for ML algorithms. It involves annotating various types of sensor data, including images, video, LiDAR, and radar data, to train and validate algorithms for perception, prediction, and decision-making.

Data labeling is crucial for developing and refining the models that enable the vehicle to perceive and understand its environment. Labels may include pedestrians, cyclists, vehicles, road signs, lane markings, traffic lights, and other relevant features. Sensor data are collected from real-world driving scenarios or simulated environments. Human annotators or automated tools label the data with predefined categories and attributes. The labeled data undergo a review process to ensure accuracy and consistency. The annotated data are used for training ML models for object detection, semantic segmentation, and behavior prediction tasks. Stringent quality measures are needed to ensure accurate labeling. Data management processes are critical to ensure the safety, accuracy, and performance of ML models.

The labeled data are used for training, validation, and testing of ML model. These labeled data when used to train the ML models are referred to as training dataset. Then, the trained model is validated using validation dataset and the model parameters are tuned based on the model outputs. Finally, the model is tested using the test dataset for accuracy and performance.

Data Security

Data security refers to the protective measures and protocols used to safeguard digital information from unauthorized access, corruption, theft, or damage. It is crucial for maintaining the integrity, confidentiality, and availability of data generated, transmitted, and stored by the vehicle's systems.

Data security encompasses many practices and technologies to protect the vehicle's data throughout its lifecycle. This includes securing communication between the vehicle and external entities, such as cloud servers, other vehicles, and infrastructure components as well as securing the data transmitted between ECUs within the vehicle network. External devices connected to vehicles such as diagnostics tools and smartphones are potential attack vectors.

It is essential to ensure that data are encrypted in transit and at rest to prevent unauthorized access. The implementation of robust authentication mechanisms is needed to verify the identities of users and devices accessing the vehicle's data. Access management protects sensitive data based on predefined policies and user roles. Checksums, hashes, and other methods are used to ensure that data have not been tampered. Establishing protocols for detecting, responding to, and recovering from security breaches is key for data security.

Data security is crucial for ensuring the safe and reliable operation of ACVs. It protects against various threats, such as hackers gaining control of the vehicle or accessing sensitive information. Another example is safeguarding personal and operational data from being exposed or stolen. Data collected from the vehicle must also comply with privacy regulations and user expectations.

DbW (Drive-by-Wire)

DbW refers to replacing traditional mechanical and hydraulic control systems in vehicles with electronic controls (**Figure D.2**). In a DbW system, electronic actuators and sensors replace the conventional mechanical linkages between the driver and the vehicle's control systems, such as steering, braking, and throttle [6].

FIGURE D.2 Driver's architecture diagram [6].

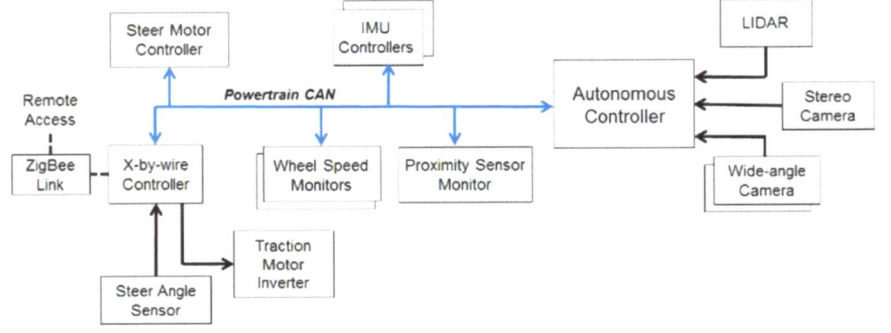

DbW technology is fundamental in ADAS and AVs. It facilitates more precise and responsive control, enabling advanced features like ACC, automated parking, and complete autonomy.

ETC replaces the mechanical throttle linkage with sensors and actuators, providing better engine response and fuel efficiency. In electronic brake control (EBC), electronic signals control braking force, allowing for features like ABS and ESC. In a full brake-by-wire (BbW) system, the hydraulic elements of the braking systems are replaced using electric motors and the motors drive the calipers to provide brake force. Steer-by-wire (SbW) eliminates the mechanical steering column, using electronic signals to control steering actuators for improved precision and integration with ADAS functions. Shift-by-wire replaces the mechanical gear shifter with electronic controls, enabling smoother gear changes and integration with ADS.

DDT (Dynamic Driving Task)

The DDT encompasses all real-time operational and tactical functions required to operate a vehicle in on-road traffic, except the strategic functions like determining destinations and waypoints. DDT includes steering, braking, accelerating, monitoring the vehicle and roadway, and responding to events (**Figure D.3**).

FIGURE D.3 View of a driving task showing DDT [7].

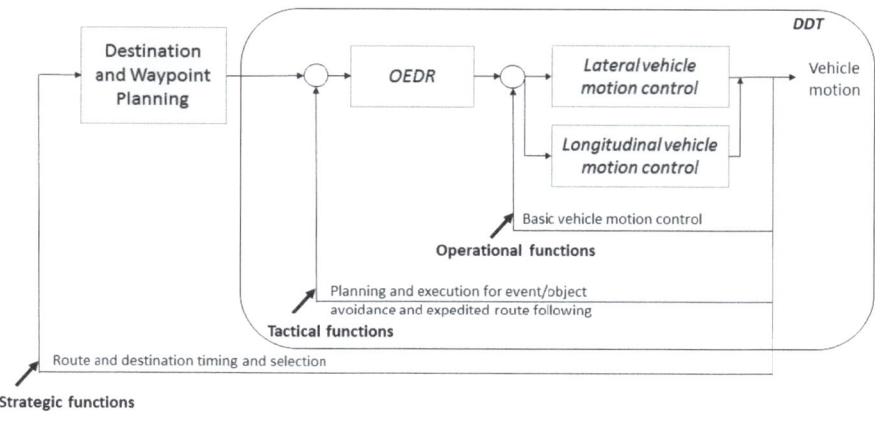

In vehicles equipped with ADS, the system may perform the entire DDT without human intervention, depending on the level of automation. The DDT for an ADS involves providing longitudinal and lateral control commands. ADAS shares the DDT with the human driver in vehicles with ADAS features, such as ACC or LKA.

Deep Learning
See **Learning**.

Digital Infrastructure
Digital infrastructure refers to the foundational digital technologies and frameworks that enable ACVs' operation, communication, and management. This includes hardware, software, networks, data centers, and communication systems essential for V2X connectivity and intelligent transportation systems.

Digital infrastructure supports various ADAS functionalities by providing a robust and interconnected environment for data exchange, real-time processing, and decision-making. It facilitates seamless communication between vehicles (V2V), vehicles and infrastructure (V2I), vehicles and pedestrians (V2P), and vehicles and networks (V2N).

High-speed, low-latency networks such as 5G and DSRC enable rapid data transfer and real-time communication between vehicles and infrastructure. Cloud computing provides scalable storage and processing power for large volumes of data generated by CVs, supporting data analysis, ML, and OTA updates. Edge computing brings data processing closer to the vehicle, reducing latency and improving response times for critical ADAS functions. Data centers and centralized facilities store, manage, and process data from CVs and infrastructure. IoT devices are sensors, cameras, and other devices embedded in infrastructure (e.g., traffic lights and road signs) that collect and transmit data to CVs.

Digital Twin

A digital twin is a virtual representation of a physical object, system, or process that mirrors its real-world counterpart in real time [8]. A digital twin involves creating a dynamic software model that simulates the behavior and performance of a vehicle, its components, or its environment (**Figure D.4**).

FIGURE D.4 Altair's digital twin platform [8].

© SAE International.

Digital twins are used to simulate and analyze vehicle performance, test new features, and optimize operations without the need for physical prototypes. They integrate real-time data from the vehicle and its environment and provide a platform for continuous monitoring and improvement.

Process:

1. Data Collection: The physical vehicle and its surroundings collect real-time data from sensors, IoT devices, and other sources.
2. Modeling: The data are used to create a detailed digital model that reflects the vehicle's current state and behavior.
3. Simulation: Various scenarios and conditions are simulated within the digital twin to predict outcomes, identify potential issues, and test new functionalities.
4. Analysis and Optimization: The simulation results are analyzed to make informed decisions on improving vehicle performance, safety, and efficiency.
5. Feedback Loop: Insights gained from the digital twin are fed back into the physical vehicle, enabling real-time adjustments and updates.

DMS (Driver Monitoring System)

A DMS is an ADAS that uses sensors and cameras to monitor the driver's state and behavior to ensure they are attentive and capable of controlling the vehicle. This technology is critical for enhancing vehicle safety by identifying driver fatigue, distraction, or impairment and taking appropriate actions to alert the driver or intervene if necessary (**Figures D.5** and **D.6**).

FIGURE D.5 Basic block diagram of emotion recognition system [9].

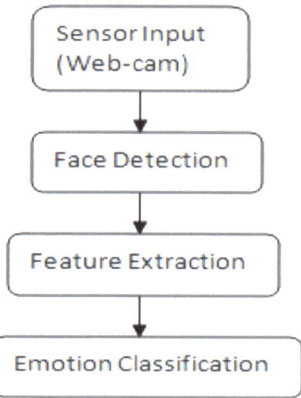

- Attention Monitoring: This method detects signs of driver distraction or drowsiness by analyzing eye movement, blink rate, and gaze direction.
- Behavior Analysis: Assesses whether the driver's actions reflect normal driving behavior or suggest impairment.
- Alert System: This system issues auditory, visual, or haptic alerts to refocus the driver's attention on the road.
- Vehicle Control Handover: Initiates protocols to safely transfer control from autonomous systems to the driver or vice versa.

FIGURE D.6 Real-time emotion detection using facial expressions [9].

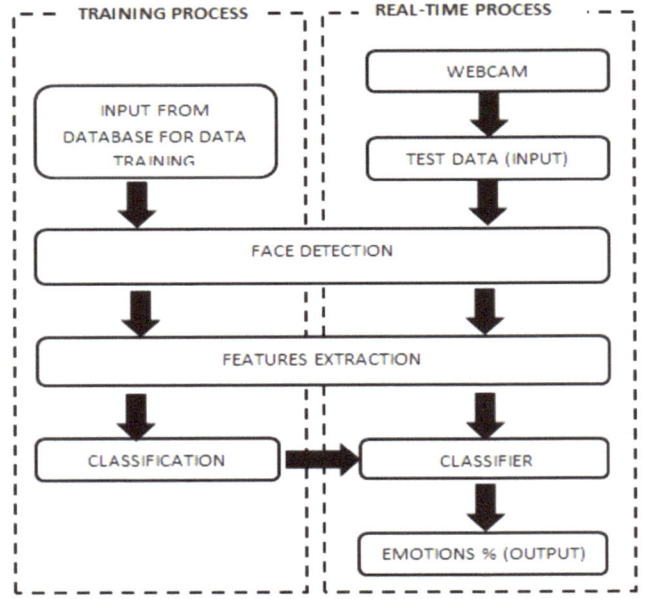

Driver Alertness Monitoring

When vehicles are driven manually, the driver performs longitudinal and lateral control by accelerating, decelerating, and steering. This is termed a DDT. The driver also actively monitors the surroundings and responds to the immediate driving situation using propulsion, steering, and braking. In other words, the driver is highly involved in the DDT. This is termed object and event detection and response (OEDR). In SAE Level 2 and Level 3 vehicles, DDT and OEDR transition between the driver and autonomy system [10].

While driving in autonomous mode, where the driver is expected to take over control when requested by the system, the driver must be engaged, even when the driver is not actively driving the vehicle, for a safe handover. Driver alertness monitoring function checks the driver engagement while the vehicle is driving in autonomous mode [11]. Based on how engaged the driver is, provide audible, visual, or haptic alerts to the driver (**Figure D.7**).

FIGURE D.7 Endogenous and exogenous factors that impact driver alertness [11].

Driver Assistance Technologies

Driver assistance technologies refer to a suite of advanced features and systems integrated into modern vehicles to assist drivers in various aspects of driving, enhancing safety, convenience, and comfort. These technologies leverage sensors, cameras, radar, and AI to provide real-time information, warnings, and automated interventions to help drivers navigate traffic, avoid collisions, and mitigate risks on the road.

- ACC: ACC adjusts the vehicle's speed to maintain a safe following distance from the vehicle ahead, automatically accelerating or decelerating as traffic conditions change.
- Lane Departure Warning (LDW) and LKA: LDW alerts drivers when the vehicle drifts out of its lane without signaling, while LKA actively steers the vehicle back into its lane to prevent unintentional lane departures [10].
- AEB: AEB detects imminent collisions with vehicles, pedestrians, or obstacles and applies the brakes automatically to mitigate or avoid accidents.
- BSM and Rear Cross Traffic Alert (RCTA): BSM warns drivers of vehicles in their blind spots, while RCTA alerts them to approaching vehicles when reversing out of parking spaces or driveways.

- Parking Assistance Systems: These systems assist drivers in parking maneuvers, including automatic parallel parking, perpendicular parking, angular parking, and parking space detection.
- Traffic Sign Recognition (TSR): TSR identifies and displays traffic signs, including speed limits, stop signs, and lane restrictions, to help drivers stay informed and compliant with road regulations.
- FCW: FCW alerts drivers to potential frontal collisions with vehicles or obstacles, providing sufficient time to react and avoid accidents.

Driver Behavior Analysis

Driver behavior analysis is a data-driven process that involves monitoring, assessing, and analyzing vehicle operators' behavior to gain insights into their driving habits, patterns, and tendencies. This analysis utilizes data collected from various sensors and onboard systems within CVs to evaluate acceleration, braking, cornering, speed, and adherence to traffic laws. Driver behavior analysis aims to improve safety, efficiency, and performance on the road by identifying areas for improvement and implementing targeted interventions or feedback mechanisms [9].

- Data Collection: Driver behavior analysis relies on data from sensors embedded within the vehicle, including accelerometers, GPSs, cameras, and onboard computers.
- Behavioral Metrics: Various metrics assess driver behavior, such as harsh acceleration, hard braking, harsh cornering, speeding, idling time, and adherence to traffic signals and speed limits.
- Pattern Recognition: Advanced algorithms and ML techniques recognize patterns and trends in driver behavior over time, allowing for more accurate analysis and prediction of future behavior.
- Feedback Mechanisms: Insights derived from driver behavior analysis can provide feedback to drivers in real time through dashboard displays, mobile applications, and visual, haptic, or audible alerts, encouraging safer and more efficient driving practices.
- Risk Assessment: Driver behavior analysis enables the identification of high-risk behaviors and situations, such as distracted driving, aggressive driving, or fatigue, which pose increased risks to road safety.
- Performance Monitoring: Fleet operators and insurance companies utilize driver behavior analysis to monitor the performance of individual drivers or entire fleets, identify top performers, and implement targeted training or incentive programs [12].

Driver Fatigue Monitoring

Driver fatigue monitoring is a safety technology used in ADAS to detect and mitigate the effects of driver fatigue on vehicle operation [11]. This system utilizes sensors and algorithms to monitor signs of drowsiness or fatigue, providing alerts and interventions to ensure the driver remains alert or suggests taking a break.

Components:
Cameras and sensors are mounted inside the vehicle; these monitor the driver's eye movements, blink rate, and head positioning. Sensors placed in the steering wheel are used to monitor the angular corrections applied by the driver to maintain the vehicle in the center of the lane. The angular movement by the driver is interpreted to determine the attentiveness of the driver. Physiological sensors may include devices to measure heart rate variability or skin conductance, and indicators of drowsiness. Software algorithms analyze the collected data to detect patterns indicative of fatigue.

Functions:

- Detection of Fatigue Signs: Identifies symptoms of fatigue such as frequent blinking, yawning, or prolonged eye closure.
- Alert Generation: The driver may receive audible, visual, or haptic alerts to regain attention or suggest a break.
- Vehicle Control: In more advanced systems, if the driver does not respond to alerts, the system may take control of the vehicle to navigate to a stop safely.

Driverless Car Regulations

Regulations are specific rules, laws, and guidelines that govern a specific sector. They are created and enforced by government agencies, and their requirements vary across different geographies. Examples include Federal Motor Vehicle Safety Standards (FMVSS) in the United States, UNECE in Europe, Japanese Industrial Standards (JIS), and similar entities.

Driverless car regulations encompass the legal framework and standards established to govern the design, testing, deployment, and operation of fully autonomous or driverless vehicles. These regulations are designed to ensure vehicle safety, reliability, and interoperability in autonomous driving scenarios. They address various aspects, including vehicle certification, road testing, liability, cybersecurity, and data protection.

Driverless Transportation

Driverless transportation refers to the operation of vehicles without human drivers. These vehicles, also known as autonomous or self-driving vehicles, use a combination of sensors, cameras, AI, and ML to navigate and operate safely in various environments [13].

- Public Transportation: Autonomous shuttles and buses designed to operate in urban environments, reducing the need for human drivers and potentially increasing transportation efficiency.
- Freight and Delivery Services: Driverless trucks and delivery vehicles that can operate longer hours than human-driven vehicles and potentially reduce operating costs.
- Personal Transportation: Fully autonomous cars that provide mobility without requiring active human control, enhancing accessibility for those unable to drive.

The development of driverless transportation is expected to transform various aspects of society, including urban planning, mobility, and the economy. Ongoing technological advancements and further refinement of standards will play critical roles in facilitating the safe and effective integration of driverless vehicles into everyday life.

DRO (Dynamic Route Optimization)

DRO refers to using advanced algorithms and real-time data to update and optimize vehicle routing continuously. This process is designed to improve transportation efficiency, reduce travel times, and address dynamic factors such as traffic conditions, road closures, and delivery schedules [12].

Applications:

- Logistics and delivery services
- Fleet management
- Passenger vehicles

The vehicle uses several systems, including GPS tracking, Geographic Information Systems (GISs), real-time traffic data, and cloud computing to adjust routes dynamically. Additionally, the incorporation of ML algorithms learns from historical data to predict traffic patterns and optimize future routing decisions.

Drowsiness Detection

Drowsiness detection refers to technologies and systems designed to identify signs of driver fatigue or drowsiness in real time, typically to enhance safety by preventing accidents caused by sleep-deprived driving. This technology uses various sensors and algorithms to monitor the driver's physical and behavioral cues that indicate fatigue.

Applications:
- Safety Systems in Vehicles: Integrated into ADAS, drowsiness detection systems can alert drivers through auditory, visual, or haptic signals to take a break or be more vigilant when signs of drowsiness are detected.
- Commercial and Industrial Fleets: Used extensively in commercial vehicles to ensure drivers adhere to safe driving practices, thus reducing the risk of accidents related to fatigue [12].

Technologies Used:
1. Physiological Monitoring: Involves sensors that can detect changes in heart rate, eye movement [percentage of eye closure (PERCLOS)], or even brain wave patterns that indicate fatigue [9].
2. Behavioral Monitoring: This analyzes changes in driving behavior, such as steering patterns or lane discipline, that might indicate that the driver is becoming drowsy.
3. Facial Recognition Systems: Utilize cameras to observe signs of sleepiness, such as frequent blinking, yawning, or drooping eyelids.

See also **Driver Alertness Monitoring**.

DSRC (Dedicated Short-Range Communication)

DSRC is a wireless communication technology specifically designed to enable secure, low-latency, and high-reliability data exchange between vehicles (V2V) and between vehicles and infrastructure (V2I) [14]. Operating in the 5.9-GHz frequency band, DSRC supports safety-critical applications for ADAS and CV ecosystems, enhancing traffic efficiency, situational awareness, and crash prevention.

Key Features:
- **Low Latency:** Ensures near-instantaneous communication for time-sensitive safety messages.
- **High Reliability:** Maintains robust communication even in challenging environments, such as high-speed scenarios and dense urban areas.
- **Standardization:** Governed by IEEE 802.11p and ETSI ITS-G5 standards, ensuring global interoperability.
- **Range:** Covers approximately 300 m, suitable for short-range communications.

Applications in ADAS and CVs:

1. **Collision Avoidance:** Real-time warnings for potential hazards such as sudden braking, intersection crashes, or blind spot collisions.
2. **Traffic Signal Priority:** Enables V2I communication to optimize signal timings for emergency vehicles or public transit.
3. **Platooning:** Supports V2V coordination for maintaining safe distances and improving fuel efficiency in convoys.
4. **Work Zone Warnings:** Alerts drivers to construction zones, lane closures, or other road hazards.

Advantages:

- DSRC's dedicated spectrum minimizes interference from other wireless systems, ensuring dependable communication for critical applications.
- No dependency on cellular networks or third-party infrastructure, making it a cost-effective solution for public safety.

REFERENCES

1. Silva, J., Silva, L., Barros, M., Santos, T. et al., "Data Analytics Applied to Automotive Wiring Harness (Electrical Distribution Systems)," SAE Technical Paper 2023-36-0058 (2023), doi:https://doi.org/10.4271/2023-36-0058.
2. Steingrimsson, B., Yi, S., Jones, R., Kisialiou, M. et al., "Big Data Analytics for Improving Fidelity of Engineering Design Decisions," SAE Technical Paper 2018-01-1200 (2018), doi:https://doi.org/10.4271/2018-01-1200.
3. Vamosi, R., Lassnig, M., and Schikuta, E., "Data Allocation Service ADAS for the Data Rebalancing of ATLAS," SAE Publishing, Warrendale, PA, 2019.
4. Li, F., Wu, Z., Zhu, Y., and Lu, K., "Radar and Smart Camera Based Data Fusion for Multiple Vehicle Tracking System in Autonomous Driving," SAE Technical Paper 2022-01-7019 (2022), doi:https://doi.org/10.4271/2022-01-7019.
5. García, F., Escalera, A., and Armingol, J.M., "Enhanced Obstacle Detection Based on Data Fusion for ADAS Applications," in *16th International IEEE Conference on Intelligent Transportation Systems*, The Hague, The Netherlands, 2013.
6. Alzu'bi, H., Dwyer, B., Nagaraj, S., Pischinger, M. et al., "Cost Effective Automotive Platform for ADAS and Autonomous Development," SAE Technical Paper 2018-01-0588 (2018), doi:https://doi.org/10.4271/2018-01-0588.

7. Society of Automotive Engineers, "SAE J3016 Surface Vehicle Recommended Practice: Taxonomy and Definitions for Terms Related to Driving Automation Systems for On-Road Motor Vehicles," SAE Publishing, Warrendale, PA, 2018.

8. Thukaram, P. and Mohan, S., "Digital Twins for Prognostic Profiling," SAE Technical Paper 2019-28-2456 (2019), doi:https://doi.org/10.4271/2019-28-2456.

9. Nandyala, S., Gayathri, K., Bhushan, C., Gandi, V. et al., "Emotion Analytics for Advanced Driver Monitoring System," SAE Technical Paper 2019-26-0025 (2019), doi:https://doi.org/10.4271/2019-26-0025.

10. Society of Automotive Engineers, "J2808 Lane Departure Warning Systems: Information for the Human Interface," SAE Publishing, Warrendale, PA, 2024.

11. Society of Automotive Engineers, "J3198 Driver Drowsiness and Fatigue in the Safe Operation of Vehicles - Definition of Terms and Concepts," SAE Publishing, Warrendale, PA, 2020.

12. Chominsky, W. and Quigley, J.M., *SAE International's Dictionary of Commercial Vehicles* (Warrendale, PA: SAE International Publishing, 2024).

13. Society of Automotive Engineers, "J3206 Taxonomy and Definition of Safety Principles for Automated Driving System (ADS)," SAE Publishing, Warrendale, PA, 2021.

14. Society of Automotive Engineers, "J2945 Dedicated Short Range Communication (DSRC) Systems Engineering Process Guidance for SAE J2945/X Documents and Common Design Concepts," SAE Publishing, Warrendale, PA, 2017.

EBA (Emergency Brake Assistance)

EBA refers to a system within a vehicle that detects when the driver initiates a sudden brake effort and automatically increases the braking force to help prevent or reduce the severity of a collision, along with a reduction of collateral damage (pedestrians and other vehicles) (Figures E.1 and E.2) [1, 2].

FIGURE E.1 Example of vehicle behavior without stop assist [1].

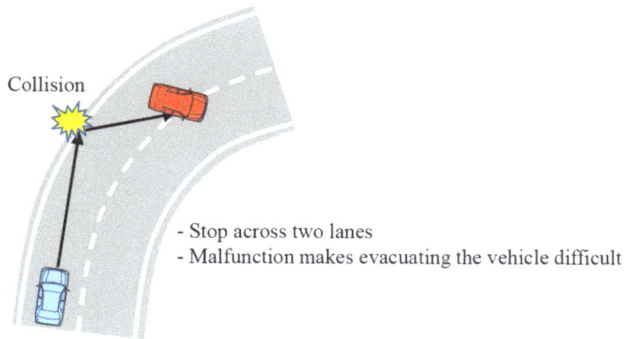

- Stop across two lanes
- Malfunction makes evacuating the vehicle difficult

FIGURE E.2 Example of vehicle behavior with stop assist [1].

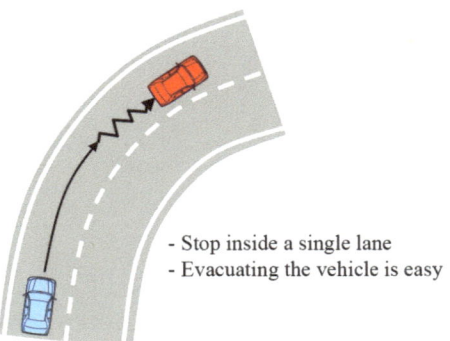

EBD (Electronic Brakeforce Distribution)

EBD is an advanced vehicle safety technology that optimizes the distribution of braking force between the front and rear wheels according to road conditions, vehicle load, and driving dynamics. EBD is designed to enhance the effectiveness of braking while maintaining vehicle stability during deceleration.

EBD consists of several sensors and actuators, beginning with wheel speed sensors that detect the speed of each wheel, which is critical for determining the necessary brake force distribution. Actuators are mechanisms within the brake system that adjust the pressure applied to each wheel. A control module is an electronic unit that processes sensor inputs to execute real-time adjustments to brake pressures.

Optimized brake force application automatically adjusts the brake force to each wheel, preventing wheel lockup and ensuring the best possible braking performance [3]. The system is responsible for maintaining vehicle stability during braking and reducing the risk of skidding or losing control, especially in emergency braking situations. This optimization includes improved wear distribution by balancing brake wear across different wheels, reducing maintenance frequency and costs.

For more information, see **J2561** and **J2909** [4, 5].

EBS (Emergency Braking System)

An EBS is an advanced automotive safety technology primarily utilized in ACVs. It is also referred to as AEB. This system uses sensors, cameras, and radar technology to detect impending collisions and automatically apply the brakes if the driver fails to respond in time. EBS is designed to either prevent accidents or reduce the severity of crashes by lowering the vehicle's speed as much as possible before impact.

EBS works with a vehicle's dynamic control systems and is an integral part of its overall safety. The system continuously monitors the vehicle's surroundings and evaluates the distance and relative speed of objects in its path. Upon detecting an imminent collision, the EBS calculates the required braking force to avoid the collision or mitigate its effects and autonomously engages the brakes.

- **Sensors:** These include radar, LiDAR, and cameras that provide real-time data on the vehicle's environment.
- **Control Unit:** A CPU that analyzes sensor data, assesses collision risks, and commands braking actions.
- **Braking System:** Enhanced braking components that can be activated automatically and rapidly to ensure maximum deceleration.

For more information, see **FMVSS 126**, **FMVSS 135**, and **FMVSS 127** [6, 7, 8].

EBW (Emergency Brake Warning)

A system within a CV that transmits a real-time warning to surrounding vehicles when an emergency braking event is detected. The warning is intended to alert nearby drivers and connected infrastructure to potential sudden decelerations, enhancing situational awareness and reducing the likelihood of rear-end collisions.

This system is part of a broader range of ADAS and CV technologies that utilize V2V and V2I communications [9, 10].

In heavy traffic, a vehicle equipped with an EBW system detects a sudden and significant brake application and sends a V2X message to the following vehicles. Upon receiving the signal, these vehicles alert the drivers through visual, auditory, or haptic feedback, allowing for a quicker response to changing traffic conditions.

For more information, see **FMVSS 126** and **FMVSS 135** [6, 7].

ECN (Electric Charging Network)

An ECN is a comprehensive system of publicly accessible EV charging stations interconnected through a network infrastructure to support EV charging needs [11]. This network facilitates the deployment, operation, and management of EV charging infrastructure across various locations, enhancing the accessibility and reliability of EV charging services. The underlying network infrastructure connects individual charging stations, provides payment systems, and is managed often by a central organization (**Figure E.3**).

FIGURE E.3 There are many components to the ECN, from generation sources to distribution networks.

Physical infrastructure equipped with one or more charging ports facilitates the charging of EVs. Software systems that manage station operations handle user authentication and billing and provide real-time data on station availability [12]. Technologies ensure seamless communication between vehicles, charging stations, and central management systems.

Eco-Driving

Eco-driving is a driving behavior optimized for energy efficiency to reduce fuel consumption and minimize environmental impact. This concept encompasses various driving strategies and techniques, such as maintaining steady speeds, minimizing unnecessary acceleration, and using vehicle technologies to enhance fuel economy [13].

The automation and connectivity features of CAVs, such as ACC and real-time TMSs that adjust vehicle operation for optimal fuel efficiency, can significantly enhance eco-driving strategies. Eco-driving is also integral to traffic management strategies that reduce congestion and improve traffic flow, contributing to lower fuel consumption and emissions [14].

ECU (Electronic Control Unit)

An ECU is a microprocessor-based vehicle device that controls one or more electrical systems or subsystems (**Figure E.4**). In the context of ADAS and CVs, ECUs are crucial for interpreting data from vehicle sensors, executing software algorithms, and managing actuator responses to perform various vehicle functions such as engine management, transmission control, and safety feature activation. Modern vehicles have a central gateway (CGW) that manages off-vehicle communications [15].

FIGURE E.4 Example of system architecture [15].

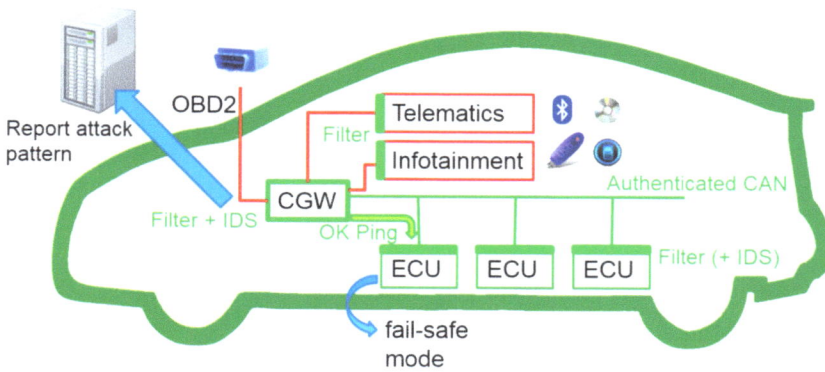

The core component processes data based on embedded software within a microprocessor or MCU. The device will include volatile and nonvolatile memory for storing operational programs and data. Input/output interfaces connect the ECU to various sensors, actuators, and other ECUs. Communication networks such as CAN [16], local interconnect network (LIN), and Ethernet enable data exchange within the vehicle.

The ECUs will manage a specific related set of vehicle functions. They will process data and compute data received from various sensors to determine appropriate responses. The ECU will exert control of tasks, managing and controlling specific vehicle functions such as braking, propulsion, steering, lighting, and other functions. The ECU will provide local diagnostic and communication to the vehicle visual interfaces via off-vehicle networks, making communication with diagnostic tools for maintenance and troubleshooting possible. Inter-ECU communication coordinates with other ECUs to ensure harmonious vehicle operation.

For more information, see **J1699** and **J3061** [17, 18].

Edge Computing

Edge computing in the context of CVs refers to the computational processes performed near the source of data generation, i.e., within or close to the car's area of interest (AOI) rather than some centralized computing center [19]. This technology enables faster processing and response times and reduces latency by handling and controlling data locally.

Edge computing is critical for applications requiring immediate decision-making, such as autonomous driving and real-time vehicle system monitoring. Edge computing enhances V2X communications by minimizing delays and improving the reliability of message exchanges between vehicles and infrastructure. By processing sensitive data locally, edge computing can enhance data security and privacy by limiting its exposure to external networks [20].

Edge Devices

In CVs, edge devices refer to hardware located physically at the "edge" of the vehicle network, close to the data sources (sensors and actuators) [20]. These devices typically perform data processing tasks locally, reducing latency and bandwidth usage by processing data near the source rather than transmitting large volumes of raw data to centralized servers [18].

Edge devices can process data from vehicle sensors in real time, supporting critical functions such as collision avoidance systems, lane keeping, and lane change assistance systems, blind spot assistance, and dynamic route adjustments. These devices can temporarily store and analyze data to optimize vehicle performance and maintenance, eliminating the need for constant cloud connectivity.

The role of edge devices is expected to grow with advancements in vehicle and high-performance computing technologies, particularly in autonomous and semi-AVs, where quick data processing is critical. Ongoing developments in IoT and AI will further enhance the capabilities of edge devices in automotive applications.

Edge Security

Edge security in the context of CVs refers to the protective measures and technologies implemented at the network's edge—where vehicles operate as nodes—to safeguard data integrity, ensure privacy, and prevent cyber threats that target vehicle systems and their communication interfaces [21].

Data protection ensures that data transmitted between vehicles and infrastructure, or between vehicles themselves, is secure from interception or manipulation [22]. Restricting access to vehicle networks and systems to authorized entities only prevents unauthorized access and manipulation. Edge security enables rapid identification and response to potential threats at the edge itself, before they can infiltrate the broader network.

Edge Sensors

In the CAV, there are sensors such as cameras, radars, LiDARs, and IMUs that process data in real time and perform automated driving or driver assistance. These are edge sensors, located physically within the vehicle network. They are often software-defined sensors and are configurable and updatable through software. This allows them to adapt to new requirements or changes in their operational environment without needing hardware alterations. This concept is part of the broader trend toward software-defined vehicles (SDVs), where software, rather than hardware, controls more features and functionalities.

EEBL (Emergency Electronic Brake Light)

The EEBL is a safety technology in CVs that communicates critical braking information between nearby vehicles. This feature is designed to alert drivers to severe braking events in vehicles ahead, even when these vehicles are not directly visible (e.g., obscured by other vehicles, curves, or weather conditions). The EEBL uses V2V communication protocols to send and receive real-time information, enhancing driver response time and helping to prevent collisions.

SAE J2945/1: This standard specifies the minimum performance requirements for V2V communications in light-duty vehicles, including aspects relevant to the operation of the EEBL feature. It outlines the communication messages, data elements, and minimum performance criteria necessary for effectively deploying this safety technology.

FMVSS 150: This focuses on V2V communication and mandates certain features and capabilities that support technologies like EEBL [23]. It outlines requirements for transmitting and receiving V2V messages to improve roadway safety through enhanced situational awareness.

E-Horizon (Electronic Horizon)

E-Horizon refers to ADAS technology that uses detailed digital map data, localization, and sensor information to provide a predictive view of the road ahead. This enables vehicles to anticipate road conditions, adjust vehicle dynamics, and optimize driving strategies in real time.

Predictive safety features use road attribute data such as curvature, slopes, and traffic signs to enhance safety functions like predictive cruise control and collision avoidance systems. Optimizing powertrain performance based on anticipated road conditions improves fuel economy and reduces emissions. Autonomous driving provides vehicles with critical foresight about the route ahead, enhancing decision-making processes and driving efficiency.

ELK (Emergency Lane Keeping)

ELK is an ADAS designed to prevent accidents by automatically correcting the vehicle's steering if it drifts out of its lane without a turn signal being activated. This system is particularly active in situations where a lane departure could result in a collision, such as when a vehicle inadvertently moves toward an adjacent lane occupied by another vehicle.

ELK systems utilize sensors and cameras to monitor lane markings and the position of other vehicles. When the system detects an unintentional lane departure that might lead to a collision, it intervenes by providing steering inputs to maintain the vehicle within its lane, thereby enhancing road safety.

Embedded Software

Embedded software in vehicles refers to software designed to run onboard computers and MCUs within the vehicle (**Figure E.5**). This software controls various vehicle functions, ranging from engine management systems to infotainment systems and, increasingly, ADAS and AV operations. The function under the control of the embedded software will have a range of ideal response times. This is the benefit of embedded software over software designed for computers at large [24].

FIGURE E.5 Real-time spectrum [24].

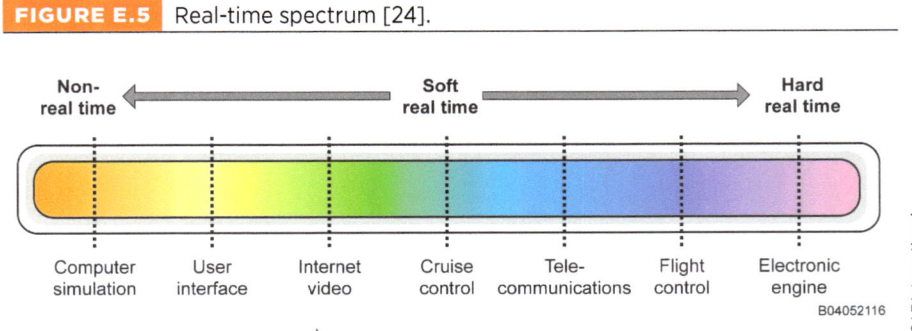

Embedded software is crucial for operating ECUs that manage various vehicle systems, such as braking, steering, and engine performance. The software manages connectivity functions and provides UIs for infotainment and vehicle management systems. This software implements functionalities of systems such as AEB, LKA, and ACC.

Emergency Assist

Emergency assist is a suite of vehicle technologies designed to enhance driver safety and vehicle control during or after an emergency through the use of dynamic driving test (DDT) [10]. These systems may include automatic braking, emergency call (eCall) services, and systems that help maintain vehicle control (Figure E.6).

FIGURE E.6 Overview of the emergency stop assist system [1].

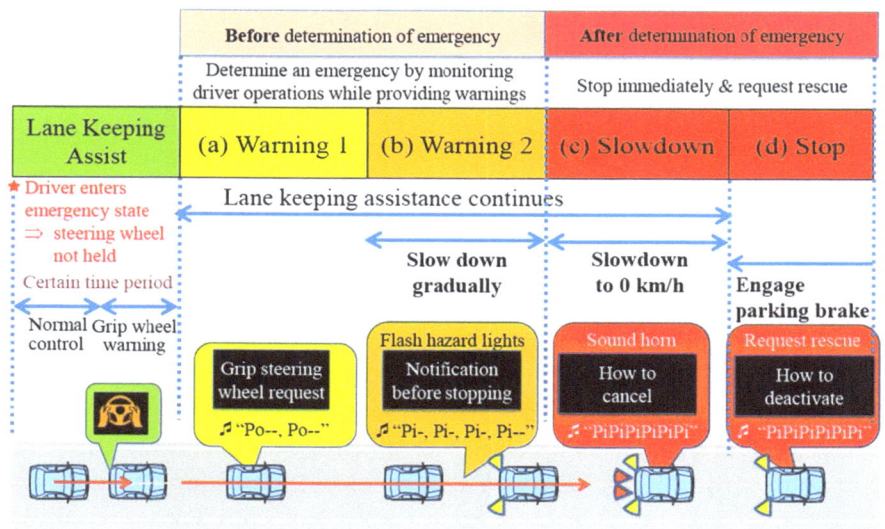

AEB detects an impending forward crash with another vehicle in time to avoid the collision or mitigate its severity. eCall systems automatically send a call for help to emergency services in the event of a serious road accident and provide essential data such as the vehicle's location. Postcrash braking helps reduce the risk of secondary collisions by automatically applying brakes when a primary collision is detected.

Advances in vehicle communication technologies, such as V2X, are expected to enhance the effectiveness of emergency assist systems by enabling vehicles to communicate with each other and with traffic infrastructure to prevent accidents before they occur.

For additional information, see **J3016** [10].

Emergency Brake Lights

Emergency brake lights, or emergency stop signals (ESSs), are specialized lighting systems equipped on ACVs to alert drivers of sudden or severe braking maneuvers. These lights are typically more intense and conspicuous than standard brake lights and may activate differently, such as by flashing or showing a different color, to differentiate from regular braking [25].

In ACVs, emergency brake lights are often integrated with the vehicle's ADAS. They are triggered automatically when the vehicle's sensors and algorithms detect a condition that requires rapid deceleration. Factors influencing activation include the speed of deceleration, the relative speed to other vehicles, and the likelihood of a collision if immediate action is not taken.

In CV ecosystems, emergency brake activation messages can be transmitted to nearby vehicles through V2V communication technologies [9]. This allows other vehicles in the vicinity, even those that may not have direct line of sight, to be aware of emergency braking events and prepare accordingly.

Emergency Services Integration

Emergency services integration in CVs refers to advanced vehicular technologies that facilitate real-time electronic communication between vehicles and emergency response services. By leveraging V2X communication systems, this integration aims to improve response times and outcomes during emergencies.

This technology enables vehicles to automatically send alerts to nearby emergency services when an accident occurs, providing precise location data, the nature of the accident, and potential hazards. It also allows for the transmission of critical health data from medical monitoring devices within the vehicle to emergency medical personnel en route.

SAE J2945/1: This standard by SAE International defines the minimum performance requirements for V2X communications in light vehicles, including aspects related to emergency service response capabilities [14].

Integrating emergency services with CV technologies significantly enhances public safety by reducing the time emergency responders take to arrive at the scene of an accident. It also improves the efficiency of the response by providing emergency personnel with advanced information about the situation, allowing for better preparedness.

Emission Monitoring

Emission monitoring refers to utilizing onboard systems to continuously monitor and report vehicle emissions data [26]. This is increasingly relevant under newer regulations like the Euro 7 standards, emphasizing the need for real-world, continuous emission data collection to ensure compliance with environmental regulations.

Key components of emission monitoring include onboard emission monitoring (OBM) and onboard fuel/energy consumption monitoring (OBFCM) [27]. These systems are integrated into modern vehicles, ranging from conventional internal combustion engines (ICEs) to hybrid and PHEVs. They record parameters such as fuel consumption, distance traveled, nitrogen oxide (NOx) emissions, and exhaust flow rates, providing a detailed picture of a vehicle's environmental impact under actual driving conditions.

The accuracy of these systems is generally high, with deviations in fuel consumption reported under 5% in test cycles and slightly higher in real-world conditions. The systems are crucial for identifying discrepancies between laboratory test results and real-world performance, helping to pinpoint areas for technological improvement and regulatory compliance.

Energy Management

Energy management refers to the integrated approach and technology used in CVs to monitor, control, and optimize the use and distribution of energy resources [28]. This concept is pivotal in enhancing efficiency, sustainability, and operational performance of electric and hybrid vehicles within connected transportation ecosystems (**Figure E.7**).

FIGURE E.7 EV energy transfer system physical context [28].

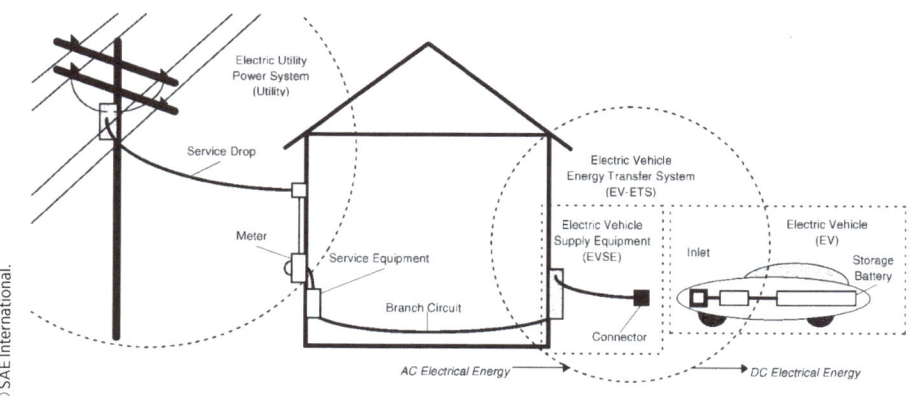

Energy management in CVs encompasses the systems and processes that manage the vehicle's energy storage (battery), generation (regenerative braking), distribution, and usage (powertrain management) to improve fuel efficiency, reduce emissions, and extend battery life and range. Advanced energy management systems leverage connectivity to optimize energy usage based on real-time data, including traffic conditions, route information, and V2V, and V2I communications [13].

Energy Storage

In the context of connected and EVs, the term "energy storage system" refers to any setup that stores electrical energy for later use in the vehicle. These systems are critical for the operation of HEVs, EVs, and PHEVs, providing necessary energy for propulsion and other vehicle functions.

The **SAE J2464** standard, "Electric and Hybrid Electric Vehicle Rechargeable Energy Storage System (RESS) Safety and Abuse Testing," is a primary guideline documenting these systems' safety and testing procedures [29]. This standard details a variety of tests to assess the durability and safety of RESS under conditions beyond their normal operational range, including extreme environmental or operational stresses.

In CV ecosystems, energy storage systems supply power, interact with V2G systems, support advanced vehicle functionalities, and enhance overall vehicle efficiency and reliability.

Energy-Efficient Driving

Energy-efficient driving refers to various practices and technologies to reduce the energy consumption and environmental impact of vehicles in motion. These practices are designed to optimize fuel usage and decrease emissions, contributing to overall vehicle efficiency and sustainability.

Techniques such as smooth acceleration and braking, maintaining steady speeds, using gears efficiently, and minimizing idling can significantly reduce fuel consumption and emissions. Regular maintenance to ensure vehicles operate at peak efficiency, including tire pressure checks, engine tuning, and recommended motor oils, is required to maintain energy efficiency. Energy-efficient driving includes optimized maintenance times based on system performance and predictive analytics. Implementation of advanced systems like automatic start–stop, cruise control, and predictive driving aids can adjust driving patterns for better fuel efficiency.

Energy-efficient driving is enhanced by integrating advanced telematics and V2X communications [9]. These technologies provide drivers with real-time data on traffic conditions, road gradients, and optimal speed, further aiding in reducing energy consumption.

See also **Energy-Efficient Routing**.

Energy-Efficient Routing

Energy-efficient routing in the context of CVs refers to using advanced algorithms and vehicular communication technologies to determine the most fuel or energy-efficient paths for vehicles to travel. This routing method considers traffic conditions, selecting type of road, road topology, weather conditions, and vehicle-specific characteristics (like fuel consumption patterns and EV battery status) to optimize routes that minimize energy usage and environmental impact.

Route dynamic adjustment is based on real-time data (e.g., traffic or accidents) to enhance fuel efficiency. Integrating smart city traffic systems to reduce idle times and stop-and-go traffic significantly increases fuel consumption. These systems can include V2V and V2I communications [9, 10].

Environmental Sensors

Environmental sensors in CVs refer to various sensor types that monitor external environmental conditions and internal vehicle states to enhance vehicle performance, safety, and connectivity. These sensors include, but are not limited to, temperature, humidity, air quality, light sensors, and sensors that monitor road conditions.

CVs use environmental sensors as part of active perception systems to detect and interpret their surroundings. These sensors are integral to technologies like radar, LiDAR, and vision-based systems, which help in navigation and obstacle detection. They also contribute to sensor fusion systems, where data from various sources are combined to provide a more comprehensive understanding of the vehicle's environment. Environmental sensors significantly contribute to vehicle safety systems by providing critical data that influence decisions in ADAS and AV operations. Environment sensors enable the determination of the conditions that limit the performance of the sensors such as radars, LiDARs, and cameras used by the perception systems of the CAVs. If such limitations are present due to environmental conditions, the CAVs transition to a degraded operating mode in which they can function safely.

ESC (Electronic Stability Control)

ESC is a vehicle safety system that improves vehicle stability by detecting and reducing loss of traction (skidding) [6]. ESC automatically applies the brakes to individual wheels and may reduce engine power to help steer the vehicle in the direction intended by the driver, thereby preventing understeer or oversteer (**Figure E.8**).

FIGURE E.8 Collection of acronyms [30].

Antilock Brake Systems	Traction Control Systems	Electronic Stability Control Systems
ABS (Antilock Brake System)	ASC (Automotive Stability Control)	ABC (Active Brake Control)
RWAL (Rear Wheel Antilock)	ASR (Automatic Stability Regulation)	AH (Active Handling System)
SCS (Stop Control System)	Brake Only Traction	Active Safety
RABS (Rear Antilock Brake System)	ETS (Enhanced Traction System)	AdvanceTrac
	TCS (Traction Control System)	ASMS (Automotive Stability Management System)
	TCB (Traction with Brake Intervention)	ASTC (Active Skid and Traction Control)
	TRAC (Traction Control System)	CBC (Cornering Brake Control)
	EDS (Electronic Differential-lock System)	DSC (Dynamic Stability Control)
		DSTC (Dynamic Stability and Traction Control)
		ESC (Electronic Stability Control)
		ESP (Electronic Stability Program)
		ICCS (Integrated Chassis Control System)
		IVD (Integrated Vehicle Dynamics)
		PCS (Precision Control System)
		PSM (Porsche Stability Management)
		SCS (Stability Control System)
		Stabilitrak
		STC (Stability and Traction Control System)
		Traxxar
		VDC (Vehicle Dynamics Control)
		VSA (Vehicle Stability Assist)
		VSC (Vehicle Stability Control)
		VSES (Vehicle Stability Enhancement System)
		YCS (Yaw Control Stability)

© SAE International.

ESC systems typically include wheel speed, gyroscopes, and steering angle sensors. These monitor the vehicle's speed, lateral acceleration, and the angle at which the steering wheel is turned. A control unit processes the data from various sensors to determine if the vehicle is moving in a different direction than the steering wheel position indicates. If a discrepancy is detected, the ESC system intervenes to correct the vehicle's path. Braking intervention functions apply brakes selectively on individual wheels to control oversteer or understeer, helping maintain vehicle stability (Figure E.9).

FIGURE E.9 The four categories of an ESC system [30].

	Systems with Engine Control	Systems Without Engine Control
Four Wheel Systems	4-Wheel ESC with Engine Control a) This system must have the means to apply all four brakes individually and a control algorithm which utilizes this capability. b) The system must have an algorithm to modify engine torque, as necessary, to assist the driver in maintaining control of the vehicle. c) The system must be operational during all phases of driving including acceleration, coasting, and deceleration (including braking). d) The system must stay operational when ABS or Traction Control are activated.*	4-Wheel ESC without Engine Control a) This system must have the means to apply all four brakes individually and a control algorithm which utilizes this capability. b) The system must be operational during all phases of driving including acceleration, coasting, and deceleration (including braking). c) The system must stay operational when ABS or Traction Control are activated.*
Two Wheel Systems	2-Wheel ESC with Engine Control a) This system must have the means to apply the brakes individually at the front wheels of the vehicle and a control algorithm that utilizes this capability. b) The system must have an algorithm to determine the need, and a means to modify engine torque, as necessary, to assist the driver in maintaining control of the vehicle. c) The system must be operational during all phases of driving including acceleration, coasting, and deceleration (including braking). d) The system must stay operational when ABS or Traction Control are activated.*	2-Wheel ESC without Engine Control a) This system must have the means to apply the brakes individually at the front wheels of the vehicle and a control algorithm that utilizes this capability. b) The system must be operational during all phases of driving including acceleration, coasting, and deceleration (including braking). c) The system must stay operational when ABS or Traction Control are activated.*

© SAE International.

e-Sim (Subscriber Identity Module)

e-SIM refers to an embedded SIM in CAVs that enables secure and remote management of cellular network subscriptions, facilitating seamless connectivity and data exchange.

e-SIM technology allows CAVs to connect to cellular networks without physically swapping SIM cards. This remote connectivity management enables vehicles to switch networks, update subscriptions, and maintain continuous connectivity for various services. e-SIM enables OTA updates for software, firmware, and configuration settings in CAVs. This OTA capability ensures that vehicles stay updated with the latest features, security patches, and network configurations without requiring manual intervention.

e-SIMs support multiple cellular network profiles, including second-generation (2G), third-generation (3G), 4G LTE, and emerging 5G networks. This multi-network compatibility ensures reliable connectivity and seamless handover between network technologies based on coverage and performance. e-SIMs employ secure authentication mechanisms, such as mutual authentication and encryption protocols, to safeguard communication between CAVs and cellular networks. This ensures data privacy, integrity, and protection against unauthorized access.

ESS (Emergency Stop Signal)

The ESS is a vehicular safety feature that activates the brake lights to flash under conditions of hard and sudden braking rapidly, alerting following drivers to the potential of a rapid slowdown or stop. This feature is designed to reduce the risk of rear-end collisions by providing an earlier warning than static brake lights.

The ESS is referenced in automotive safety standards, which outline the performance requirements and test procedures for such safety systems. These standards ensure that emergency signaling systems meet consistent and reliable safety measures, enhancing roadway safety for all users.

Documents such as **SAE J1432** focus on the requirements and guidelines for high-mounted stop lamps and signal lamps, which are relevant to the broader context of emergency signaling systems, including the ESS [31]. ESS, while not exclusively a CV technology, complements broader V2X communications by enhancing immediate safety responses in urgent scenarios. It aligns with the goals of improving vehicular communication and safety as outlined in various SAE standards related to CV technologies.

SAE International provides guidelines and standards related to the implementation and performance of automotive safety features, including the ESS. SAE standards specify the parameters such as the flashing frequency, visibility, and activation conditions for ESS to ensure consistency and effectiveness across different vehicle brands and models.

The National Highway Traffic Safety Administration (NHTSA) outlines safety regulations under FMVSS in the United States. These standards may include specific requirements for signaling devices, including those used in emergency braking situations, to ensure they meet minimum safety and operational criteria.

Like the FMVSS, Canadian Motor Vehicle Safety Standards (CMVSSs) are the regulatory standards set by Transport Canada concerning the safety and performance of motor vehicles on Canadian roads. CMVSS includes provisions for vehicular lighting and signaling systems, potentially covering the specifications and usage of ESS.

The adoption of ESS technology is part of broader efforts to enhance road safety through advanced vehicular communication systems. By integrating ESS, vehicles contribute to a safer driving environment, potentially lowering the incidence of rear-end collisions, which are common during abrupt traffic halts.

EV Range Optimization

EV range optimization refers to techniques and technologies designed to maximize the distance an EV can travel on a single charge. This involves improving the efficiency of various vehicle subsystems, including the BMS, drivetrain, thermal management, and the integration of ADAS to enhance overall vehicle performance and reduce energy consumption.

BMSs optimize battery performance through temperature control and charge management. Energy-efficient drivetrains use advanced motor and transmission technologies to reduce energy wastage. Aerodynamic design features vehicle designs that minimize drag and enhance energy efficiency. Regenerative braking systems convert braking energy back into battery power, extending driving range.

For more information, see **J2954, J3072,** and **J2847** [32, 33, 34].

EVC (Emergency Vehicle Communication)

EVC refers to using dedicated communication systems in emergency vehicles such as ambulances, fire trucks, and police vehicles. These systems enable direct communication between emergency vehicles and other road users, infrastructure, connected intersections, and dispatch centers, improving response times and enhancing road safety during emergencies.

The primary purpose of EVC is to ensure that emergency vehicles can navigate traffic more efficiently and safely, particularly when responding to emergencies. This is achieved by transmitting real-time data about the vehicle's location and emergency status to surrounding vehicles, connected intersections, and TMSs.

EVC leverages technologies within the framework of CV technology, including DSRC, cellular communication (such as C-V2X), and other wireless communication technologies. These technologies facilitate V2X communication, enabling emergency vehicles to send alerts to other CVs and infrastructure about their presence and intended actions.

In practice, EVC systems can be used to control traffic lights to create a clear path for emergency vehicles, send warnings to vehicles on potential collision paths, and improve coordination during large-scale emergency responses.

EVP (Emergency Vehicle Priority)

EVP is a feature in the CV environment that allows traffic control systems to recognize and prioritize the movement of emergency vehicles such as ambulances, fire trucks, and police vehicles through traffic signals and other control devices. The primary goal of EVP is to reduce response times for emergency services and improve safety by minimizing delays at intersections and along routes.

EVP systems typically use a combination of V2I and V2V communications technologies [9]. These systems enable emergency vehicles to send signals to TMSs, which then adjust the traffic signals in real time to create a congestion-free corridor. This adjustment can include extending green lights or shortening red lights to expedite the emergency vehicle's progress.

Implementing EVP requires coordination between vehicle manufacturers, infrastructure providers, and regulatory bodies. It involves integrating advanced communication systems into both vehicles and traffic management infrastructure, training emergency personnel, and conducting public awareness campaigns to inform drivers about the expected behavior in response to EVP systems.

EVs (Electric Vehicles)

EVs are powered entirely or partially by electricity, utilizing one or more electric motors for propulsion. They differ from traditional ICE vehicles in that they rely on rechargeable battery packs as their primary energy source. EVs include BEVs, HEVs, and PHEVs.

Electric motor(s) provide the mechanical power for vehicle propulsion. A battery pack stores the electrical energy required to operate electric motors. Power electronics control the flow of electrical power between the battery, motor, and other systems. A charging system manages recharging the battery from an external power source. The EV battery is charged during a drive cycle by utilizing the braking energy to charge the vehicle battery.

EV charging technologies are categorized by their speed and the charging station types. Level 1 charging uses standard household outlets and offers slow charging. Level 2 charging requires a higher-voltage power source and provides faster charging. Direct current (DC) fast charging (DCFC) provides the fastest charging speeds, ideal for rapid recharging during longer trips [35].

EVs frequently incorporate ADAS and are integral to developing CV technologies. Their electrical architecture enables the easier integration of sensors, controllers, and communication devices, facilitating features such as automated driving, V2G communications, and enhanced fleet management solutions.

For more information, see **J2954** and **J2911** [32, 36].

EVTM (Emergency Vehicle Traffic Management)

EVTM refers to the systems, protocols, and technologies used in CV environments to enhance the operational efficiency and safety of emergency vehicles, including ambulances, fire engines, and police vehicles. EVTM leverages vehicle-to-everything (V2X) communication technologies to facilitate priority movement through traffic and enhance situational awareness among all road users [9].

V2V communication enables emergency vehicles to transmit their location and movement intentions to surrounding vehicles, allowing other vehicles to adjust their routes or yield the right of way [9]. V2I communication involves the interaction between emergency vehicles and road infrastructure, such as traffic signals, to create clear paths and reduce response times [37].

EVTM systems significantly enhance the efficiency of emergency response operations by reducing response times and increasing the safety of emergency personnel and the general public. EVTM plays a crucial role in modern traffic management and emergency services in urban environments by improving communication and coordination among vehicles and traffic infrastructure.

EVTSP (Emergency Vehicle Traffic Signal Priority)

EVTSP refers to a system integrated within the broader framework of ITS and CV technologies. EVTSP systems enable emergency vehicles, such as ambulances, fire trucks, and police vehicles, to communicate with traffic signal control (TSC) systems to facilitate the preemption or prioritization of traffic signals. This ensures a congestion-free traffic corridor for these vehicles during emergencies, reducing response times and improving the overall efficiency of emergency services.

Through V2I communication technologies, EVTSP systems transmit a signal from an approaching authorized emergency vehicle to the TMS, which then alters the traffic signal phases. This alteration can halt cross traffic earlier than scheduled or extend the green light phase, allowing the emergency vehicle to pass through the intersection with minimal delay [9].

Implementing EVTSP involves various challenges, including widespread infrastructure upgrades, interoperability among diverse systems, and ensuring privacy and security in communication channels. Additionally, consistent and clear regulations are required to guide the deployment and operation of such systems across different jurisdictions.

REFERENCES

1. Takano, M., Morimoto, K., Takagi, M., Oda, T. et al., "Development of an Emergency Stop Assist System," SAE Technical Paper 2019-01-1025 (2019), doi:https://doi.org/10.4271/2019-01-1025.

2. Department of Transportation, "FMVSS 128: Automatic Emergency Braking Systems for Light Vehicles," Washington, DC, 2023.

3. National Highway Traffic Safety Administration, "FMVSS No. 135, Light Vehicle Brake Systems," US Department of Transportation, Washington, DC, 2004.

4. Society of Automotive Engineer, "J2561 Bluetooth Wireless Protocol for Automotive Applications," SAE Publishing, Warrendale, PA, 2016.

5. Society of Automotive Engineers, "J2909 Light Vehicle Dry & Wet Stopping Distance Test Procedure," SAE Publishing, Warrendale, PA, 2018.

6. Department of Transportation, "FMVSS 126: Electronic Stability Control System," Washington, DC, 2007.

7. Department of Transportation, "FMVSS 135 Laboratory Test Procedure for Light Vehicle Brake Systems," Washington, DC, 2005.

8. U.S. Department of Transportation, National Highway Traffic Safety Administration, "Federal Motor Vehicle Safety Standards No. 126, 135, and 138. Title 49, Code of Federal Regulations, Part 571," 2024.

9. Society of Automotive Engineers, "J2735 V2X Communications Message Set Dictionary," SAE Publishing, Warrendale, PA, 2023.

10. Society of Automotive Engineers, "J3016 Taxonomy and Definitions for Terms Related to Driving Automation Systems for On-Road Motor Vehicles," SAE Publishing, Warrendale, PA, 2021.

11. Society of Automotive Engineers, "J1772 SAE Electric Vehicle and Plug-In Hybrid Electric Vehicle Conductive Charge Coupler," SAE Publishing, Warrendale, PA, 2024.

12. Society of Automotive Engineers, "J3068 Electric Vehicle Power Transfer System Using a Three-Phase Capable Coupler," SAE Publishing, Warrendale, PA, 2022.

13. Society of Automotive Engineers, "J2735 V2X Communications Message Set Dictionary™ Set," SAE Publishing, Warrendale, PA, 2023.

14. Society of Automotive Engineers, "J2945 Minimum Requirements for Road Geometry and Attributes Definition," SAE Publishing, Warrendale, PA, 2024.

15. Wasicek, A. and Weimerskirch, A., "Recognizing Manipulated Electronic Control Units," SAE Technical Paper 2015-01-0202 (2015), doi:https://doi.org/10.4271/2015-01-0202.

16. Society of Automotive Engineers, "J1939-71 Vehicle Application Layer," SAE Publishing, Warrendale, PA, 2020.

17. Society of Automotive Engineers, "J1699 OBD-II Communications Anomaly List," SAE Publishing, Warrendale, PA, 2021.

18. Society of Automotive Engineers, "J3061 Cybersecurity Guidebook for Cyber-Physical Vehicle Systems," SAE Publishing, Warrendale, PA, 2021.

19. Wang, H., Xie, J., and Magboul Ali Muslam, M., "FAIR: Towards Impartial Resource Allocation for Intelligent Vehicles with Automotive Edge Computing," *IEEE Transactions on Intelligent Vehicles* 8, no. 2 (2023): 1971-1982.

20. Ibn-Khedher, H., Laroui, M., Ben Mabrouk, M., Moungla, H. et al., "Edge Computing Assisted Autonomous Driving Using Artificial Intelligence," in *2021 International Wireless Communications and Mobile Computing (IWCMC)*, Harbin City, China, 2021.

21. Society of Automotive Engineers, "J3101 Hardware Protected Security Environment - Application Programming Interface Analysis - Information Report," SAE Publishing, Warrendale, PA, 2024.

22. Ward, D. and Wooderson, P., *Automotive Cybersecurity: An Introduction to ISP/SAE 21434* (Warrendale, PA: SAE Publishing, 2022).

23. Department of Transportation, "FMVSS 150 V2V Communications," Washington, DC, 2017.

24. Society of Automotive Engineers, "J2640 General Automotive Embedded Software Design Requirements," SAE Publishing, Warrendale, PA, 2008.

25. Department of Transportation, "FMVSS 108 Federal Motor Vehicle Safety Standards; Lamps, Reflective Devices, and Associated Equipment," Washington, DC, 2018.

26. Society of Automotive Engineers, "J1978 OBD-II Scan Tool," SAE Publishing, Warrendale, PA, 2022.

27. Franco, V., Dilara, P., Hennig, N., Manara, D. et al., "On-Board Monitoring of Emissions in the Future Euro 7 Standard," SAE Technical Paper 2023-24-0111 (2023), doi:https://doi.org/10.4271/2023-24-0111.

28. Society of Automotive Engineers, "J2293 Energy Transfer System for Electric Vehicles - Part 2: Communication Requirements and Network Architecture," SAE Publishing, Warrendale, PA, 2014.

29. Society of Automotive Engineers, "J2464 (R) Electric and Hybrid Electric Vehicle Rechargeable Energy Storage System (RESS) Safety and Abuse Testing," SAE Publishing, Warrendale, PA, 2021.

30. Society of Automotive Engineers, "J2564 (R) Automotive Stability Enhancement Systems," SAE Publishing, Warrendale, PA, 2023.

31. Society of Automotive Engineers (SAE), "SAE J1432 – High-Mounted Stop, Tail, and Turn Signal Lamps for Passenger Cars and Multipurpose Passenger Vehicles Less Than 2032 mm in Overall Width," SAE International, Warrendale, PA, 2015.

32. Society of Automotive Engineers, "J2954 Wireless Power Transfer for Light-Duty Plug-In/Electric Vehicles and Alignment Methodology," SAE Publishing, Warrendale, PA, 2022.

33. Society of Automotive Engineers, "J3072 Interconnection Requirements for Onboard, Grid Support Inverter Systems," SAE Publishing, Warrendale, PA, 2021.

34. Society of Automotive Engineers, "J2847-1 Communication Between Plug-In Vehicles and Off-Board DC Chargers," SAE Publishing, Warrendale, PA, 2023.

35. Society of Automotive Engineers, "J1773 SAE Electric Vehicle Inductively Coupled Charging," SAE Publishing, Warrendale, PA, 2014.

36. Society of Automotive Engineers, "J2911 Procedure for Certification that Requirements for Mobile Air Conditioning System Components, Service Equipment, and Service Technician Training Meet SAE J Standards," SAE Publishing, Warrendale, PA, 2020.

37. Society of Automotive Engineers, "J3216 Taxonomy and Definitions for Terms Related to Cooperative Driving Automation for On-Road Motor Vehicles," SAE Publishing, Warrendale, PA, 2016.

Fail-Operational Systems

Fail-operational systems are designed to maintain the functionality of a vehicle's functions when there are failures present in the vehicle systems. Modern vehicles have several electrical components. Electrical technologies are used for many "by-wire" systems such as DbW, BbW, and SbW. In some "by-wire" systems, the mechanical backups are replaced entirely by electrical components and elements. Such systems also must be fail-operational to ensure the safety of occupants and other traffic participants. Fail-operational systems are also pivotal for higher levels of vehicle automation (SAE Levels 3 to 5), where the vehicle must continue to operate safely without human intervention (Figure F.1).

 Levels of autonomous driving systems [1].

As vehicles progress toward full automation, the inability of a human driver to intervene promptly in the event of a system failure necessitates that the vehicle's systems themselves can manage or contain failures effectively and operate safely. This capability is essential to ensure safety and robustness in automated driving environments.

Implementing fail-operational systems often involves using redundant architectures—both hardware and software—that allow the vehicle to continue operating despite failures. This might include redundant sensors, actuators, or control units that can take over if one fails. The goal is to avoid any safety-critical impact on the vehicle's operation. Redundancy could be homogeneous or heterogeneous. In homogeneous redundancy, similar hardware and software elements are used. For ensuring genuine redundancy and fail-operational characteristics, this might not be sufficient. Homogeneous redundancy will not be able to address standard failure modes or common cause failures. Therefore, diversity in software and hardware design and implementation is required to ensure proper redundancy. This is referred to as heterogeneous redundancy, and it enables actual fail-operational behavior.

The design and implementation of such systems need to address the increased complexity and potential for higher energy consumption and costs due to redundancy. The systems must also meet stringent safety integrity levels as outlined by automotive safety standards such as ISO 26262 [2] and ISO 21448 [3].

Fail-operational architectures are critical in ADAS and autonomous driving systems. They ensure that functions such as steering and braking can continue to operate safely, even if one part of the system fails, thereby increasing the overall safety and reliability of the vehicle.

Fail-Safe Systems

Fail-safe systems are designed to transition to a safe state, degraded state, or minimal risk condition in the event of a failure within the system. These systems are critical in automotive applications where reliability and safety are paramount, including braking, steering, and power systems. The primary objective of safety mechanisms is to prevent accidents and mitigate any hazardous outcomes arising from system malfunctions.

Modern vehicles incorporate multiple ECUs that perform safety-critical functions such as airbag deployment, braking systems, chassis control, and engine management. Fail-safe tasks in these systems ensure that, even in the case of a malfunction such as hardware failures, software errors, or unexpected environmental influences, the system remains in or switches to a safe state to prevent catastrophic outcomes.

International safety standards like ISO 26262 guide fail-safe systems in vehicles [2], which outline the requirements for safety integrity levels that these systems must meet. These standards help ensure that the fail-safe systems are robust, reliable, and effective across all levels of vehicle operation. In addition, fail-safe systems are also influenced by regulations such as the FMVSS in the United States and the CMVSSs, which set forth the necessary vehicle safety requirements.

FCW (Forward Collision Warning)

FCW systems are ADASs designed to warn drivers of an impending collision with a vehicle or object ahead. These systems use sensors such as radar, LiDAR, and cameras to detect potential obstacles, providing the driver with visual, auditory, or haptic alerts to prevent or mitigate collisions. FCW systems detect vehicles and objects in the vehicle's path using onboard sensors (**Figure F.2**). When the system detects a potential collision threat, drivers receive alerts through various modalities (visual, auditory, and haptic) [4].

FIGURE F.2 Overhead view of the stationary target test [5].

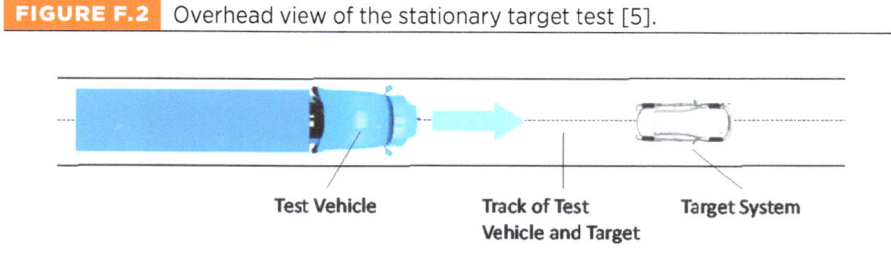

Test Vehicle · Track of Test Vehicle and Target · Target System

© SAE International.

Firmware

Firmware is a specific class of computer software that provides low-level control for a device's specific hardware [6]. This software controls and manages the electronic systems and modules that facilitate vehicle connectivity and autonomous functions.

Firmware is often stored in a device's read-only memory (ROM) or flash memory, making it a permanent part of its hardware. There are several instances of firmware in vehicles; often, each ECU will have its firmware. It provides the necessary instructions for communicating with other hardware components. This includes booting the device, controlling basic functions, and managing data flow. Firmware is explicitly designed for the hardware it runs on, ensuring compatibility and optimal performance for the device [7]. Firmware can often be updated to fix bugs, improve performance, or add new features. The device manufacturer usually provides special update tools for these updates (**Figure F.3**).

FIGURE F.3 Critical expertise of a firmware update management system [7].

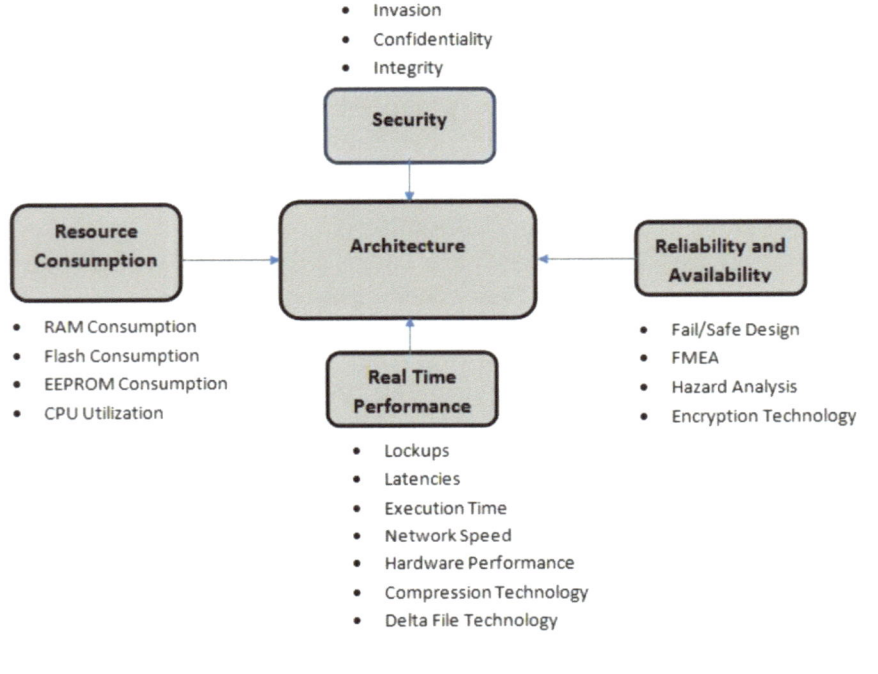

Firmware in CVs interacts with various vehicle systems to ensure communication and interoperability among vehicle technologies, including telematics, infotainment systems, and ADAS. It is critical to the functionality, safety, and security of CVs.

Fleet Management

Fleet management in the context of CVs refers to using telematics systems and software solutions to monitor, manage, and optimize vehicle fleets [8]. This technology is integrated into commercial vehicles to enhance operational efficiency, improve safety, reduce costs, and support compliance with regulatory requirements (**Figure F.4**).

FIGURE F.4 System overview [8].

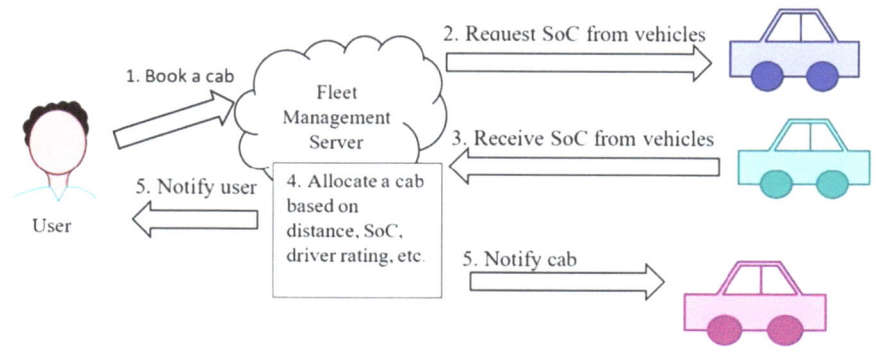

Fleet management systems use GPS and onboard diagnostics (OBD) to provide real-time vehicle location, speed, and route history data. Vehicle systems monitor driver behavior, analyzing data on driving patterns to promote safer driving behaviors and reduce risk. The system monitors fuel use to identify wasteful practices and suggests efficiency improvements. On- and off-board systems use predictive analytics to schedule maintenance, preventing breakdowns and extending vehicle life.

Fleet Management Software

Fleet management software is a specialized tool designed to assist in managing, coordinating, and optimizing vehicle fleets within an organization. It centralizes various operational aspects such as vehicle maintenance, driver management, fuel tracking, and compliance with transportation regulations, facilitating better logistical planning and decision-making.

Fleet Route Optimization

Fleet route optimization uses various technologies and algorithms to determine the most efficient routes for a fleet of vehicles [9]. This process aims to minimize operational costs, enhance fuel efficiency, reduce travel time, and improve overall service quality in transporting goods or passengers. The approach incorporates real-time data on traffic conditions, vehicle status, weather, and other environmental factors (**Figure F.5**).

FIGURE F.5 Basis for fleet target breakdown [9].

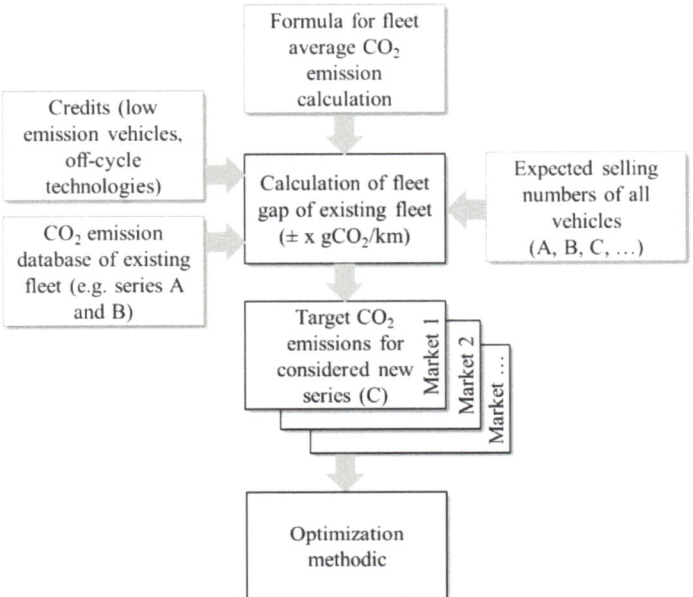

DSRC Message Set Dictionary provides a standardized set of data elements for V2V and V2I communication, essential for real-time route optimization [10]. On-Board System Requirements for V2V Safety Communications lays down the requirements for V2V communication systems, ensuring that vehicles can exchange information reliably to support dynamic routing decisions [11]. Ensuring the data used for routing decisions are accurate and updated in real time is crucial for the effectiveness of fleet route optimization. Robust security protocols must be implemented to protect the data exchange from cyberthreats [12].

Systems should be compatible with various vehicle makes and models and standardized for broad implementation across different fleets (**Figure F.6**).

FIGURE F.6 Consideration of the influence of CO_2 emission improvement measures for other driving cycles [9].

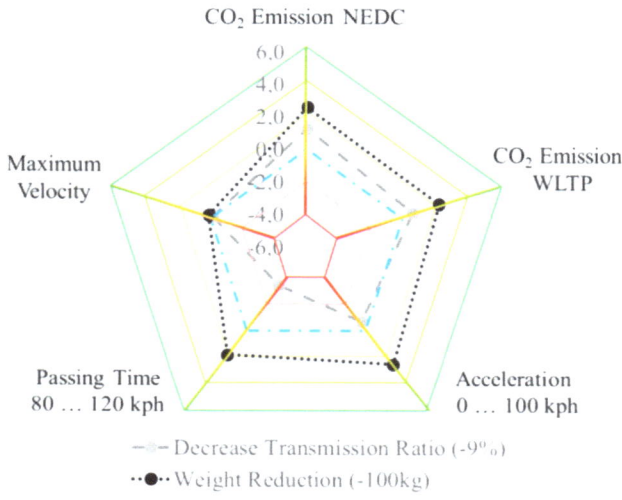

Fleet Safety Management

Fleet safety management refers to the comprehensive approach and strategies employed by CAVs operating in fleet settings to ensure the safety of passengers, pedestrians, and other road users.

CAVs in fleet management incorporate ADAS technologies such as LDW, ACC, AEB, and blind spot detection to enhance safety and mitigate collision risks. Using telematics systems, CAV fleets gather and analyze vehicle performance, vehicle data, driver behavior, route optimization, and real-time monitoring data. This data-driven approach enables proactive safety measures and performance improvements. The gathered data could be used for evaluating the state of components in vehicles and predicting and performing proactive maintenance (**Figure F.7**).

CAV fleets integrate DMSs that use sensors and cameras to assess driver attentiveness, fatigue, distraction, and compliance with safety protocols. Real-time feedback and alerts help prevent accidents caused by human error [14]. Fleet safety management relies on V2V and V2I communication standards, which include sharing collision warnings, traffic conditions, and emergency notifications.

FIGURE F.7 Block diagram of road scanning of static hazards [13].

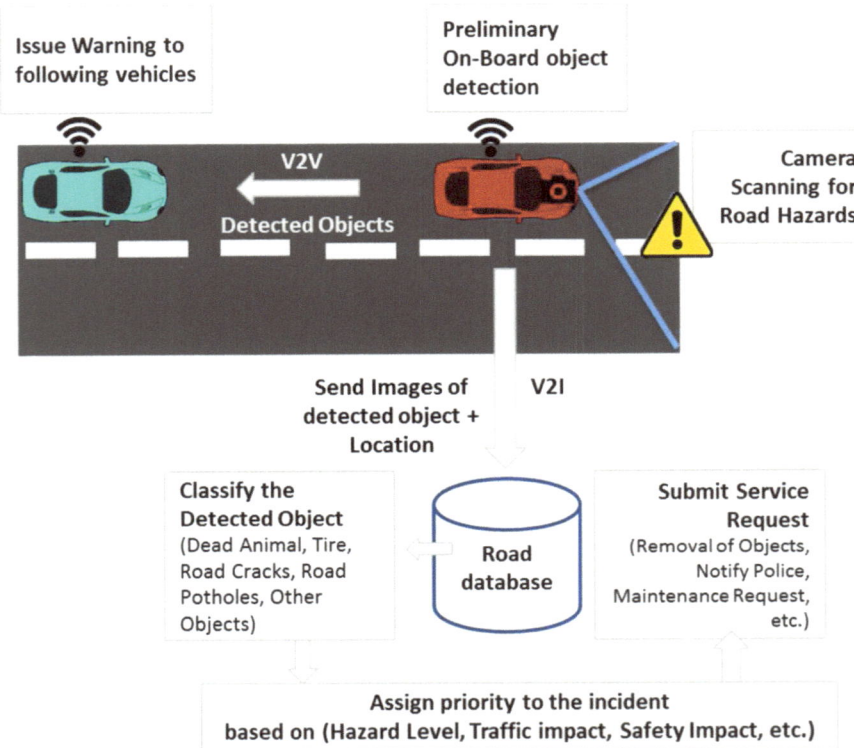

Fleet Telematics

Fleet telematics refers to the use of telecommunications and informatics to enhance the management of fleet vehicles. It involves the integration of technologies such as GPS tracking, OBD, and mobile communications to monitor, control, and optimize various aspects of fleet operations. This includes vehicle maintenance, fuel management, route optimization, and driver behavior monitoring.

GPS technology enables real-time location tracking of fleet vehicles, facilitating efficient route management and rapid response to emergent situations. Telematics systems can monitor driver behavior, including speed, braking patterns, and adherence to traffic laws, contributing to improved safety and reduced operational costs. The telematics system can also limit the areas of vehicle application, excluding some operating areas (geofencing). OBD can predict vehicle maintenance needs and schedule services proactively, minimizing downtime and extending vehicle life. Telematics data help identify fuel wastage and optimize routes, significantly reducing fuel consumption and environmental impact.

Fleet Tracking

Fleet tracking is a system within CV technologies that utilizes GPS and other navigation tools to monitor fleet vehicles' real-time location and status. This system is pivotal for optimizing route planning, enhancing productivity, ensuring safety, and supporting decision-making processes in fleet management.

Vehicle telematics are installed in vehicles to collect data on location, speed, and driving patterns. Communication systems enable the transmission of data between vehicles and central management systems, often facilitated by cloud-based platforms.

Modern fleet tracking integrates advanced data analytics and ML to predict delivery time and dates, anticipate vehicle maintenance needs, and optimize fleet performance based on historical data. Integration with broader smart city infrastructure is also an evolving trend, enhancing overall urban mobility management.

FoV (Field of View)

FoV refers to the total observable area visible to a vehicle's sensor systems at any given moment. This includes the vision ranges provided by cameras, LiDAR, radar, and other sensory technologies equipped in a vehicle. FOV is critical for ensuring vehicles can detect and respond to environmental variables, obstacles, and other road users [15].

A comprehensive FOV is essential for the safe operation of CAVs. It impacts the vehicle's ability to navigate and interact effectively with its surroundings. Proper FOV ensures that autonomous systems can detect obstacles and recognize pedestrians and other critical safety maneuvers (**Figure F.8**).

FOV is influenced by several factors, including sensor placement, the technology type, and integrating multiple sensor inputs through processes like sensor fusion. The optimal FOV varies based on the vehicle's design and operational needs, particularly in how these vehicles detect peripheral objects and events, which are critical for DDTs.

FIGURE F.8 Direct horizontal FoV. The binocular obstruction is seen by both eyes. The monocular obstruction is seen only by the left eye [15].

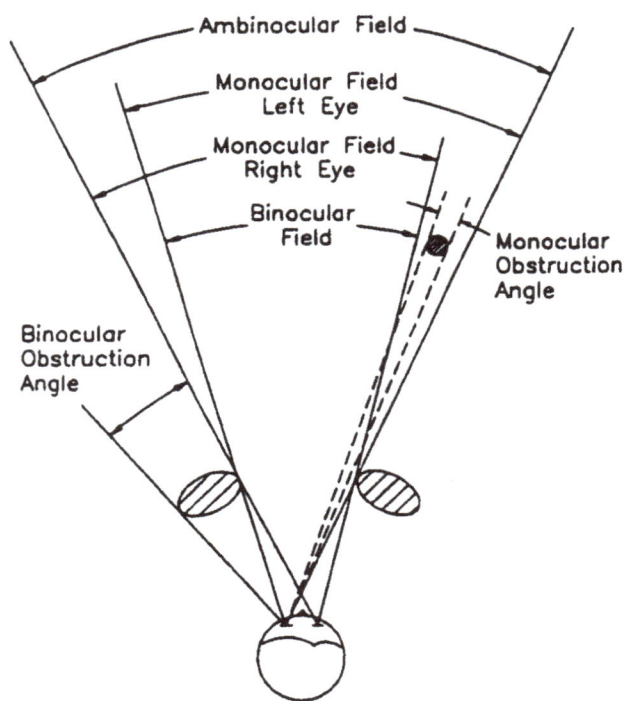

FOW (Forward Obstruction Warning)

FOW systems are ADAS designed to alert drivers of potential obstacles in their path, including vehicles, pedestrians, and other hazards. This system utilizes sensors and cameras to detect objects in front of the vehicle and provides visual, auditory, or haptic warnings to prevent collisions.

FOW systems typically use radar, LiDAR, or camera-based sensors to detect obstacles. Depending on the vehicle model and system design, the warnings can be auditory (alarms), visual (dashboard lights or heads-up displays), or haptic (vibration in the steering wheel or seat) [4]. The collision avoidance system primarily enhances safety by preventing front-end collisions. Driver assistance acts as an additional layer of security for drivers, particularly in complex driving environments like urban centers or congested highways.

REFERENCES

1. Society of Automotive Engineers, "SAE Standards News: J3016 Automated-Driving Graphic Update," January 7, 2019, accessed July 23, 2024, https://www.sae.org/news/2019/01/sae-updates-j3016-automated-driving-graphic.

2. ISO, "ISO 26262 Road Vehicles - Functional Safety Part 1-12, Second Edition," Geneva, 2018.

3. ISO, "ISO 21448:2022 Road Vehicles - Safety of the Intended Functionality," Geneva, 2022.

4. Department of Transportation, "FMVSS 101 Controls and Displays," Washington, DC, 2000.

5. Society of Automotive Engineers, "J3029 Forward Collision Warning and Automatic Emergency Braking Test Procedure and Minimum Performance Requirements - Truck and Bus," SAE Publishing, Warrendale, PA, 2023.

6. Society of Automotive Engineers, "J2640 General Automotive Embedded Software Design Requirements," SAE Publishing, Warrendale, PA, 2008.

7. Silva, J., "Advanced Firmware Device Manager for Automotive: A Case Study," *SAE Int. J. Passeng. Cars - Electron. Electr. Syst.* 5, no. 1 (2012): 34-45, doi:https://doi.org/10.4271/2012-01-0013.

8. Muneer, M. and Haseena, V., "An ML-Based Fleet Management System for Electric Vehicles," SAE Technical Paper 2022-28-0300 (2022), doi:https://doi.org/10.4271/2022-28-0300.

9. Michael, M., Eichberger, A., and Dragoti-Cela, E., "Optimization Approach to Handle Global CO_2 Fleet Emission Standards," SAE Technical Paper 2016-01-0904 (2016), doi:https://doi.org/10.4271/2016-01-0904.

10. Society of Automotive Engineers, "J2735 V2X Communications Message Set Dictionary™ Set," SAE Publishing, Warrendale, PA, 2023.

11. Society of Automotive Engineers, "J2945 Performance Requirements for Cooperative Adaptive Cruise Control (CACC) and Platooning," SAE Publishing, Warrendale, PA, 2023.

12. Society of Automotive Engineers, "J3061 Cybersecurity Guidebook for Cyber-Physical Vehicle Systems," SAE Publishing, Warrendale, PA, 2021.

13. Horani, M., Al-Refai, G., and Rawashdeh, O., "Towards Video Sharing in Vehicle-to-Vehicle and Vehicle-to-Infrastructure for Road Safety," SAE Technical Paper 2017-01-0076 (2017), doi:https://doi.org/10.4271/2017-01-0076.

14. Department of Transportation, "FMVSS 111 Rear Visibility," Washington, DC, 2019.

15. Society of Automotive Engineers, "J1050 Describing and Measuring the Driver's Field of View," SAE Publishing, Warrendale, PA, 2009.

GNSS (Global Navigation Satellite System)

GNSS refers to a constellation of satellites providing signals from space that transmit positioning and timing data to GNSS receivers. The receivers then use these data to determine location.

GNSS is a standard term for satellite navigation systems that provide autonomous geospatial positioning with global coverage. This term encompasses systems such as the United States' GPS, Russia's GLONASS, Europe's Galileo, and China's BeiDou. GNSS uses multiple satellites to enable a receiver to determine its position accurately.

In ACV technologies, GNSS is crucial for accurate positioning, navigation, vehicle tracking, and ADAS [1]. These systems are not standalone but rather rely on the integration of GNSS data with other sensors and data sources to operate safely and efficiently, enhancing location accuracy and reliability. Differential GNSS uses additional correction information transmitted by reference stations and provides high-accuracy positioning (Figure G.1).

FIGURE G.1 An example of the system of satellites around the earth.

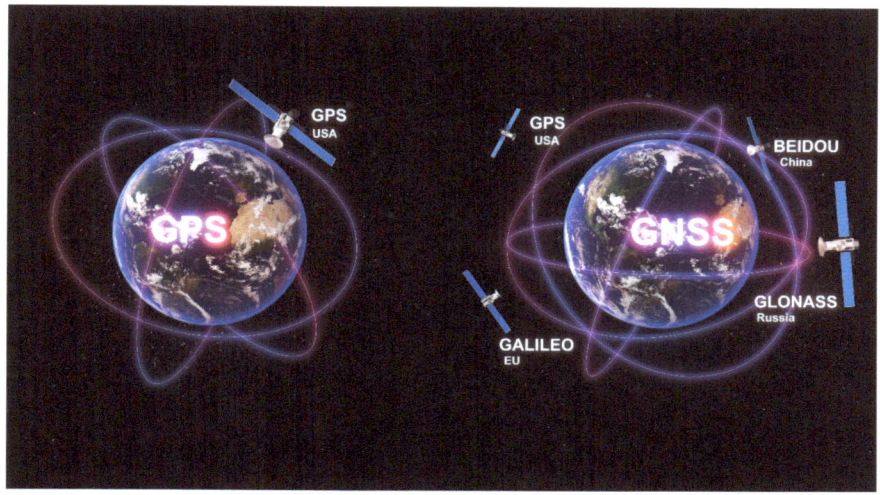

GPS (Global Positioning System)

GPS is a satellite-based navigation system composed of satellites that orbit the earth and transmit precise signals, allowing GPS receivers to calculate and display accurate location, speed, and time information to users anywhere on the globe (see GNSS). Originally developed by the US Department of Defense for military navigation, GPS has since become indispensable in various civilian, commercial, and scientific applications.

GPS is fundamental for operating location-based services, navigation, and mapping in ACVs. With minimal human intervention, these vehicles can determine their location and navigate to destinations. GPS data are often combined with other sensors like LiDAR, radar, and cameras to enhance the vehicle's understanding of its environment, contributing to safer and more efficient operation. This integration is crucial for path planning, obstacle avoidance, and maintaining real-time connectivity with TMSs to optimize routes and reduce congestion. GPS recordings are also used for accident reconstruction efforts [2].

For more information, see **J2735** and **J2945** [3, 4].

Green Driving

Green driving refers to driving practices, vehicle technologies, and transportation policies aimed at reducing the environmental impact of vehicles, particularly those that are autonomous (self-driving) and connected (interconnected via networks with other vehicles and infrastructure) [5]. This concept is integral to the development of sustainable mobility solutions.

Key Features:

1. **Eco-Friendly Routing:** AVs utilize algorithms to select the most fuel-efficient routes, considering factors like traffic, road conditions, and elevation changes to minimize energy consumption.
2. **Optimal Speed Management:** These vehicles can adjust their speed in real time to maintain optimal fuel efficiency, reducing unnecessary acceleration and idling.
3. **V2V Communication:** CVs share information about traffic conditions and road hazards, allowing for cooperative and synchronized driving patterns that can lower fuel use and reduce emissions.
4. **V2I Communication:** Interaction with traffic signals and other infrastructure components to maintain steady traffic flow and reduce stop-and-go driving.
5. **Advanced Energy Management:** Integration of systems that monitor and control energy use, including hybrid and electric powertrains that are managed to maximize efficiency and battery life.

For more information, see **J3016** [6].

Green Route Planning

Green route planning in CVs involves optimizing travel routes based on criteria such as reducing fuel consumption, lowering emissions, and minimizing the environmental impact of vehicle operations. This strategy utilizes advanced route algorithms and real-time data analytics to guide vehicles along the most environmentally friendly routes. Techniques like adaptive route planning consider dynamic variables, including traffic congestion and vehicle density, to enhance transport efficiency and environmental benefits. This form of route planning is particularly crucial in urban areas, where reducing congestion and emissions is a top priority.

Green Transportation

Green transportation encompasses systems and methods that minimize environmental impacts through reduced emissions and energy consumption. This term is particularly significant in CVs, where advanced technologies and data analytics are employed to optimize vehicle performance and fuel efficiency, thus contributing to environmental sustainability.

CVs generate vast data on vehicle performance, including fuel consumption and emission rates. These data can be analyzed to identify inefficiencies and develop more environmentally friendly driving patterns or vehicle features. Integrating AI and other intelligent technologies in CVs supports eco-driving practices. These technologies can optimize routes, improve traffic flow, and enhance vehicle operations, contributing to lower fuel consumption and reduced emissions. Connected EVs, including hybrid and fully electric models, leverage connectivity to enhance battery management and driving strategies, thus improving energy efficiency and reducing dependence on fossil fuels. Advanced research in this area explores the dynamics of energy management and predictive analytics to reduce energy consumption further.

REFERENCES

1. Kitano, H., Kazuo, H., and Tanaka, H., "An Estimation Method of Vehicle Position for Automated Driving with GNSS," SAE Technical Paper 2016-01-0166 (2016), doi:https://doi.org/10.4271/2016-01-0166.

2. Bortolin, R., Hrycay, J., and Golden, J., "GPS Device Comparison for Accident Reconstruction," *SAE Int. J. Passeng. Cars - Electron. Electr. Syst.* 5, no. 1 (2012): 343-357, doi:https://doi.org/10.4271/2012-01-0997.

3. Society of Automotive Engineers, "J2735 V2X Communications Message Set Dictionary," SAE Publishing, Warrendale, PA, 2023.

4. Society of Automotive Engineers, "J2945 Performance Requirements for Cooperative Adaptive Cruise Control (CACC) and Platooning," SAE Publishing, Warrendale, PA, 2023.

5. Marques, N., Campos, G., and Cavalcante, M., "Driving towards a Sustainable Future: Leveraging Connected Vehicle Data for Effective Carbon Emission Management," SAE Technical Paper 2023-36-0145 (2023), doi:https://doi.org/10.4271/2023-36-0145.

6. Society of Automotive Engineers, "J3016 Taxonomy and Definitions for Terms Related to Driving Automation Systems for On-Road Motor Vehicles," SAE Publishing, Warrendale, PA, 2021.

HAVs (Hybrid Autonomous Vehicles)

HAVs refer to vehicles that combine autonomous driving features and traditional manual control systems. Unlike fully AVs, which operate without human intervention, HAVs allow drivers to switch between autonomous and manual driving, offering flexibility based on driving conditions or personal preference.

HAVs are equipped with systems that enable both self-driving capabilities and manual control by the driver. This dual functionality allows the vehicle to operate autonomously in safe and efficient situations, while also providing the option for human control when needed or desired.

These vehicles integrate advanced sensors, cameras, and AI to navigate roads, detect obstacles, and make driving decisions in autonomous mode. When switched to manual mode, the control reverts to human input, although supportive technologies such as ADAS may remain active.

HAVs are designed with a focus on safety, featuring systems that can intervene or alert the driver in critical situations. Switching to manual driving enables drivers to handle complex scenarios that AI may not yet be fully capable of managing.

As HAVs include both autonomous and manual driving capabilities, they must comply with regulatory standards for both types of vehicles. This includes passing safety inspections and meeting the legal requirements for autonomous technology and manual vehicle operation.

HAVs are particularly useful in transitioning periods where full autonomy is not yet practical or legally permitted across all environments. They offer a practical solution for consumers and fleets to benefit from autonomous technologies while retaining the option to drive manually when preferred or necessary. This dual capability makes them a versatile choice in the evolving landscape of automotive technology.

Highway Assist

Highway assist is an ADAS designed to assist the driver in maintaining longitudinal (propulsion and braking) and lateral (steering) vehicle control. This system combines ACC, lane-keeping assist, and sometimes other functionalities to provide semi-autonomous driving capabilities. Highway assist systems aim to improve safety and comfort during long-distance highway travel.

ACC automatically adjusts the vehicle's speed to maintain a safe distance from the vehicle ahead. LKA provides steering assistance or alerts if the vehicle begins to drift without signaling. LCC maintains the vehicle in the center of the lane when able, together with ACC.

Highway Pilot/Highway Chauffeur

Highway pilot, also known as highway chauffeur, refers to an advanced autonomous driving mode in CAVs designed to operate on highways with minimal human intervention. This mode allows the vehicle to navigate highways, manage lane changes, maintain safe distances from other vehicles, and adapt to traffic conditions autonomously.

Highway pilot-equipped CAVs have a sophisticated sensor suite, including lidar, radar, cameras, and ultrasonic sensors. These sensors provide a 360-degree perception of the vehicle's surroundings, enabling it to detect other vehicles, obstacles, lane markings, and road signs. Advanced perception, planning, and control systems, powered by AI algorithms, process sensor data in real time to make driving decisions. These systems control steering, acceleration, braking, and lane changes to navigate the highway safely and efficiently.

Highway pilot systems may utilize V2V and V2I communication to exchange relevant data with other vehicles and TMSs. This communication enhances situational awareness and enables cooperative driving functionalities.

Highway pilot systems adhere to safety standards such as those specified by SAE International, FMVSS, and CMVSS. These standards ensure that autonomous driving functions are reliable, safe, and compliant with regulatory requirements.

HMI (Human-Machine Interface)

HMI refers to the system or set of systems that interact between a human operator and a machine, vehicle, or device [1]. In the context of CVs, HMI encompasses the technologies and methods that enable drivers and passengers to communicate with vehicle systems and access information. This includes visual displays, touchscreens, auditory commands and feedback, gesture recognition, and other user input and output mechanisms (**Figure H.1**).

FIGURE H.1 Driver as part of the control structure [1].

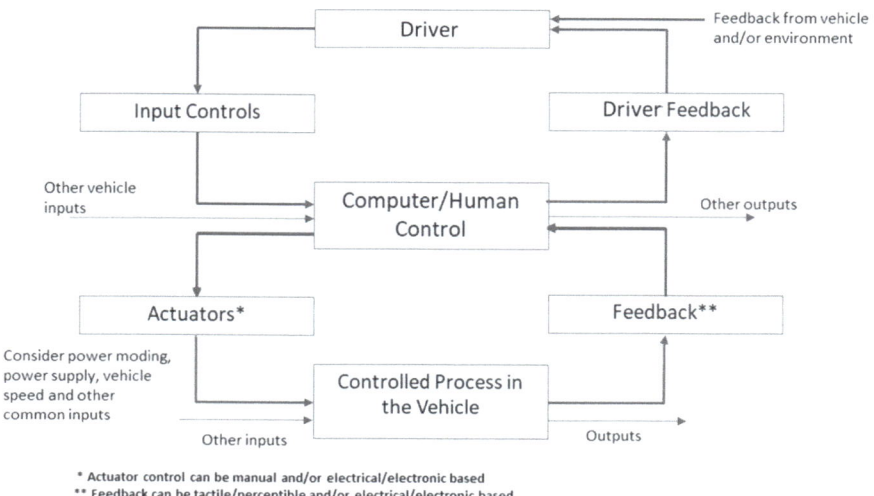

* Actuator control can be manual and/or electrical/electronic based
** Feedback can be tactile/perceptible and/or electrical/electronic based

© SAE International.

In CVs, HMI is crucial for facilitating the efficient, intuitive, and safe interaction between the human user and the sophisticated technologies embedded within modern vehicles. It plays a critical role in how information from V2V and V2I communications is conveyed to the driver, influencing decision-making and vehicle control.

Driver information systems display real-time information about vehicle status, navigation, traffic conditions, and environmental alerts [2]. Infotainment systems integrate audio, video, Internet applications, and telecommunications with vehicle functions.

Homologation

Homologation is the process of certifying that a vehicle, its components, and its systems meet the regulatory standards and requirements of various countries or regions before it can be legally sold or used [3]. This process ensures that vehicles conform to internationally recognized safety, environmental, and performance standards [4].

Safety certification ensures vehicles meet safety standards concerning crashworthiness, materials, and system integrity [5]. Environmental compliance certifies that vehicles comply with emissions and environmental impact standards. Performance verification and testing verify that vehicles perform according to manufacturers' specifications [6].

Navigating the differences in regulations across countries and regions can be complex and costly for manufacturers (**Figure H.2**). Keeping up with rapid advancements in vehicle technology, such as electric powertrains and autonomous driving systems, may outpace existing regulations. State-of-the-art standards published by organizations such as the SAE and the ISO help bridge the gap between technological advancements and regulatory frameworks (**Figure H.3**).

FIGURE H.2 Process scheme of the test management for homologation including the attached interaction with market sales programs and the resulting drag coefficients in those markets [6].

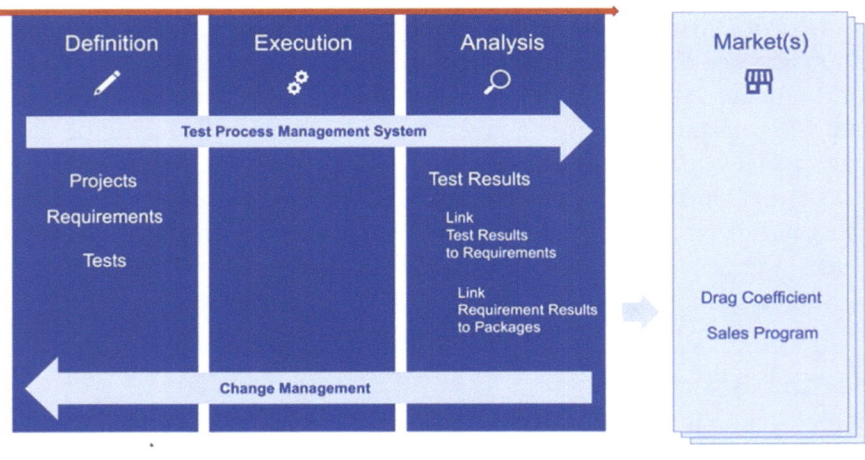

FIGURE H.3 The Homologation 4.0 Framework [7].

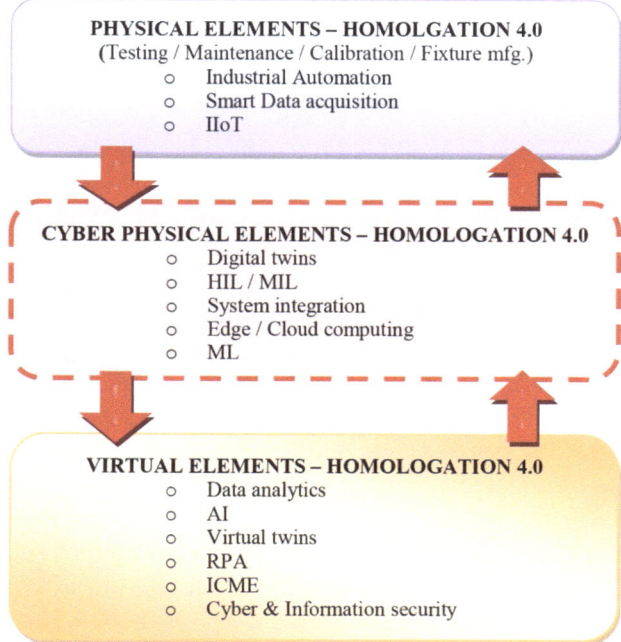

HUD (Head-Up Display)

A HUD is an electronic system used in automotive vehicles to present critical data within the driver's line of sight. This technology projects information, usually on the windshield, allowing drivers to focus on the road ahead. HUD systems typically display speed, navigation directions, and safety warnings.

The primary function of a HUD is to increase driver safety and provide a more comfortable driving experience by minimizing the need to look away from the road to check instruments and controls. Modern HUD systems can also interface with external systems and sensors to provide real-time information such as traffic conditions, vehicle status, and ADAS alerts.

HUD systems utilize a combination of optics, projectors, and reflective materials to display information directly in the driver's field of vision. Technology advancements have expanded HUD capabilities to include AR, integrating real-time digital information with the external environment to significantly enhance decision-making processes and boost driver confidence.

HUD systems are subject to regulatory standards that focus on safety, including guidelines for driver distraction and visibility requirements under various lighting conditions [2].

Human Factors

Human factors refer to the study and application of psychological, physiological, and ergonomic principles to optimize human well-being and overall system performance. In the context of CAVs, human factors involve analyzing and designing vehicle systems that facilitate intuitive interactions between humans and technology, enhance comfort, ensure safety, and improve usability.

The design of interfaces in CAVs is crucial for effective communication between the vehicle and the driver or passengers. This includes the layout of dashboard displays, auditory signals, and tactile feedback systems, which must be clear, intuitive, and accessible.

The physical environment of CAVs is adjusted to fit human physical abilities and limitations. Seat design, spacing, and control accessibility are tailored to enhance comfort and minimize strain during interaction with vehicle technologies.

CAVs must manage and present information in a way that does not overwhelm the user. The systems should aid in decision-making without causing distraction or cognitive overload, considering the limits of human attention and information processing.

In CAVs, maintaining high levels of situation awareness is essential. The vehicle should provide the human operator with relevant information about its operation and the external environment, keeping them informed and ready to take control if necessary.

The overall experience of interacting with CAVs should be positive, fulfilling user expectations and needs. This encompasses ease of use, satisfaction, and the ability to foster trust and reliability in the technology. Systems should be flexible to accommodate various user capabilities and preferences. Adaptive interfaces that adjust to individual user profiles can enhance safety and comfort.

REFERENCES

1. Society of Automotive Engineers, "J3187 System Theoretic Process Analysis (STPA) Recommended Practices for Evaluations of Safety-Critical Systems in Any Industry - Appendix: STPA and Human Machine Interactions (HMIs)," SAE Publishing, Warrendale, PA, 2023.

2. Department of Transportation, "FMVSS 101 Controls and Displays," Washington, DC, 2000.

3. Pries, K.H. and Quigley, J.M., *Testing Complex and Embedded Systems* (Boca Raton, FL: CRC Press, Taylor & Francis Group, 2011).

4. Quigley, J.M., *Dictionary of Testing, Verification and Validation* (Warrendale, PA: SAE Publishing, 2024).

5. Quigley, J.M. and Shenoy, R., *Project Management for Automotive Engineers: A Field Guide* (Warrendale, PA: SAE Publishing, 2016).

6. Jacob, J., "Software-Supported Processes for Aerodynamic Homologation of Vehicles," SAE Technical Paper 2024-01-3004 (2024), doi:https://doi.org/10.4271/2024-01-3004.

7. Thipse, Y., "Genesis of the 'Automotive Homologation 4.0' Framework for India," SAE Technical Paper 2024-26-0360 (2024), doi:https://doi.org/10.4271/2024-26-0360.

I

I2V (Infrastructure-to-Vehicle) Communication
See **V2I (Vehicle-to-Infrastructure)**.

IMU (Inertial Measurement Unit)
An IMU is an electronic device that combines multiple inertial sensors—typically accelerometers and gyroscopes—to measure a vehicle's acceleration and rotational dynamics along three orthogonal axes—namely yaw, pitch, and roll, as well as longitudinal, lateral, and vertical movements; see the following equation [1]. IMUs provide essential data for vehicle motion tracking, stability control, navigation, and ADAS, especially in scenarios where GPS or external sensors are unavailable or unreliable:

$$g_x^2 + g_y^2 + g_z^2 = |g|^2.$$

When GPS data are unavailable or unreliable, the last known good GPS data are used to estimate motion parameters by integrating them with the IMU data. This is known as "dead reckoning."

Key Features:
- **Six-Degree-of-Freedom Sensing:** Measures three-axis acceleration (linear movement) and three-axis angular velocity (rotational movement), enabling full six-degree (6D) motion capture.
- **Sensor Fusion:** IMUs work with other sensors (GPS, LiDAR, radar, and cameras) to enhance vehicle localization, navigation, and object/event detection, especially for autonomous and ADAS-equipped vehicles.
- **Reliability and Redundancy:** Triple-redundant IMU architectures are used in safety-critical applications to ensure reliability and fault tolerance by cross-checking data from multiple independent IMUs.
- **AI and Low Power:** Newer automotive IMUs integrate AI cores for on-sensor event detection, classification, and ultra-low power operation, supporting always-on use cases such as theft prevention, telematics, and motion-activated functions.

In-Car Adaptive Lighting

In-car adaptive lighting refers to advanced automotive lighting systems that automatically adjust the vehicle's headlights (high and low beams), fog lamps, and corner lamps [2]. This includes adaptive forward lighting to accommodate curves [3]. In some cases, interior lighting is based on real-time driving conditions, vehicle dynamics, and environmental factors (**Figure I.1**). These systems are a key feature of ADASs and connected car platforms, enhancing visibility, safety, and comfort for drivers and other road users [4].

FIGURE I.1 Maximum horizontal swivel angle [3].

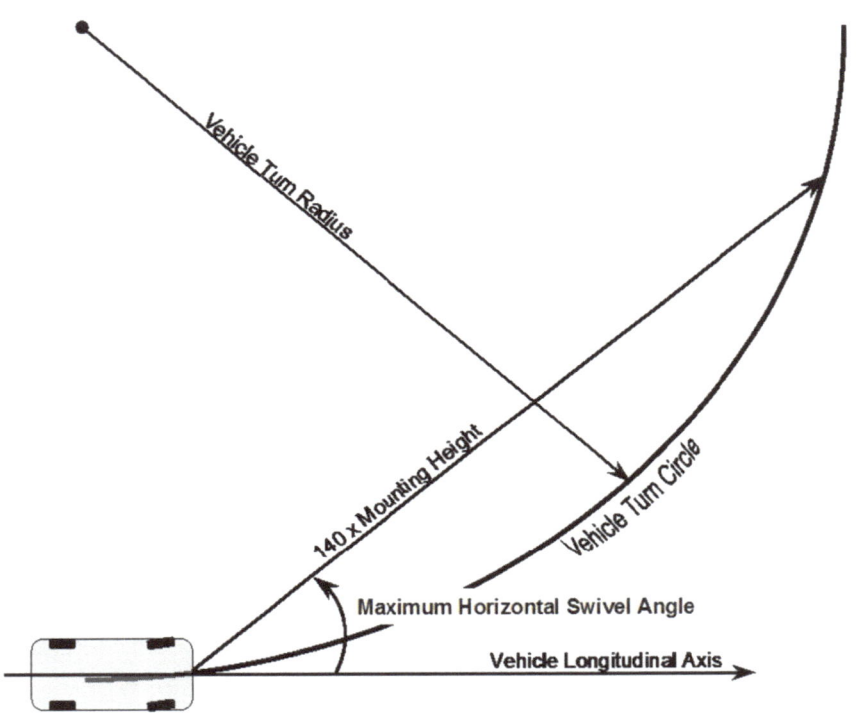

Key Features:

- **Adaptive Front Lighting System (AFS):** Adjusts the direction and intensity of headlights based on steering angle, vehicle speed, road curvature, and ambient lighting. This allows for better lighting around bends and corners, especially at night or in low-light conditions [3].

- **ADB:** Dynamically controls the headlight beam pattern to maximize forward illumination while minimizing glare for oncoming or preceding vehicles. ADB systems use sensors and cameras to detect other road users and selectively dim or redirect portions of the beam [5].
- **AHB Control:** Switches between high and low beams based on traffic, ambient lighting, type of road, and the presence of other vehicles, reducing the need for manual intervention (**Figure I.2**).
- **Sensor Integration:** Utilizes data from cameras, steering angle sensors, speed sensors, and sometimes radar or LiDAR to inform lighting adjustments in real time.

FIGURE I.2 CIE 1931 chromaticity diagram [6].

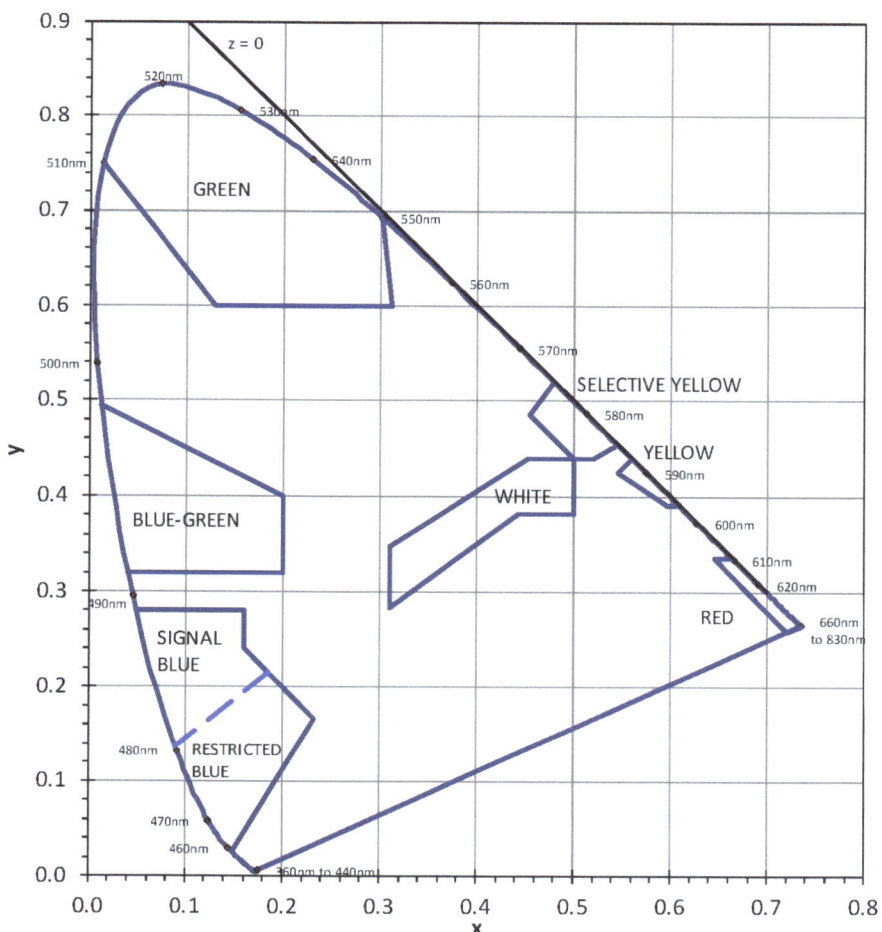

In-Car Biometric Authentication

In-car biometric authentication refers to the use of unique biological traits, such as fingerprints, facial features, iris patterns, and voice recognition, to identify and authenticate drivers and passengers within a vehicle. This technology replaces or supplements traditional key-based or Personal Identification Number (PIN)-based access, providing secure, personalized, and seamless interaction with vehicle systems and connected services [7].

Key Features:
- **Vehicle Access and Ignition:** Biometric systems, such as fingerprint readers or facial recognition cameras (often placed on the B-pillar or dashboard), enable keyless entry and ignition, ensuring only authorized users can unlock or start the vehicle.
- **Personalization:** Upon authentication, the vehicle automatically adjusts settings (seat position, climate control, infotainment, mirrors) according to the recognized user's profile, enhancing comfort and convenience.
- **Secure Access Control:** Restricts access to sensitive vehicle functions and data based on user identity and authorization level, protecting privacy and preventing unauthorized use.
- **In-Car Payments and Services:** Enables secure transactions for fuel, charging, tolls, and in-car purchases by authenticating the user before authorizing payments.
- **Car Sharing and Fleet Management:** Facilitates secure, user-specific access in shared mobility scenarios, ensuring only registered users can operate the vehicle and automatically loading their preferences.

Types of Biometric Modalities:
- **Physical Biometrics:** Fingerprints, facial recognition, iris/retina scans.
- **Behavioral Biometrics:** Voice recognition, gait analysis, keystroke dynamics.
- **Multimodal Systems:** Combine two or more biometric methods for enhanced security and reliability.

In-Car Breathalyzer

An in-car breathalyzer—an ignition interlock device (IID) or breath-based alcohol detection system—is an integrated automotive safety feature that measures a driver's **br**eath **a**lcohol **c**oncentration (BrAC) [8]. This includes before and/or during vehicle operation. If the detected BrAC exceeds a preset legal or programmed threshold, the system prevents the vehicle from starting or may intervene while the vehicle is in motion. The detection may be from a mouthpiece or passive devices. This technology is designed to prevent alcohol-impaired driving and is increasingly mandated or integrated as standard equipment in new vehicles. The integration of in-car breath analyzers helps minimize accidents due to driving under the influence, which is one of the significant causes of road accidents.

Key Features [8]:
- **Breath-Based Detection:** The most common approach requires the driver to provide a breath sample, which is analyzed for alcohol content. Modern systems may use passive sensors to detect alcohol from ambient breath, eliminating the need for a mouthpiece.
- **Ignition Interlock:** The system is linked to the vehicle's ignition; if BrAC exceeds the threshold, the engine will not start.
- **Rolling Retests:** Many devices require periodic breath samples while the vehicle is in motion to ensure ongoing sobriety. If a retest fails or is missed, the system may trigger alarms or log violations, but it does not stop the engine for safety reasons.
- **Data Logging:** Devices record test results, compliance, and tampering attempts, which authorities or program administrators can review.
- **Continuous and Passive Monitoring:** Advanced systems can continuously monitor blood alcohol concentration (BAC) through sensors embedded in the cabin, steering wheel, or dashboard, providing seamless and nonintrusive operation.

In-Car Child Safety Alerts

In-car child safety alerts—often referred to as child presence detection (CPD) or rear occupant alert (ROA) systems—are advanced safety features designed to detect the presence of children (or other vulnerable occupants such as pets) left unattended in a vehicle. These systems provide escalating notifications and interventions to alert drivers, caregivers, or emergency services, helping prevent heatstroke and other dangers associated with children being left in hot cars.

Key Features:
- **Occupant Detection:** Uses sensors such as radar, ultrasonic, weight sensors, or cameras to detect the presence and even vital signs (like breathing) of children in the vehicle, including those asleep or hidden in footwells.
- **Alert Escalation:** Triggers a series of warnings—audible alarms, dashboard messages, flashing lights, and push notifications to

smartphones or connected devices—if a child is detected after the driver exits or locks the vehicle.
- **Connected Alerts:** Some systems integrate with mobile apps or cloud services to send alerts to driver and/or emergency contacts and may even override smartphone silent modes for critical notifications.
- **Climate and Vehicle Controls:** Advanced systems may restrict remote climate control changes or software updates if a child is detected, maintaining a safe cabin environment.
- **Rolling and Continuous Monitoring:** Systems remain active for a defined period (e.g., 15 min post-ignition off) and can detect children who enter an unlocked vehicle after the driver has left.
- **Integration with Vehicle Systems:** May connect with GPS, voice assistants, and other vehicle infrastructure for enhanced safety and UX.

In-Car Cognitive Computing

In-car cognitive computing integrates advanced AI, ML, NLP, and data analytics within vehicles to enable real-time perception, reasoning, learning, and decision-making [9]. These systems mimic human cognitive abilities to enhance automation, safety, personalization, and efficiency in driver-operated and AVs.

Key Features:
- **Perception and Sensing:** Cognitive computing platforms process data from various vehicle sensors (cameras, LiDAR, radar, microphones, etc.) to interpret the environment, recognize objects, and assess driving conditions.
- **Contextual Understanding:** Systems analyze and comprehend complex scenarios—including traffic, weather, and driver behavior—enabling context-aware responses and dynamic adaptation.
- **Natural Language Interaction:** This technology enables conversational interfaces, allowing drivers and passengers to interact with vehicles using natural language for commands, queries, and feedback.
- **Personalization:** Learns driver preferences, habits, and routines to tailor infotainment, navigation, and comfort settings and provide proactive recommendations.
- **Predictive Analytics:** Anticipates hazards, maintenance needs, and route optimizations by learning from historical and real-time data.

- **Collaborative Intelligence:** Vehicles can share insights and learnings with other vehicles, infrastructure, and cloud platforms, contributing to collective intelligence for safer and more efficient transportation.
- **Continuous Learning:** Cognitive systems update their models and improve performance over time by learning from new data and user interactions.

In-Car Driver Profiles

In-car driver profiles are digital systems that store and automatically recall a wide range of personalized vehicle settings and preferences for individual users. These profiles enhance comfort, convenience, safety, and personalization by tailoring the driving experience to each recognized driver in private vehicles, shared cars, or fleet environments.

Key Features:

- **Personalized Settings:** Driver profiles save preferences such as seat position, steering wheel and mirror adjustments, climate control, lighting, infotainment settings, and even HUD configurations.
- **Automatic Recognition:** Vehicles identify drivers using key fobs, smartphones, biometric authentication (e.g., facial recognition and fingerprints), or cloud-based accounts and automatically load the corresponding profile upon entry.
- **Cloud and Cross-Vehicle Sync:** Many systems store profiles in the cloud, allowing settings to follow drivers across multiple vehicles of the same brand or within a shared mobility ecosystem.
- **Security and Privacy:** Some systems offer PIN protection or guest profiles to prevent unauthorized access or changes, especially when the vehicle is being handed over for servicing or sharing.
- **Behavioral and Usage Data:** Advanced driver profiling can include analysis of driving habits, trip patterns, and behavioral data to provide insights, recommendations, or even adaptive ADAS features.
- **Special Modes:** Profiles can be created for specific scenarios, such as "Easy Entry" for easier ingress/egress, "Loading" for cargo space, or "Camping" for comfort during stationary use.

In-Car Entertainment

In-car entertainment, or in-vehicle infotainment (IVI), refers to the integrated hardware and software systems in automobiles that provide drivers and passengers with audio, video, connectivity, gaming, and information services. Initially limited to radios and compact disc (CD) players, modern in-car entertainment systems now encompass navigation, smartphone integration, Internet access, conferencing, and multimedia features. See infotainment.

In-Car Facial Recognition

In-car facial recognition is a biometric technology that uses cameras and AI algorithms to analyze drivers' and passengers' facial features to identify and authenticate them. This technology is increasingly integrated into modern vehicles to enhance security, personalization, driver monitoring, and convenience.

Key Features:

- **Keyless Access and Ignition:** Facial recognition enables vehicles to automatically unlock doors and start the engine when an authorized user is detected, eliminating the need for traditional keys or fobs. Cameras are often integrated into the B-pillar or invisibly behind the driver display for a seamless UX.

- **Personalization:** Upon recognizing a driver or passenger, the system can automatically adjust seat position, climate control, mirrors, infotainment settings, and more, based on stored user profiles. This creates a tailored in-cabin experience for each individual.

- **Driver Monitoring and Safety:** Facial recognition systems, often combined with DMSs, track the driver's gaze, eyelid movement, and head position to detect drowsiness, distraction, or signs of impairment. Real-time alerts can prompt the driver to refocus or take a break, and advanced systems may even prevent the vehicle from starting if impairment is detected.

- **Enhanced Security:** Only authorized users can access or start the vehicle, reducing the risk of theft or unauthorized use. Liveness detection and anti-spoofing features ensure that photos or videos cannot easily fool the system.

- **In-Car Payments and Services:** Facial recognition can authenticate users for secure in-car payments, such as for fuel, parking, or tolls, without the need for physical cards or devices.

- **Car Sharing and Multi-User Management:** The technology simplifies car sharing by automatically identifying different users, applying their preferences, and managing access permissions.

How It Works:
1. **Enrollment:** Drivers and passengers enroll by having their facial features captured and stored securely by the vehicle's system.
2. **Recognition:** When approaching or entering the vehicle, cameras scan faces and match them against stored profiles.
3. **Authentication:** If a match is found, the system grants access, starts the vehicle, and loads personalized settings. Access is denied or limited if not, and the owner may be notified of unauthorized attempts.
4. **Monitoring:** During driving, the system continuously monitors the driver for signs of fatigue, distraction, or impairment, issuing alerts or taking action as needed.

In-Car Gesture Recognition

In-car gesture recognition is a technology that enables drivers and passengers to control vehicle functions by simply glancing at the display [10]. Hand or finger movements are interpreted by sensors and AI algorithms. Integrated into ADAS and connected car platforms, it reduces the need for physical interaction with buttons or touchscreens, thereby enhancing safety and convenience [9].

Key Features:
- **Touchless Control:** Adjust infotainment (volume and navigation), climate settings, calls, and lighting via predefined gestures (e.g., swipes, circles, and pinches).
- **Sensor Integration:** This technique combines infrared (IR) cameras, 3D depth sensors, radar (e.g., millimeter wave), or capacitive electric field sensors to detect gestures.
- **AI and ML:** Algorithms like 3D convolutional neural networks (3DCNNs) and long short-term memory networks (LSTMs) analyze motion data to improve accuracy and adapt to user-specific gestures.
- **Multimodal Interaction:** This feature works alongside voice commands and touchscreens, allowing drivers to choose the least distracting input method.
- **Privacy-Focused:** Unlike optical cameras, radar and IR systems avoid capturing detailed visuals.

How It Works:
- **Detection:** Sensors (e.g., IR cameras in roof modules and millimeter-wave radar) capture hand movements within a defined 3D zone.

- **Processing:** Trained AI models (e.g., CNNs) translate gestures into commands.
- **Execution:** Commands adjust various functions (e.g., muting audio and zooming maps) or trigger connected services (e.g., smart home controls).

In-Car Health Analytics

In-car health analytics refers to the integration of biometric sensors, AI, and data analytics within vehicles to monitor, analyze, and respond to the physiological and cognitive state of drivers and occupants. These systems aim to enhance safety, wellness, and UX by detecting health-related risks, providing real-time alerts, and enabling proactive interventions.

Key Features:
- **Biometric and Physiological Monitoring:** Sensors embedded in seats, seatbelts, steering wheels, or dashboards can measure vital signs such as heart rate, electrocardiogram (ECG), respiration, skin temperature, and blood oxygen levels. Some systems also monitor behavioral cues such as eyelid movement, gaze, and steering patterns to assess fatigue, distraction, or drowsiness.
- **Real-Time Health Assessment:** AI algorithms analyze sensor data to detect signs of fatigue, drowsiness, stress, distraction, or medical emergencies (e.g., heart attack or asthma attack). Systems can issue alerts, suggest breaks, or adjust vehicle settings to support driver wellness.
- **Cognitive and Emotional State Detection:** Advanced DMSs use cameras, steering-mounted capacitive sensors, and AI to evaluate cognitive load, emotional state, and alertness, enabling tailored safety interventions.
- **Integration with Vehicle Safety Systems:** Health analytics can trigger preventive actions, such as adjusting cruise control, initiating emergency stops, or alerting emergency services via V2X communication, if a critical health event is detected.
- **Occupant Health and Wellness:** Interior sensing solutions extend health analytics to all occupants, supporting features such as CPD and personalized comfort adjustments.
- **Remote Diagnosis and Data Sharing:** Health data can be stored locally or in the cloud, enabling remote diagnostics, integration with healthcare providers, and swifter emergency medical response.

For additional information, see **JA6268** *Design and Run-Time Information for Health Ready Components* [11].

In-Car Mood Lighting

In-car mood lighting, also known as ambient lighting, refers to the use of customizable, often color-changing light-emitting diode (LED) lighting systems within a vehicle's cabin to enhance the atmosphere, comfort, and overall driving experience [12]. Modern systems can dynamically adjust color, brightness, and animation effects in response to user preferences, driving conditions, or vehicle events, thereby enhancing the overall driving experience.

Key Features:

- **Customizable Atmosphere:** Drivers and passengers can select from a wide spectrum of colors and brightness levels to match their mood or the occasion, creating a personalized and welcoming environment.
- **Dynamic and Intelligent Effects:** Advanced mood lighting systems can animate light (e.g., moving, pulsing, or "breathing" effects), synchronize with music, or react to vehicle events such as door opening, incoming calls, or safety alerts.
- **Integration with Smart Technology:** Mood lighting can be linked to infotainment systems, smartphones, and vehicle sensors, allowing for interactive and context-aware lighting scenarios (e.g., changing color based on driving mode or temperature).
- **Human-Centric Lighting:** Some systems use specific color temperatures and high color rendering to support occupant well-being, reduce eye strain, and improve nighttime visibility.
- **Safety and Communication:** Beyond aesthetics, ambient lighting can provide visual cues for warnings, guidance, or interaction, such as highlighting controls or alerting occupants to hazards.

In-Car Payments

In-car payments can initiate and complete financial transactions directly from a vehicle's onboard systems, typically via the infotainment interface. This technology transforms the car into a secure, connected commerce platform, allowing drivers and passengers to pay for goods and services, such as fuel, EV charging, parking, tolls, food, and subscriptions, without leaving the vehicle or using a separate device.

Key Features:

- **Infotainment Integration:** Payments are managed through the car's central display, where users can select services, enter or store payment details, and authorize transactions.

- **Contactless and Seamless:** This type of payment uses wireless technologies and secure authentication (e.g., PIN and biometrics) to enable frictionless, often contactless, transactions.
- **Mobile Wallet and Card Support:** This feature supports integration with credit/debit cards, digital wallets, and sometimes direct links to bank accounts.
- **Connected Services:** Enables OTA purchases of vehicle features (e.g., ADAS upgrades and subscriptions), digital content, and third-party services via integrated marketplaces.
- **Personalization and User Profiles:** Payment preferences and transaction history can be linked to individual driver profiles, allowing personalized offers and streamlined recurring payments.
- **Fleet and Expense Management:** For commercial vehicles, in-car payments can automate expense tracking and integrate with fleet management systems, reducing administrative overhead.

In-Car Personal Assistant

An in-car personal assistant—also known as a virtual personal assistant (VPA) or digital voice assistant is an AI-powered system embedded in a vehicle that enables drivers and passengers to interact with vehicle functions, infotainment, navigation, and connected services using natural language voice commands or, in some cases, multimodal inputs. These assistants leverage AI, ML, and NLP to provide a hands-free, personalized, and context-aware driving experience. They are integrated with other vehicle systems such as driver monitoring, navigation, and driver assistance systems.

Key Features:
- **Voice-Activated Control:** Allows users to operate navigation, communication, entertainment, climate, and vehicle settings by speaking naturally (e.g., "Hey vehicle, navigate to the nearest coffee shop").
- **Personalization:** Learns user preferences for music, routes, climate, and seat positions, adapting over time to deliver a tailored experience.
- **Proactive Assistance:** Offers predictive suggestions and reminders, such as suggesting a break if the driver appears tired, recommending alternate routes to avoid traffic, or reminding about appointments and vehicle maintenance.
- **Contextual Awareness:** Understands complex, multi-step requests and follows up with relevant questions or actions (e.g., finding a restaurant, comparing prices, and booking a table).

- **Integration with Connected Services:** Can make payments, book parking, order food, and access digital marketplaces, extending the assistant's utility beyond the vehicle.
- **Emergency and Support Functions:** This position assists in emergencies, such as contacting roadside support, scheduling service, or giving instructions during critical situations.
- **Continuous Learning and Updates:** Improves with every interaction and can receive OTA updates to expand capabilities and enhance performance.

For additional information, see **J2396** [10].

In-Car Personalization

In-car personalization refers to the use of advanced technologies, such as AI, ML, IoT, sensors, and cloud connectivity, to automatically adapt and customize a vehicle's settings, features, and services to the unique preferences, behaviors, and routines of individual drivers and passengers. This transforms the vehicle from a generic mode of transportation into a highly individualized and interactive mobility experience.

Key Features:
- **User Profiles:** Store and recall personalized settings for seat position, steering wheel, mirrors, climate control, ambient lighting, infotainment preferences, and more. Profiles can be linked to smart keys, biometric authentication, or mobile devices and are often portable across vehicles and platforms.
- **AI-Driven Adaptation:** AI and ML analyze real-time data from sensors, cameras, and user interactions to learn habits and anticipate needs, such as preferred routes, music, or even driving modes, adapting the vehicle dynamically to the user.
- **Contextual and Predictive Personalization:** The system responds to contextual cues (weather, time of day, location, and traffic) and predicts user needs, such as suggesting alternate routes, activating seat warmers in cold weather, or recommending nearby amenities based on past behavior.
- **Hyper-Personalization:** This goes beyond basic settings to include mood detection, stress monitoring, and well-being interventions, using biometric sensors and interior cameras to adjust the in-vehicle environment for comfort and safety.

- **Routine Automation:** Users can define routines—such as "morning commute" or "family trip"—that trigger multiple vehicle functions automatically (e.g., seat and climate adjustments, navigation, and playlist selection) based on time, location, or detected user.
- **Integration with Connected Services:** Personalized recommendations and services, such as maintenance alerts, in-car payments, and digital assistant interactions, are tailored to the user's preferences and driving patterns.
- **Portability across Platforms:** Preferences and profiles can be transferred across different vehicles or shared mobility platforms, ensuring a consistent experience regardless of ownership or usage model.

In-Car Privacy Controls

In-car privacy controls are features and settings within connected and modern vehicles that allow drivers and passengers to manage, restrict, or opt out of collecting, sharing, and processing their personal and behavioral data. These controls are essential as vehicles increasingly function as data hubs, gathering information ranging from location and driving habits to voice recordings and biometric identifiers [13].

Key Features:
- **Data Collection Management:** Users can review and limit what data the vehicle collects, such as location, driving behavior, voice interactions, and biometric data. These are often integrated with the manufacturer's mobile app. Some systems offer granular controls for different types of data [14].
- **Opt-Out and Deletion Requests:** Many automakers now allow consumers to submit requests to:
  - Opt out of data sharing with third parties (e.g., insurers and data brokers).
  - Limit use and disclosure of sensitive personal information (e.g., geolocation and biometrics).
  - Delete personal data from automaker and third-party records, with some exceptions for legal or operational requirements.
- **Mobile App and Infotainment Controls:** Privacy settings are often accessible via a connected mobile app or the vehicle's infotainment system.

Users can toggle data sharing, disable location tracking, or manage consent for specific features.

- **Transparency and Consent:** Modern privacy controls require explicit user consent before collecting or using personal data, in line with regulations like General Data Protection Regulation (GDPR) and state-level privacy laws. Users are informed about what data are collected and how they will be used, with options to review and correct stored data.
- **Data Minimization and Purpose Limitation:** Vehicles are increasingly designed to collect only the data necessary for specific functions, reducing unnecessary storage of sensitive information. Data are used strictly for specified purposes, such as safety features or user-requested services.
- **Encryption and Security:** Vehicle data is encrypted both in transit and at rest, ensuring robust protection against unauthorized access, tampering, and data breaches. This end-to-end encryption framework safeguards sensitive information such as location, user, biometrics, and behavioral data across cloud services, mobile apps, and in-vehicle systems. It forms a critical layer in the vehicle's cybersecurity architecture, supporting compliance with privacy regulations and reinforcing consumer trust in connected mobility platforms.

In-Car Safety Notifications

In-car safety notifications are real-time alerts and messages delivered to drivers and passengers via the vehicle's display, audio, haptic feedback, or connected devices [15]. These notifications are generated by ADAS and connected car platforms to inform users of potential hazards, system status, or required actions, thereby enhancing situational awareness and road safety [9].

Key Features:

- **Hazard Alerts:** Warn drivers of immediate dangers such as collisions, lane departures, blind spot intrusions, sudden braking, or pedestrian detection.
- **System Status Notifications:** Inform users about the operational state of ADAS features (e.g., ACC, lane-keeping assist, and emergency braking) and any malfunctions or limitations [16].
- **Driver Attention and Wellness:** Provide alerts for drowsiness, distraction, or health events detected by monitoring systems, prompting corrective action.

- **V2X Integration:** Receive safety notifications from other vehicles, infrastructure, or cloud services, such as road hazard warnings, emergency vehicle proximity, road congestion and route change, or weather alerts.
- **Escalation and Redundancy:** Use multimodal delivery (visual, auditory, haptic) to ensure critical safety messages are noticed and acted upon, even if one channel is missed.
- **Personalization and Prioritization:** Tailor notification type, urgency, and modality to the driver's preferences, context, and current workload, minimizing distraction while maximizing effectiveness.

In-Car Shopping

In-car shopping, also known as in-car commerce or connected commerce, enables drivers and passengers to browse, purchase, and pay for goods and services directly from a vehicle's infotainment system or through voice assistants. This technology leverages the car's connectivity, user profiles, and payment integration to offer a seamless, hands-free shopping experience while prioritizing safety and minimizing driver distraction.

Key Features:
- **Voice-Driven Commerce:** This technology enables users to make purchases using natural language commands, keeping their hands on the wheel and eyes on the road. Voice assistants are the preferred input method for in-car shopping, especially for drivers.
- **Integrated Marketplaces:** Dashboards and HUDs can host marketplaces where users can order fuel, EV charging, parking, food, and other services. Some systems also support feature-on-demand (FOD), allowing users to unlock vehicle features or subscriptions with a purchase.
- **Personalization:** Shopping recommendations and offers can be tailored to individual driver profiles, trip context, or past purchase history to enhance convenience and relevance.
- **Connected Payments:** Secure payment processing is integrated into the vehicle, supporting credit/debit cards, digital wallets, or direct billing through the automaker's platform.
- **Passenger Participation:** Passengers, not just drivers, can shop during trips—expanding the use case to ridesharing, family travel, and more.

In-Car Sleep Monitoring

In-car sleep monitoring refers to advanced systems integrated into vehicles that continuously assess the driver's state of alertness and detect signs of drowsiness or impending sleep. These systems, a core part of ADAS and connected car safety features, use a combination of sensors, cameras, and AI algorithms to monitor physiological and behavioral cues, issuing timely alerts or interventions to prevent fatigue-related accidents.

Key Features:

- **DMS:** These utilize cameras (often IR), facial recognition, and embedded hardware to track eyelid movements, blink rates, gaze direction, head position, and facial expressions—key indicators of drowsiness or distraction.
- **Physiological and Behavioral Sensing:** Some systems incorporate additional sensors to measure heart rate, steering patterns, and lane-keeping behavior to enhance detection accuracy.
- **Real-Time Alerts:** When early or acute signs of drowsiness (such as microsleeps, prolonged eye closure, or nodding) are detected, the system issues visual, audible, or haptic alerts to prompt the driver to take corrective action.
- **Adaptive and Personalized:** Advanced solutions use ML to adapt to individual drivers, refining detection thresholds and minimizing false alarms over time.
- **Integration with Vehicle Controls:** In higher levels of automation (SAE Level 3+), if the driver fails to respond, the system can initiate a minimal risk maneuver, such as slowing down or stopping the vehicle safely [9].

In-Car Temperature Control

In-car temperature control refers to the systems and UIs within vehicles that regulate the cabin climate, encompassing heating, ventilation, and air conditioning (HVAC), to maintain occupant comfort and safety (**Figure I.3**). These systems can operate manually, automatically, or remotely and are increasingly integrated with ADAS and connected car platforms for enhanced personalization and efficiency.

FIGURE I.3 An example of temperature controls.

Cherkas/Shutterstock.com.

Key Features:

- **Automatic Climate Control:** Uses temperature sensors to monitor cabin and ambient conditions, adjusting airflow, temperature, and humidity automatically to maintain user-set or optimal comfort levels.

- **Remote and Connected Operation:** Modern systems allow users to precondition the cabin via smartphone apps or key fobs, and some can automatically maintain safe temperatures for unattended vehicles (e.g., when pets or sensitive cargo are inside).

- **Personalization:** Driver profiles can store individual temperature and airflow preferences and automatically apply them when a specific user is recognized.

- **Safety and Health Integration:** Systems may integrate with occupant detection, child safety alerts, or health analytics to adjust the climate based on passenger needs or medical conditions.

- **Energy Efficiency:** Advanced controls optimize compressor and fan operation to balance comfort with reduced energy consumption, which is essential for EVs.

In-Car Touchscreen Controls

In-car touchscreen controls are interactive displays embedded in vehicle dashboards that enable drivers and passengers to operate various vehicle functions, including infotainment, navigation, climate control, ADAS features, and connectivity, through direct touch, swipes, and gestures. These systems are designed for ease of use, accessibility, and integration with modern digital lifestyles.

Key Features:

- **Centralized Interface:** Touchscreens consolidate multiple controls (audio, navigation, climate, vehicle settings) into a single, customizable interface, reducing dashboard clutter and physical buttons [10].
- **Intuitive Operation:** Operate much like smartphones, responding to finger presses and swipes, with large icons and simplified menus for safer use while driving [16].
- **Personalization:** Displays can be customized to show preferred functions, layouts, and themes, enhancing comfort and UX.
- **Integration with Smartphones:** Support for Apple CarPlay and Android Auto enables seamless access to phone functions, apps, and content directly from the touchscreen.
- **Advanced Features:** Touchscreens often display feeds from reversing cameras, provide instant access to EV charging and battery status, and integrate with voice controls for hands-free operation.

In-Car Voice Assistant

In-car voice assistants are AI-powered systems embedded in vehicles that allow drivers and passengers to control vehicle functions, access information, and interact with connected services using natural language voice commands. These assistants can manage navigation, entertainment, communication, climate control, and more, providing a hands-free, eyes-on-road experience.

For more information, see **In-Car Personal Assistant**.

In-Car Voice Biometrics

In-car voice biometrics refers to the use of advanced voice recognition and speaker identification technologies to authenticate and identify drivers and passengers within a vehicle. This enables secure, personalized, and hands-free interaction with vehicle systems and connected services. Unlike standard voice command systems, voice biometrics analyze unique vocal characteristics to determine precisely who is speaking, not just what is being said [17].

In-Car Voice Command

In-car voice command uses spoken language to control vehicle functions, access information, and interact with connected services through embedded or cloud-based voice recognition systems [17]. This technology is a core feature of modern ADAS and CVs. It enables hands-free operation of navigation, infotainment, climate, communication, and other systems for improved safety and convenience.

In-Car Weather Updates

In-car weather updates are real-time notifications and data services delivered to drivers and vehicle systems, providing current and forecasted weather and road conditions along a vehicle's route. These updates are integrated into ADAS navigation and connected car platforms to enhance the safety, comfort, and operational decision-making of both human and automated vehicles [16].

In-Car Wi-Fi

In-car Wi-Fi refers to the integration of wireless Internet connectivity within a vehicle, allowing drivers and passengers to connect multiple devices, such as smartphones, tablets, and laptops, to the Internet via an onboard hotspot. This onboard hotspot is enabled by the SIM integrated into modern vehicles. This feature is a core component of modern CVs, supporting entertainment, productivity, navigation, OTA updates, and seamless access to cloud-based services (**Figure I.4**).

FIGURE I.4 Concretization of the environment model [18].

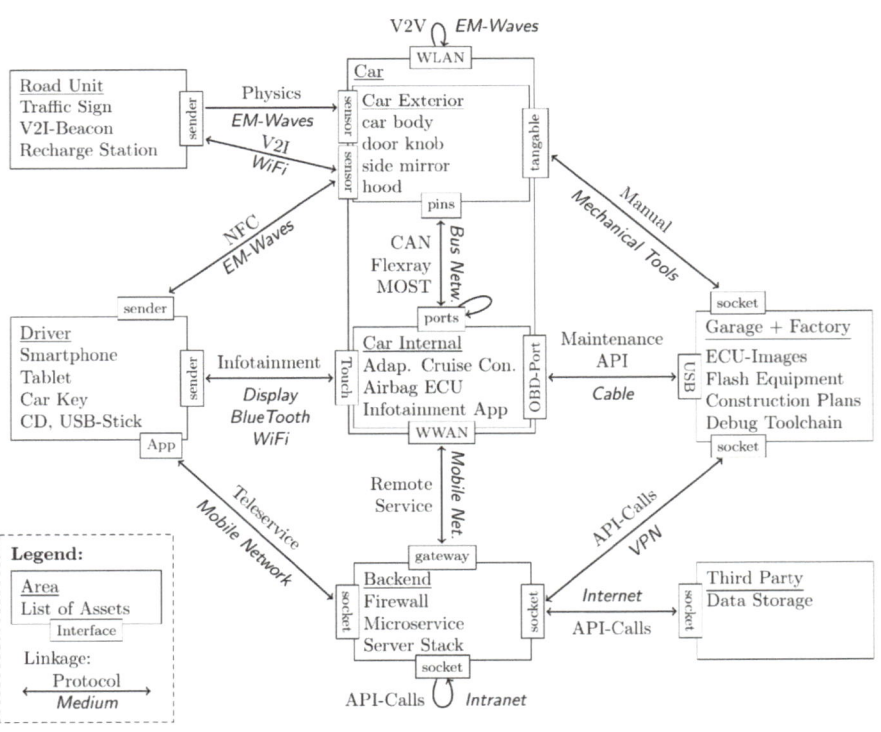

In-Car Wireless Charging

In-car wireless charging refers to technologies that enable electronic devices (such as smartphones and wearables) and EVs without physical cables. For personal devices, this is typically used with inductive charging pads built into the vehicle's interior. For EVs, wireless charging involves transferring power from a ground-based pad to a receiver on the vehicle using electromagnetic resonance or inductive coupling, allowing for efficient, automatic charging simply by parking over the pad (**Figure I.5**).

FIGURE I.5 Predefined AOIs and an illustration of driver fixation and saccadic movements between and within an AOI [10].

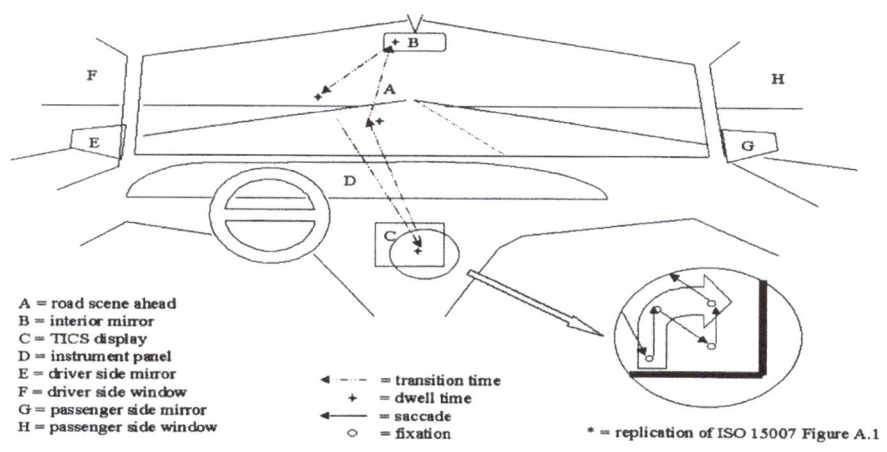

A = road scene ahead
B = interior mirror
C = TICS display
D = instrument panel
E = driver side mirror
F = driver side window
G = passenger side mirror
H = passenger side window

◄---- = transition time
+ = dwell time
◄—— = saccade
○ = fixation

* = replication of ISO 15007 Figure A.1

© SAE International.

Infotainment

Infotainment is a portmanteau of "information" and "entertainment," referring to the integrated digital system in a vehicle that centrally manages a wide range of functions, including multimedia, navigation, connectivity, and vehicle controls (**Figure I.1**). It is also referred to as in-vehicle information systems (IVISs) [19]. Infotainment systems are typically accessed via a touchscreen but may also use physical buttons, rotary controllers, voice commands, or gesture controls [16].

TABLE I.1 Example of information items [16].

Appropriate information items
Turn right onto Poplar in 0.1 mile
Heavy fog on planned route
Toll cost: $1.25
School bus stopped ahead
Construction delays on Route 12
Engine overheating
Icy bridge ahead

© SAE International.

Key Features:

- **Audio and Video Playback:** Access to radio, CD/digital video disc (DVD), Universal Serial Bus (USB), Bluetooth, streaming services, and sometimes video for rear-seat passengers.
- **Navigation:** GPS-based navigation with real-time traffic updates, turn-by-turn directions, and POIs.
- **Smartphone Integration:** Seamless smartphone connectivity using Apple CarPlay, Android Auto, or MirrorLink for music, calls, messaging, and apps.
- **Hands-Free Communication:** Bluetooth-enabled hands-free calling and messaging to reduce driver distraction.
- **Internet and App Access:** In-car Wi-Fi, Internet browsing, and access to downloadable apps for news, weather, music, and more.
- **Vehicle Controls:** Centralized control of climate, seat positions, lighting, trailer management, and other comfort or safety features.
- **Camera Integration:** Display for rearview, 360°, or parking cameras to enhance safety.
- **Telematics:** Display vehicle performance data, diagnostics, and status messages.
- **Voice and Gesture Controls:** Voice recognition and, in some advanced systems, gesture-based controls for safer, hands-free operation.

The system is powered by a CPU running a specialized operating system, often displayed on a dashboard-mounted touchscreen (**Figure I.6**). Modern cars typically feature a single display that is divided into an instrument cluster (IC) and an infotainment display. The IC part of the display is non-touch-activated, and the infotainment display is a touch screen. The IC displays several safety-critical data points that inform the driver about the vehicle's current state, such as its speed. Engine rpm, cruise control state, warnings related to vehicle components, and ADAS. The infotainment software also has stringent requirements as it displays safety-related information such as rear-view camera images. This makes modern infotainment systems complex in terms of software and hardware.

FIGURE I.6 Information item filter [16].

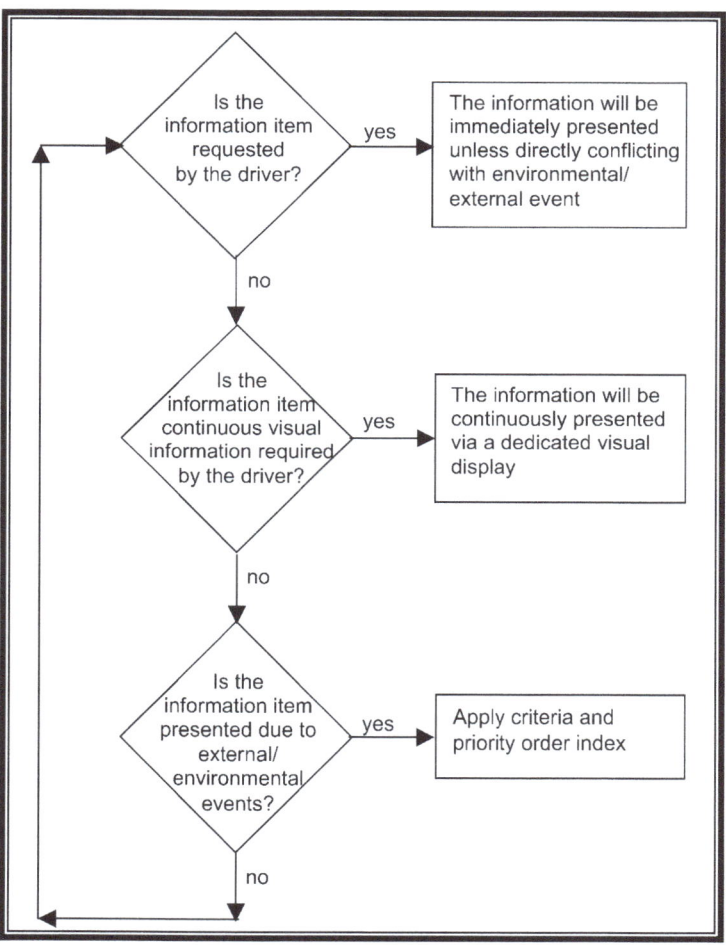

Infotainment systems connect to other devices via Bluetooth, USB, High-Definition Multimedia Interface (HDMI), or Wi-Fi, enabling integration with smartphones, tablets, and cloud services. Data from vehicle sensors, GPS, and external networks are processed to provide real-time information and personalized experiences. Users interact with the system through touch, physical controls, voice, or gestures, depending on the vehicle make and model (**Figure I.2**).

TABLE I.2 Examples of types of driver information requirements for key IVIS message sources [19].

IVIS message sources	Information types/functions
Routing and navigation	Trip planning
	Pre-drive route and destination selection
	Off-route recalculation
	Route guidance
	Route navigation
	Automated toll collection
	Route scheduling
	Post-trip summary
Motorist services	Broadcast services/attractions
	Services/attractions directory
	Destination coordination
	Delivery-related information
Augmented signage	Guidance sign information
	Notification sign information
	Regulatory sign information
Safety/warning	Immediate hazard warning
	Road condition information
	Automatic aid request
	Manual aid request
	Vehicle condition monitoring
	Driver monitoring devices
	Sensory augmentation
Collision avoidance and vehicle control	Forward collision avoidance
	Road departure collision avoidance
	Lane change and merge collision avoidance
	Intersection collision avoidance
	Railroad crossing collision avoidance
	Backing aid
	Advanced cruise control
Driver comfort, communication, and convenience	Real-time communication (e.g., electronic road signs)
	Asynchronous communications
	Contact search and history
	Entertainment and general information
	Interactive entertainment
	Heating, ventilation, air conditioning, and noise
	Automatic system configuration

© SAE International.

Infrastructural Connectivity

Infrastructural connectivity refers to the integration and real-time communication between vehicles and the surrounding physical and digital infrastructure. This connectivity enables vehicles to interact with roadways, TMSs, and other infrastructure elements to enhance safety, efficiency, automation, and the overall mobility experience.

See **V2V (Vehicle-to-Vehicle)** and **V2I (Vehicle-to-Infrastructure)**.

Intelligent Parking Solutions

Intelligent parking solutions, also known as smart parking systems, refer to integrated technologies and platforms that automate, optimize, and enhance the process of finding, reserving, and managing parking spaces [20]. These systems utilize IoT sensors, real-time data analytics, cloud computing, mobile applications, and AI to enhance operational efficiency, mitigate congestion, and deliver a seamless UX in urban and private parking environments [21].

Key Features:

- **IoT-Based Sensors:** Sensors embedded in parking spaces detect vehicle presence in real time, transmitting data to centralized platforms for accurate monitoring and management.
- **Digital Signage and Guidance:** Car parking facility displays and wayfinding systems guide drivers to available spaces, reducing search time and congestion.
- **Automated Parking and Retrieval:** Advanced systems use robotics or autonomous mobile robots (AMRs) to park and retrieve vehicles autonomously, maximizing space utilization and minimizing human intervention.
- **License Plate Recognition (LPR):** Cameras equipped with LPR automate access control, track occupancy, and enable ticketless entry/exit, improving security and convenience.
- **Cloud-Based Management:** Centralized software collects, processes, and analyzes data from sensors and cameras, supporting dynamic pricing, reporting, and efficient resource allocation.
- **Dynamic Pricing and Reservations:** Systems can adjust pricing based on demand and allow for booking, optimizing revenue, and user satisfaction.
- **Environmental and Energy Benefits:** By reducing the time spent searching for parking, these systems lower vehicle emissions and fuel consumption, contributing to sustainability goals.

Intelligent Transportation Infrastructure
See **V2I (Vehicle-to-Infrastructure)**.

Intersection Collision Warning
See **V2I (Vehicle-to-Infrastructure)** and **V2V (Vehicle-to-Vehicle)**.

Intersection Management
See **V2I (Vehicle-to-Infrastructure)** and **V2V (Vehicle-to-Vehicle)**.

Intersection Movement Assist
See **V2I (Vehicle-to-Infrastructure)** and **V2V (Vehicle-to-Vehicle)**.

In-Vehicle Software Platforms
In-vehicle software platforms are comprehensive software frameworks that serve as the backbone for modern vehicles, enabling the integration, management, and continuous evolution of vehicle functions, applications, and connectivity (Figure I.7). These platforms underpin the shift to SDVs, where software, not hardware, drives feature deployment, system updates, and UXs throughout the vehicle's life cycle [22].

FIGURE I.7 Software is ubiquitous in ADAS vehicles with interfaces that extend beyond the vehicle.

Gorodenkoff/Shutterstock.com.

Key Features [23]:
- **Centralized and Modular Architecture:** Modern platforms use centralized or zonal computing architectures, which allow high-performance processors to manage multiple vehicle domains (infotainment, ADAS, powertrain, etc.) and enable flexible, scalable deployment of software functions.
- **Operating System and Middleware:** The platform's core typically consists of a real-time operating system (RTOS) and middleware that manages hardware resources, provides standardized application programming interfaces (APIs), and supports application development.
- **Standardized Interfaces (e.g., AUTOSAR):** Platforms like AUTOSAR provide modular, standardized interfaces between application software and hardware, streamlining integration, supporting code reuse, and enabling easier upgrades and maintenance [22].
- **OTA Updates:** Support for remote software updates allows new features, bug fixes, and performance improvements to be delivered without dealer visits, keeping vehicles current and secure [24].
- **Security and Compliance:** Platforms are designed to comply with cybersecurity standards, ensuring robust protection against cyberthreats throughout the vehicle life cycle [13].
- **Life Cycle and Data Management:** Centralized data storage and management facilitate predictive maintenance, performance monitoring, and user personalization.

IoT (Internet of Things)

The IoT in the automotive context refers to the networked integration of sensors, devices, software, and cloud connectivity within vehicles and across transportation infrastructure. This ecosystem enables vehicles to collect, exchange, and analyze data in real time, driving advancements in safety, efficiency, automation, infotainment, predictive maintenance, and the overall UX.

IPGS (Intelligent Parking Guidance System)

An IPGS is an integrated solution that uses sensors, cameras, digital signage, and real-time data analytics to guide drivers to available parking spaces quickly and efficiently. These systems are designed to reduce congestion, minimize search time, and optimize parking facility utilization, significantly enhancing the parking experience for both drivers and operators [20].

Key Features:
- **Sensor-Based Detection:** This method utilizes ultrasonic, IR, or camera-based sensors to monitor the occupancy status of individual parking spaces, relaying real-time data to a central management system.
- **Digital Wayfinding Signage:** LED displays and digital signs at entry points, aisles, and decision points show the number and location of available spaces, guiding drivers directly to open spots.
- **Overhead Indicators:** LED lights (commonly green for available and red for occupied) above parking spaces provide immediate visual cues, reducing the time needed to find a spot.
- **Mobile App Integration:** Many systems offer mobile apps that allow users to check real-time availability, reserve spots, and navigate to the selected space, extending the guidance beyond the parking facility itself.
- **Central Management Software:** Acts as the system's brain, collecting and analyzing sensor data, generating reports, and optimizing space allocation and facility operations.
- **Automated Payment and License Plate Recognition:** Integrates with contactless payment systems and LPR for seamless entry/exit and enhanced security.

ISA (Intelligent Speed Adaptation)

ISA, also known as intelligent speed assistance, is an ADAS designed to help drivers comply with legal and safe speed limits by utilizing real-time data from TSR cameras, GPS-linked speed limit maps, and, in some cases, direct infrastructure broadcasts. One of the primary causes of road accidents is speeding. ISA aims to reduce speeding, thereby reducing road injuries and fatalities. The system monitors the vehicle's speed relative to the current road's speed limit. It provides feedback or intervention to prevent speeding, reduce accident risk, and support safer, more environmentally friendly driving [25].

Key Features:
- **Speed Limit Detection:** ISA uses a combination of TSR cameras, GPS-based speed limit databases, and sometimes V2I communications to determine the correct speed limit for the vehicle's location.
- **Driver Feedback and Alerts:** The system can provide visual, audible, or haptic warnings (open ISA) when the speed limit is exceeded, allowing the driver to take corrective action.
- **Active Speed Limiting:** More advanced ISA systems (closed or half-open) can actively limit engine power to prevent the vehicle from accelerating

beyond the speed limit or increase resistance on the accelerator pedal when the limit is exceeded. Some systems allow the driver to override the intervention temporarily for safety reasons.

- **Dynamic and Fixed Limits:** ISA can recognize static (posted) and dynamic (variable, time-dependent, or weather-dependent) speed limits, adapting to changing road conditions and signage.
- **Manual Override:** Regulations require that drivers be able to deactivate ISA, but the system may continue to display speed limit information even when deactivated.

ITS (Intelligent Transportation System)

ITSs are integrated networks of advanced technologies, including sensors, communications, AI, and real-time analytics, designed to enhance the safety, efficiency, and sustainability of transportation networks. ITS connects vehicles, infrastructure, and travelers to enable more innovative mobility, predictive traffic management, and enhanced UXs.

See **V2I (Vehicle-to-Infrastructure)** and **V2V (Vehicle-to-Vehicle)**.

IVI (In-Vehicle Infotainment)

See **In-Car Entertainment**.

REFERENCES

1. Schoenebeck, K., Melbert, J., and Weiser, F., "Motion Tracking in Crash Test Applications with Inertial Measurement Units," SAE Int. J. Passeng. Cars - Mech. Syst. 2, no. 1 (2009): 247-253, doi:https://doi.org/10.4271/2009-01-0056.

2. Society of Automotive Engineers, "J2838 Full Adaptive forward Lighting Systems," SAE Publishing, Warrendale, PA, 2020.

3. Society of Automotive Engineers, "J2591 Limited Adaptive forward Lighting System," SAE Publishing, Warrendale, PA, 2024.

4. Society of Automotive Engineers, "J1383 Performance Requirements for Motor Vehicle Headlamps," SAE Publishing, Warrendale, PA, 2018.

5. Society of Automotive Engineers, "J3069 Adaptive Driving Beam System," SAE Publishing, Warrendale, PA, 2021.

6. Society of Automotive Engineers, "J578 Chromaticity Requirements for Ground Vehicle Lamps and Lighting Equipment," SAE Publishing, Warrendale, PA, 2020.

7. Society of Automotive Engineers, "J3101-3 Hardware Protected Security Environment Management of Confidential Data," SAE Publishing, Warrendale, PA, 2024.
8. Society of Automotive Engineers, "J3214 Breath-Based Alcohol Detection System," SAE Publishing, Warrendale, PA, 2023.
9. Society of Automotive Engineers, "J3016 Taxonomy and Definitions for Terms Related to Driving Automation Systems for On-Road Motor Vehicles," SAE Publishing, Warrendale, PA, 2021.
10. Society of Automotive Engineers, "J2396 Definitions and Experimental Measures Related to the Specification of Driver Visual Behavior Using Video-Based Techniques," SAE Publishing, Warrendale, PA, 2023.
11. Society of Automotive Engineers, "J6268 Design & Run Time Information Exchange for Health Ready Components," SAE Publishing, Warrendale, PA, 2023.
12. Society of Automotive Engineers, "J2938 LED Light Sources Tests and Requirements Standard," SAE Publishing, Warrendale, PA, 2024.
13. Society of Automotive Engineers, "J3061 Cybersecurity Guidebook for Cyber-Physical Vehicle Systems," SAE Publishing, Warrendale, PA, 2021.
14. Society of Automotive Engineers, "J3201 Guidelines for Automotive Environment Cybersecurity Key Management and Credential Distribution," SAE Publishing, Warrendale, PA, 2024.
15. Society of Automotive Engineers, "J2400 Human Factors in forward Collision Warning Systems: Operating Characteristics and User Interface Requirements," SAE Publishing, Warrendale, PA, 2003.
16. Society of Automotive Engineers, "J2395 ITS In-Vehicle Message Priority," SAE Publishing, Warrendale, PA, 2002.
17. Society of Automotive Engineers, "J2988 Guidelines for Speech Input and Audible Output in a Driver Vehicle Interface," SAE Publishing, Warrendale, PA, 2015.
18. Hutzelmann, T., Banescu, S., and Pretschner, A., "A Comprehensive Attack and Defense Model for the Automotive Domain," SAE Int. J. Transp. Cyber. & Privacy 2, no. 1 (2019): 5-20, doi:https://doi.org/10.4271/11-02-01-0001.
19. Society of Automotive Engineers, "J2831 Development of Design and Engineering Recommendations for In-Vehicle Alphanumeric Messages," SAE Publishing, Warrendale, PA, 2020.
20. Society of Automotive Engineers, "J3164 Ontology and Lexicon for Automated Driving System (ADS)-Operated Vehicle Behaviors and Maneuvers in Routine/Normal Operating Scenarios," SAE Publishing, Warrendale, PA, 2023.

21. Society of Automotive Engineers, "J3063 Active Safety Systems Terms and Definitions," SAE Publishing, Warrendale, PA, 2023.

22. Aly, S., "Consolidating AUTOSAR with Complex Operating Systems (AUTOSAR on Linux)," SAE Technical Paper 2017-01-1617 (2017), doi:https://doi.org/10.4271/2017-01-1617.

23. Patel, J., "Delivering the Promise of AUTOSAR," in 2017 NDIA Ground Vehicle Systems Engineering and Technology Symposium Power & Mobility (P&M) Technical Session, Novi, MI, 2017.

24. Fuchs, A., Automotive Telematics: An Introduction into the Technical Aspects of Automotive Telematics with Reference to Business Model and User Needs (Warrendale, PA: Society of Automotive Engineers, 2002).

25. Hultkrantz, L. and Lindberg, G., "Intelligent Economic Speed Adaptation," in 10th International Conference on Travel Behaviour Research, Borlänge, Sweden, 2003.

L

Last-Mile Connectivity

In ACVs, last-mile connectivity refers to the final segment of the transportation journey, where passengers or goods are delivered from a local hub or transit center to their final destination. This phase is critical in urban and suburban settings, where the distance may be too short for conventional vehicle use but too long to walk comfortably. The vehicles used in last-mile connectivity are often EVs to reduce environmental impact around population centers.

Last-mile connectivity is considered a significant challenge within urban planning and transportation networks. It often determines a transportation system's overall efficiency and user satisfaction. Solutions in this area are crucial for reducing congestion, minimizing carbon footprints, and enhancing accessibility to transit systems.

- **Autonomous Shuttles:** Self-driving shuttles that can transport individuals or small groups within residential areas, campuses, or between transport hubs and final destinations.
- **Delivery Robots:** Autonomous or semi-autonomous robots are designed to deliver goods and packages from local distribution centers to the doorstep, optimizing the logistics chain.
- **Personal Mobility Devices:** This includes electric scooters, bikes, and other forms of personal transport that can be integrated with public transit, enhancing users' flexibility and convenience.
- **Infrastructure Integration:** Requires significant coordination with existing public transit systems and infrastructure.
- **Regulatory Issues:** Involves navigating complex legal frameworks that vary widely by locality, affecting the deployment of autonomous technologies.
- **Technological Barriers:** Ensuring reliable connectivity and safety in diverse urban environments poses ongoing technological challenges.

Last-Mile Delivery

Last-mile delivery refers to the final step of the delivery process, where goods are transported from a transportation hub to their destination, typically a personal residence or business. This delivery segment is crucial because it directly interacts with the customer and significantly affects customer satisfaction due to its impact on delivery speed and cost.

ACVs are increasingly being used to optimize last-mile delivery. These vehicles operate without direct human control and can communicate with other vehicles and infrastructure. ACVs in last-mile delivery are expected to grow as technological advances and regulatory frameworks evolve. This growth is anticipated to transform the logistics industry by making deliveries faster, safer, and more cost-effective, ultimately enhancing the overall customer experience.

LDW (Lane Departure Warning)

LDW is a safety technology in ACVs designed to alert drivers when their vehicle begins to move out of its designated lane without the use of a turn signal. This system uses cameras and sensors to monitor lane markings and detect unintentional lane departures (**Figure L.1**).

FIGURE L.1 LDW icon, from ISO 7000:2019-2682 [1].

© SAE International.

LDW systems are designed to work in tandem with the driver. They process images captured by cameras mounted on the vehicle, which scan the road for visible lane markings. Advanced algorithms interpret the imagery to determine the vehicle's position relative to lane boundaries. When the system detects that the vehicle is drifting toward or crossing a lane marking without an activated turn signal, it alerts the driver through visual, auditory, or tactile signals (such as steering wheel or seat vibrations).

The primary goal of the LDW is to enhance road safety by reducing accidents caused by unintended lane departures, which can occur due to driver inattention, drowsiness, or distraction (**Figure L.2**).

FIGURE L.2 Graphical representation of warning and control zones [2].

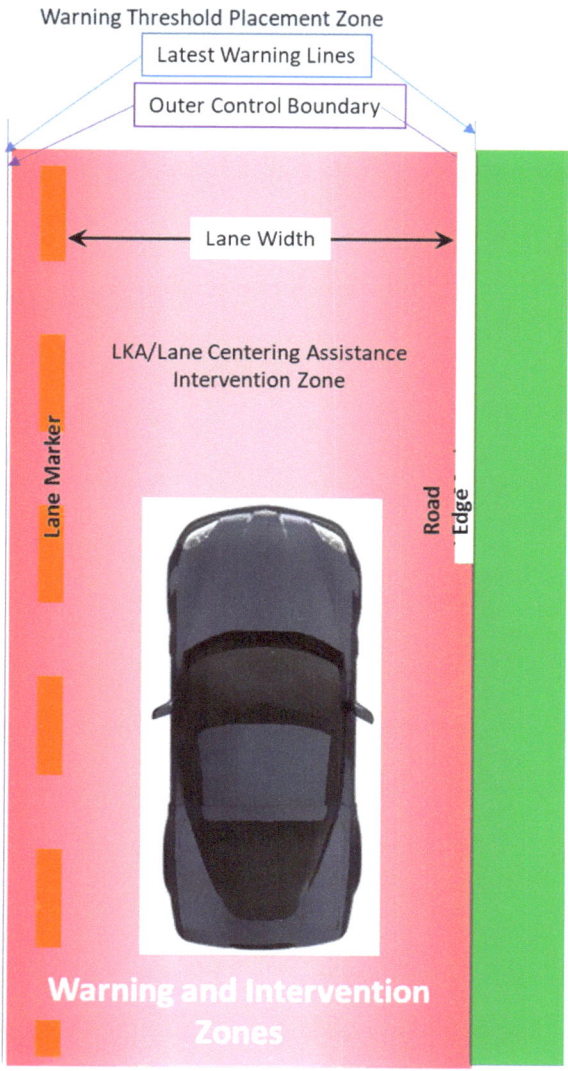

Initially developed as standalone systems, LDW technology has been increasingly integrated with other safety features, such as LKA and ACC, to provide a more comprehensive safety net. In fully AVs, LDW functions are part of a broader suite of technologies that control vehicle navigation and ensure compliance with traffic laws, thereby reducing the reliance on human driver input.

Learning

In the context of ACV technologies, system learning refers to the methodologies and processes through which an autonomous system improves its performance over time by learning from data [3]. This concept is foundational to the development and operational efficiency of autonomous driving and CV systems (**Figure L.3**).

FIGURE L.3 Block diagram representation of the proposed deep learning (DL)-based sensor fusion algorithm [3].

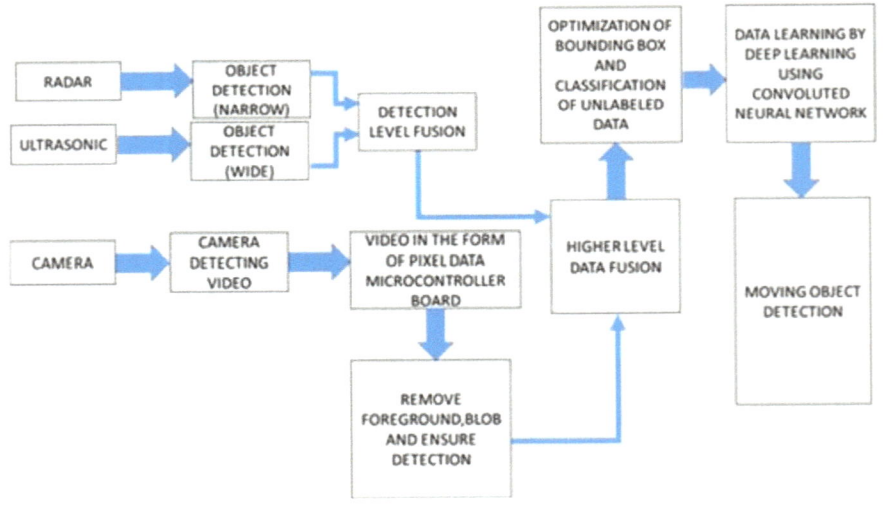

Key components that allow learning are as follows:
- **ML:** Algorithms that enable a system to learn from and make predictions or decisions based on data.
- **DL:** A subset of ML that uses neural networks with many layers (deep networks) to analyze various factors of the driving environment.

- **Reinforcement Learning (RL):** An area of ML concerned with how software agents ought to take actions in an environment to maximize some notion of cumulative reward.
- **Data Collection:** Continuous gathering of real-time data from the vehicle's sensors (such as cameras, radar, and LiDAR).
- **Data Processing:** Analyzing the collected data to extract useful features that help in understanding the environment.
- **Model Training:** Developing models using historical and real-time data to predict outcomes or make decisions.
- **Feedback Loops:** Utilizing the outcomes of the model's predictions to refine and improve the learning algorithms.

Levels of Automation

Level 0—No Automation: The vehicle has no automation features; the driver is responsible for all driving tasks, including steering, braking, accelerating, and monitoring the environment (Figure L.4).

FIGURE L.4 SAE J3016 levels of driving automation [4].

Level 1—Driver Assistance: This level includes basic automation features such as ACC and LKA. However, the driver must remain fully engaged and monitor the environment, as they are still primarily in control of the vehicle.

Level 2—Partial Automation: At this level, the vehicle can control both steering and acceleration/deceleration under certain conditions. The driver must remain alert, keep their hands on the wheel, and be ready to take over at any time, as the system can only handle certain aspects of driving.

Level 3—Conditional Automation: Vehicles at this level can make informed decisions about driving tasks and can manage most aspects of driving in specific conditions, such as highway driving. The driver must still be present and able to take over if the system requests it, but they can disengage from some driving tasks.

Level 4—High Automation: Vehicles can perform all driving tasks independently in most environments and conditions but may still require human intervention in extreme scenarios. Drivers can completely disengage from driving in certain situations, such as within geofenced areas or specific types of roads.

Level 5—Full Automation: At the highest level of automation, the vehicle is fully autonomous and requires no human intervention at any time. It can operate under all road and environmental conditions that a human driver could handle. These vehicles may not even be equipped with traditional driving controls, such as steering wheels or brake pedals.

LiDAR (Light Detection and Ranging)

LiDAR is an active sensor. It is referred to as lidar, LiDAR, and LIDAR. A LiDAR emits light/laser pulses into the surrounding environment. A receiver receives the reflected pulses, and characteristics such as distance and time of flight are measured. A processing unit in the sensor creates point clouds of objects in the surrounding environment based on the reflected beams received by the sensor continuously. This is a 3D reconstruction of the area surrounding the vehicle.

Automated vehicles use mechanical LiDARs or solid-state LiDARs. Mechanical LiDARs have a rotating element that emits laser beams in a 360° range. Solid-state LiDARs do not have moving parts. Modern LiDARs have a range of more than 200 m and very high resolution. Adverse weather situations impact their performance as the beams get scattered. LiDARs are costly sensors compared to cameras and radar.

LiDAR Mapping

LiDAR mapping is a remote sensing method used extensively in ACVs to generate precise, 3D information about the surrounding environment's shape and surface characteristics. Utilizing pulsed laser light, LiDAR sensors measure the time it takes for the reflected light to return to the sensor, thereby calculating distances.

LiDAR mapping is crucial for real-time navigation, obstacle detection, and environment recognition. Mounted on vehicles, LiDAR sensors create detailed 3D maps of the surroundings, enabling vehicles to navigate safely and efficiently in complex environments without human intervention.

LiDAR provides high-resolution environmental mapping, 3D geographical data, which is essential for creating detailed maps used by AVs to navigate roads, detect lane boundaries, and identify potential hazards [5].

FIGURE L.5 Vehicle detects the pedestrian body from different angles when the pedestrian is crossing the road and along the road [5].

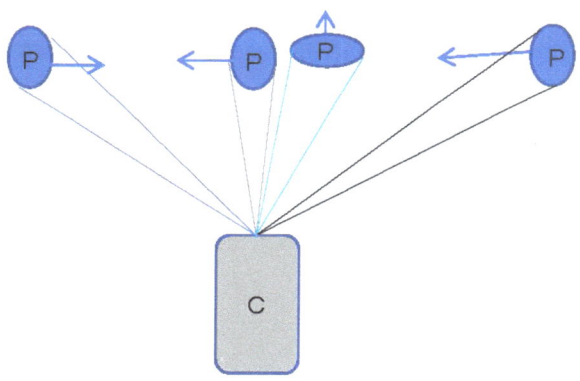

By continuously scanning the vehicle's environment, LiDAR helps detect and classify objects such as pedestrians, other vehicles, and road infrastructure, facilitating robust decision-making processes for the vehicle's control systems (**Figure L.5**). LiDAR helps AVs determine their exact location relative to their surroundings, which is critical for path planning and maneuver execution (**Figure L.6**).

FIGURE L.6 Path planning block diagram [6].

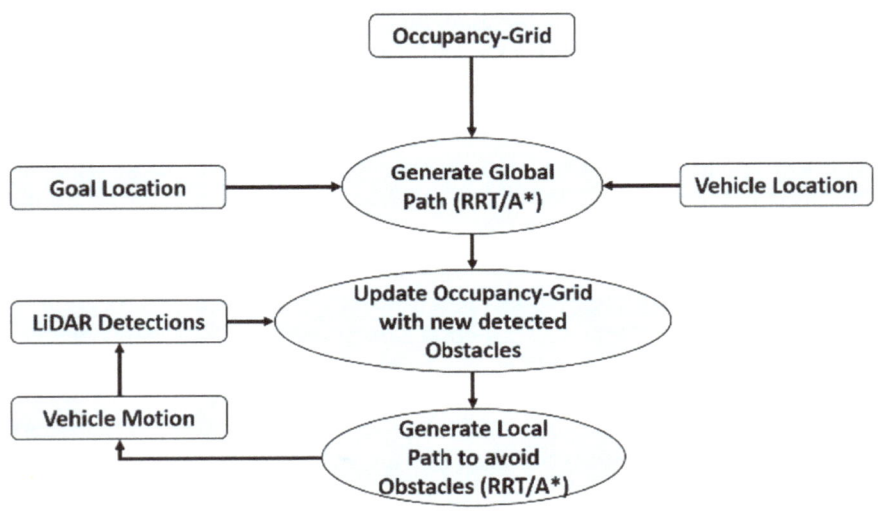

There are limitations to LiDAR: higher acquisition and maintenance costs compared to other sensing technologies, such as radar and cameras. Additionally, the sensors are sensitive to weather conditions. Performance can be affected by adverse weather conditions such as fog, rain, or snow, which can scatter laser pulses.

LKA (Lane-Keeping Assist)
See **LLC (Lane Centering Control)**.

LLC (Lane Centering Control)
LCC is an ADAS technology used in ACVs to maintain a vehicle's position within its lane on the road. Depending on the operation, this is also referred to as LKA. This feature utilizes cameras, onboard high-definition maps, and positioning sensors to continuously monitor road markings and detect the vehicle's position relative to them, providing drivers with constant reassurance of their vehicle's location. The system can automatically adjust the steering to keep the vehicle centered in its lane by processing these data, enhancing driving safety and comfort.

The system integrates inputs from various onboard sensors (**Figure L.7**), including cameras and radar, which scan the lane markings. The positioning information from the vehicles is correlated with the stored high-definition maps to determine the vehicle's position within the lane. Sophisticated algorithms interpret the data to determine the vehicle's position and trajectory.

If the system detects a deviation from the center of the lane, it automatically initiates minor steering corrections. Lane centering is particularly useful on highways and well-marked roads, providing valuable support during long drives. Reducing the need for constant steering adjustments can help alleviate driver fatigue, enhancing the overall driving experience.

FIGURE L.7 CarMaker line sensor [7].

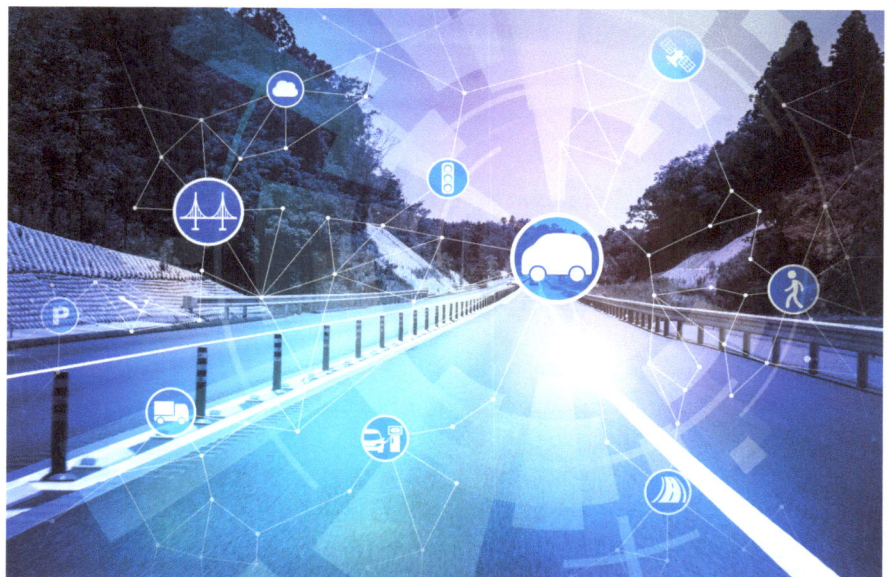

While lane centering greatly aids in driving, it has limitations, particularly when road markings are worn, obscured, or absent and when vision-based technologies alone are used for centering the vehicle in the lane (**Figure L.8**). Additionally, the technology relies heavily on the clarity of lane markings and environmental conditions; poor visibility due to weather can impair its effectiveness. When positioning data and high-definition maps are utilized, the system can overcome the weakness that affects the vision sensors. It is also essential for drivers to remain alert and ready to take control when necessary, as the system is designed to assist rather than replace human driving.

FIGURE L.8 LDW/LKA/lane centering assistance system sensitivity settings for testing [2].

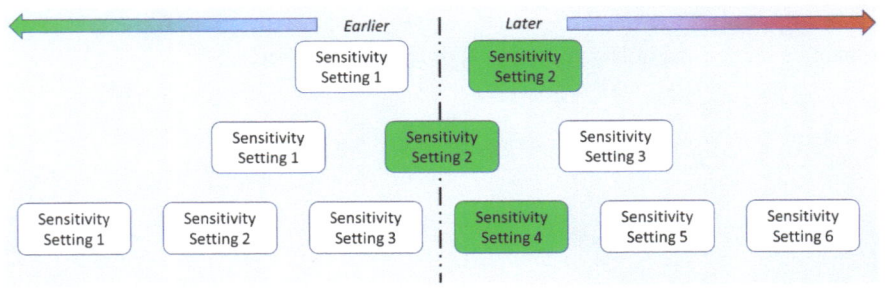

Localization

Localization is a critical function in ACVs. It refers to the ability of a vehicle to determine its precise position within an environment. This capability is essential for safe navigation and operation without human intervention.

Localization technologies allow vehicles to pinpoint their exact location on a given map, enabling them to navigate complex environments, adjust to dynamic changes, and interact safely with other vehicles and pedestrians. It relies on a combination of sensors, data, and algorithms to achieve high levels of accuracy and reliability.

The GPS provides geographical positioning with varying degrees of accuracy but is often augmented with other technologies due to limitations such as signal blockage in urban environments. LiDAR utilizes laser beams to map out the vehicle's surroundings in detail, providing high-resolution images that aid in recognizing lane markings, obstacles, and road features. Cameras capture visual information that algorithms interpret to identify road signs, traffic signals, and other critical navigational cues.

Radar uses radio waves to detect the distances and speeds of nearby objects, which is helpful in adverse weather conditions where optical sensors might fail. IMU detects a vehicle's specific movement changes (acceleration and rotation) to assist in tracking its trajectory and orientation when GPS data are unavailable. Odometry estimates a vehicle's position changes over time based on wheel rotation data, which can be particularly useful in GPS-denied environments, such as tunnels. IMU data and the most recent position fix are integrated over time, and positions can be determined when GPS connectivity is lost. This is referred to as dead reckoning.

Localization is used to navigate and operate AVs safely and enhance connectivity among vehicles (V2V communication) and between vehicles and infrastructure (V2I communication). This integration supports many applications, including traffic management, automated parking, and emergency response maneuvers.

REFERENCES

1. Society of Automotive Engineers, "J2808 Lane Departure Warning Systems: Information for the Human Interface," SAE Publishing, Warrendale, PA, 2024.

2. Society of Automotive Engineers, "J3240 Passenger Vehicle Lane Departure Warning, Lane Keeping Assistance, and Lane Centering Assistance Systems Test Procedure," SAE Publishing, Warrendale, PA, 2023.

3. Dheekonda, R.S., Panda, S., Nazmuzzaman Khan, M., Hasan, M. et al., *Object Detection from a Vehicle Using Deep Learning Network and Future Integration with Multi-Sensor Fusion Algorithm* (Warrendale, PA: SAE Publishing, 2017).

4. Society of Automotive Engineers, "Taxonomy and Definitions for Terms Related to Driving Automation Systems for On-Road Motor Vehicles J3016_202104," SAE Publishing, Warrendale, PA, 2021.

5. Society of Automotive Engineers, "J3116 Active Safety Pedestrian Test Mannequin Recommendation," SAE Publishing, Warrendale, PA, 2023.

6. Alzu'bi, H., Jarbo, A.T., Alrousan, Q., and Tasky, T., *LiDAR-Based Predictive Cruise Control* (Warrendale, PA: SAE Publishing, 2020).

7. Varunjikar, T., Awathe, A., Kushwaha, P., and Chen, C., *Simulation Study to Evaluate Robustness of the Lane Centering Feature* (Warrendale, PA: SAE Publishing, 2021).

MaaS (Mobility as a Service)

MaaS is an integrated transportation concept that combines various transport services into a single, accessible, and on-demand mobility platform. MaaS aims to provide an alternative to private vehicle ownership by facilitating the efficient and convenient use of public and private transportation options. These can include, but are not limited to, mass rapid transportation systems such as subways, buses, trams, taxis, and personal mobility solutions, including car rentals, bike-sharing, and ride-sharing services.

- **Reduced Traffic Congestion:** By optimizing route and vehicle usage, MaaS can decrease the number of vehicles on the road to reduce traffic congestion.
- **Environmental Impact:** Encourages shared and public transport options, potentially reducing carbon footprint and GHG emission reductions.
- **Cost Efficiency:** Offers flexible pricing models and subscription services that can be more economical than owning and maintaining private vehicles.
- **Accessibility:** Enhances mobility for a broader demographic, including those unable to drive.

MAP (Map Data)

MAP is a V2X message used to communicate intersection and road geometries [1]. The data elements include intersection geometry, road segment list, road curvature, and lane information. Lane geometries are specified using a sequence of nodes (based on latitude/longitude coordinates) that communicate the shape and width of the lane, speed limit information, and permitted and inhibited traffic movements. CV applications in the vehicle receive the MAP and SPaT [1] messages from signalized intersections and are used together to indicate phase change in specific lanes of an intersection. MAP and SPaT messages are used together to perform V2X functions that are associated with signalized intersections.

MEC (Multi-Access Edge Computing)

MEC is a network architecture concept that enables cloud computing capabilities and an information technology (IT) service environment at the edge of the network. In the context of ADAS, MEC plays a pivotal role in processing large volumes of data generated by vehicles and other road users in real time, close to the source of data generation.

MEC provides a low-latency, high-bandwidth, and resource-rich computing environment at the edge of the cellular network. By decentralizing computing resources and bringing them closer to the vehicle, MEC facilitates quicker data processing and decision-making, which is essential in ADAS applications. This includes object recognition, collision avoidance, navigation, and seamless V2X communications.

- **Real-Time Traffic Management:** MEC can process information from various sensors and sources on the road, providing real-time updates and guidance to AVs to optimize routes and speed based on current traffic conditions.
- **Safety Applications:** Enables immediate local processing of pedestrian detection, hazardous object recognition, and other critical safety functions, allowing vehicles to react swiftly to potential dangers.
- **Connectivity and Communication:** Facilitates efficient and reliable communication among vehicles, infrastructure, and other road users through enhanced V2X communication capabilities, supporting cooperative driving and smart city integrations.

MFM (Mobile Fleet Management)

MFM in the context of ADAS and CVs refers to the integrated approach of managing a fleet of vehicles using various technologies that enhance connectivity, automation, and real-time operations management. This system leverages ADAS capabilities to enhance safety, improve efficiency, and streamline the overall management of fleet vehicles, encompassing a range of vehicles from cars and trucks to specialized equipment.

- **Data Collection and Transmission:** MFM systems in CVs collect vast amounts of data from multiple sources, including vehicle telematics, GPS tracking, OBD, and ADAS sensors such as cameras, radar, and LiDAR. These data are transmitted in real time via cellular or satellite networks to central databases for processing and analysis.
- **Database Management:** Efficient database systems are essential for storing, retrieving, and managing the vast amounts of data generated by fleet vehicles. These systems must handle high-velocity, high-volume data streams securely and efficiently. To support scalability and accessibility,

technologies such as cloud databases, big data platforms, and advanced data warehousing solutions are common.

- **Data Security and Privacy:** Protecting sensitive information about vehicle locations, driver behavior, and proprietary operational data is critical. Robust cybersecurity measures, including data encryption, secure access protocols, and regular security audits, are crucial for safeguarding against unauthorized access and cyberthreats.
- **Data Analysis and Utilization:** Advanced analytical tools and ML algorithms interpret the collected data, providing insights into vehicle performance, maintenance needs, driver behavior, and route optimization. These analyses help fleet managers make informed decisions to improve efficiency, reduce costs, and enhance safety.
- **Integration with Other IT Systems:** MFM systems often need to integrate with other enterprise IT systems, such as human resources management systems (HRMSs), enterprise resource planning (ERP), and customer relationship management (CRM), to enable comprehensive business management. API technology facilitates this integration, allowing seamless data exchange and process automation.
- **Regulatory Compliance:** Compliance with local and international regulations regarding data protection, vehicle safety, and emissions is mandatory. MFM systems must include capabilities to monitor compliance and generate reports as required by regulatory bodies.

Micro-Mobility

Micro-mobility refers to the use of small, lightweight vehicles, typically weighing less than 500 kg, designed for individual use over short distances. This category includes vehicles such as electric scooters, bicycles, Segways, and e-bikes. In the context of ADASs and CV technologies, micro-mobility focuses on integrating these more minor transport modes into the broader transportation ecosystem, enhancing safety, sustainability, efficiency, and communication between various modes of transport.

As urban areas become more congested and the need for sustainable transport options increases, micro-mobility solutions equipped with ADAS technologies are gaining prominence. These technologies can include automated braking, collision avoidance systems, and real-time data communication between vehicles and infrastructure. This integration aims to reduce accidents, optimize traffic flow, and provide a seamless mobility experience.

Middleware

Middleware in ACVs refers to the software layer between the vehicle's operating system and its applications. This software facilitates communication and data management between automotive systems and components, such as sensors, actuators, and external networks.

The primary purpose of middleware in these vehicles is to ensure seamless integration and interoperability of diverse automotive systems and applications. It manages the vehicle's communications and operation complexities, enabling different components and software applications to function efficiently without direct interaction.

ML (Machine Learning)

ML is a subset of AI that enables computer systems to learn from data, identify patterns, and make decisions similar to those made by humans. In ADAS and AVs, ML is pivotal for enhancing vehicle intelligence and enabling autonomous functionalities [2].

Application in ADAS and AVs:
- **Perception:** ML algorithms interpret data from sensors such as cameras, radar, and LiDAR to detect and classify objects, predict their behavior, and construct a virtual model of the region around the vehicle consisting of the static and dynamic elements. This capability is crucial for functions such as object detection, lane detection, traffic sign detection, and other perception functions (**Figure M.1**)
- **Prediction:** ML models predict the actions of other road users, such as whether a vehicle will change lanes or whether a pedestrian might cross the street unexpectedly. These predictions help the autonomous system make safer driving decisions.
- **Decision-Making:** ML algorithms help AVs make complex, real-time driving decisions. These decisions include route planning, speed control, and maneuvering around obstacles or through traffic.
- **Learning and Adaptation:** Over time, ML enables vehicles to improve their performance based on accumulated experience, much like human drivers refine their skills. This adaptive learning helps in dealing with diverse and changing driving conditions [4].

FIGURE M.1 Schematic of the process for sensor-level analysis [3].

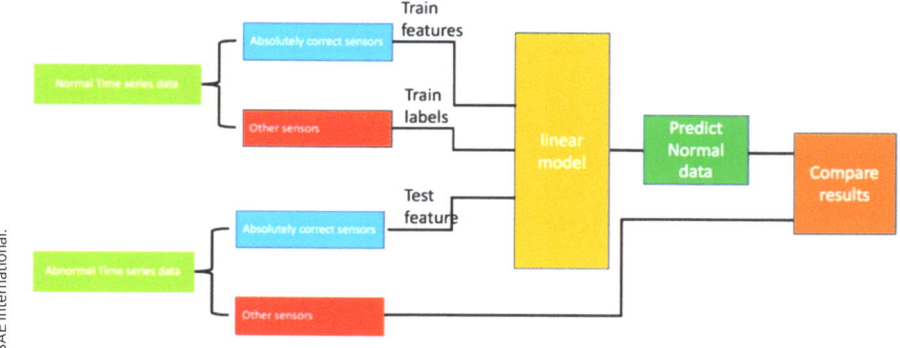

Technologies and Techniques:

- **Supervised Learning:** Used widely for regression and classification tasks such as object detection and TSR.
- **Unsupervised Learning:** Applied for clustering and association tasks, which can help in discovering patterns in driving data without labeled responses.
- **RL:** Employs strategies where the vehicle learns to make sequences of decisions by receiving feedback from its actions, optimizing long-term goals such as safety and efficiency.

Mobile App Integration

Mobile app integration in ACV systems (ADAS) refers to the technological process that allows vehicle systems to connect and interact with mobile applications. Manufacturers offer mobile applications that provide access to vehicle systems, compatible with common platforms such as iOS and Android. This integration enables drivers and passengers to control vehicle functions and access vehicle data through smartphone apps, thereby enhancing the driving experience, safety, and overall functionality. Mobile platforms such as Android and iOS have apps that integrate into the intelligent cockpit of the vehicles to seamlessly integrate media, work-related communications such as Microsoft Teams, and phone-based navigation.

Features included are as follows:

- **Remote Control:** Users can operate certain vehicle functions remotely, such as locking/unlocking doors, starting the engine, or configuring climate control, all through a mobile app.

- **Real-Time Data Access:** Integration enables direct monitoring of vehicle parameters, such as fuel levels, battery status, and tire pressure, from the user's mobile device.
- **Navigation and Traffic Updates:** Enhanced GPS functionalities that sync directly with the vehicle's navigation system, providing real-time traffic updates, route suggestions, and destination sharing.
- **Entertainment and Connectivity:** Users can manage multimedia content, including music, podcasts, and video streaming services, directly through their mobile apps via the vehicle's infotainment system.
- **Safety Alerts:** The system can send instant notifications to the user's mobile app regarding vehicle diagnostics, maintenance reminders, and ADAS features, such as proximity and collision warnings. Some applications support theft protection and allow for the remote location of stolen vehicles.

REFERENCES

1. Society of Automotive Engineers, "J2735 V2X Communications Message Set Dictionary," SAE Publishing, Warrendale, PA, 2023.
2. Borrego-Carazo, J., Castells-Rufas, D., Biempica, E., and Carrabina, J., "Resource-Constrained Machine Learning for ADAS: A Systematic Review," *IEEE Access* 8 (2020): 40573-40598.
3. Sun, Y., Xu, Z., and Zhang, T., "On-Board Predictive Maintenance with Machine Learning," SAE Technical Paper 2019-01-1048 (2019), doi:https://doi.org/10.4271/2019-01-1048.
4. Ball, J.E. and Tang, B., *Machine Learning and Embedded Computing in Advanced Driver Assistance Systems (ADAS)* (Basel: MDPI, 2019).

Navigation

Navigation refers to the system's ability to guide a vehicle from one location to another using GPS/GNSS technology, digital maps, positioning infrastructure, and onboard sensors [1]. The navigation system provides real-time directions, route planning, and traffic updates, enhancing the driver's ability to reach their destination efficiently and safely [2].

GPS integration is a crucial component that uses satellite signals to determine the vehicle's location on Earth accurately. Digital maps stored in the vehicle or on smart devices display detailed information, including roads, landmarks, traffic conditions, and POIs. Route planning calculates the best route to the destination, considering factors such as distance, travel time, and traffic conditions. Turn-by-turn directions provide step-by-step instructions, through either a visual display or voice commands (or both), to guide the driver along the selected route. Traffic updates monitor and display real-time traffic conditions, suggesting alternative routes in case of delays or road closures. POIs enable drivers to search for nearby services, including gas stations, restaurants, and hotels.

Safety enhancement through clear and accurate directions and navigation systems helps reduce the driver's cognitive load, allowing them to focus more on the road. Navigation is integrated with other ADAS features. Navigation systems often incorporate features such as ACC and LKA to optimize vehicle performance and enhance overall driving safety.

NCAP (New Car Assessment Program)

The NCAP is a government or nongovernmental organization that conducts crash tests and provides safety ratings for new vehicles [3]. These programs encourage car manufacturers to improve vehicle safety by offering consumers transparent and comparable information on vehicle crashworthiness and overall safety performance (Figure N.1).

FIGURE N.1 Map of consumer programs around the world and date of first [3].

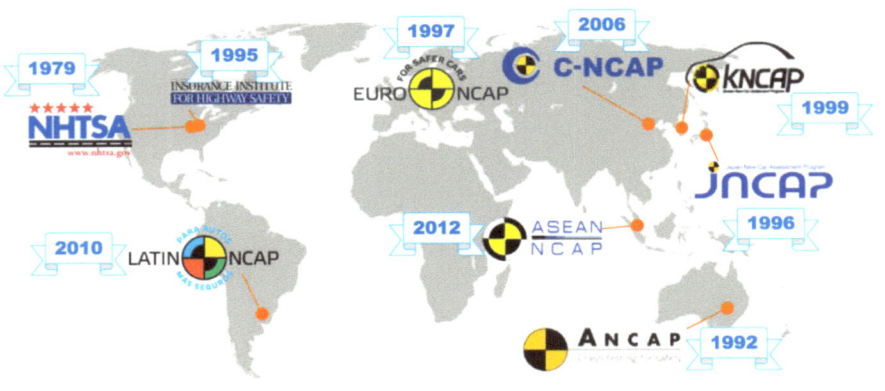

NCAP organizations assess vehicles based on various safety tests, including frontal, side, and rollover crash tests, pedestrian protection, and, increasingly, the performance of ADAS [4]. The results are typically presented as star ratings or scores, with higher ratings indicating better safety performance.

Different regions have their own NCAP organizations, such as Euro NCAP (Europe), ANCAP (Australia and New Zealand), and US-NCAP from NHTSA (United States). Each NCAP may have slightly different testing protocols and criteria tailored to regional priorities and regulations. Many NCAP programs now include the evaluation of ADAS technologies, such as AEB, LDW [4], and ACC. These systems are assessed for their effectiveness in preventing or mitigating crashes, and their performance contributes to the vehicle's overall safety rating.

Neural Networks

Neural networks are a type of ML algorithm inspired by the human brain's structure and function. They consist of interconnected layers of nodes (neurons) that process and learn from data. In ADAS, neural networks are used for object detection, classification, and decision-making tasks to enhance vehicle safety and automation (**Figure N.2**).

FIGURE N.2 Neural network algorithm flowchart describes the working flow [5].

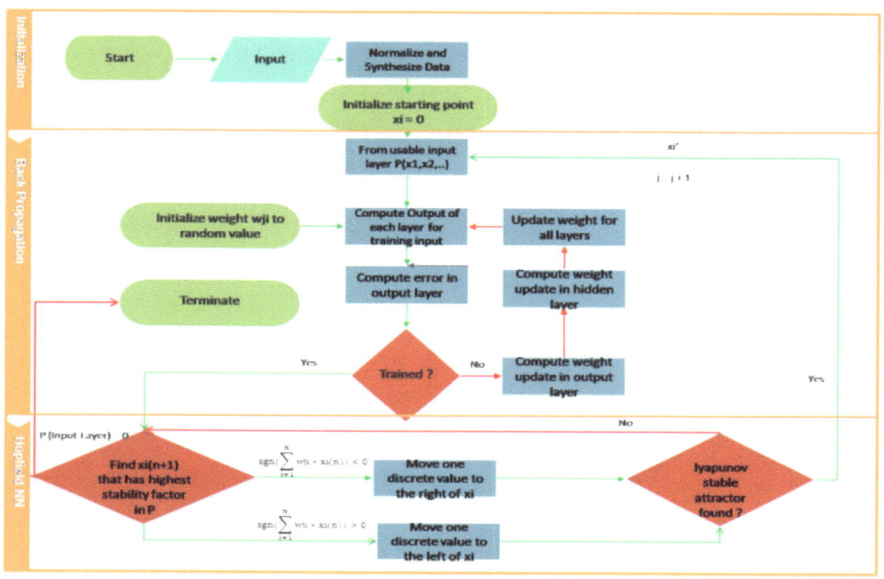

Neurons (nodes) are the fundamental units of a neural network that process inputs, apply weights, and pass the result through an activation function to produce an output. Neural networks are typically organized into layers. Those layers consist of an input layer, which receives raw data (e.g., images and sensor data); hidden layers, where data are processed and features are extracted; and intermediate layers. Lastly, the output layer produces the final prediction or decision, such as identifying a pedestrian or recognizing a stop sign. Parameters that are adjusted during training to minimize errors in forecasts are known as hyperparameters. The activation function is a mathematical function applied to the output of each neuron to introduce nonlinearity, enabling the network to solve complex problems.

In ADAS, neural networks are used to detect objects. They are trained to identify and classify objects in the vehicle's environment, such as other vehicles, pedestrians, traffic signs, and obstacles. Neural networks also assist in detecting lane markings and LKA systems. They are used to monitor driver attentiveness and detect signs of drowsiness or distraction.

Night Vision

Night vision in ADAS refers to a technology that enhances a driver's ability to see in low-light or dark conditions by detecting and displaying objects or obstacles that may not be visible to the naked eye (**Figure N.3**). This system typically uses IR sensors or thermal imaging cameras to capture and interpret the heat emitted by objects, such as pedestrians, animals, or vehicles. Then, it projects this information onto a display within the vehicle, such as a HUD, infotainment display unit, or instrument panel. The primary objective of night vision systems (NVSs) is to improve safety by increasing a driver's situational awareness and reaction time during nighttime driving or in poorly lit environments.

FIGURE N.3 Night vision improves visibility and reduces risk.

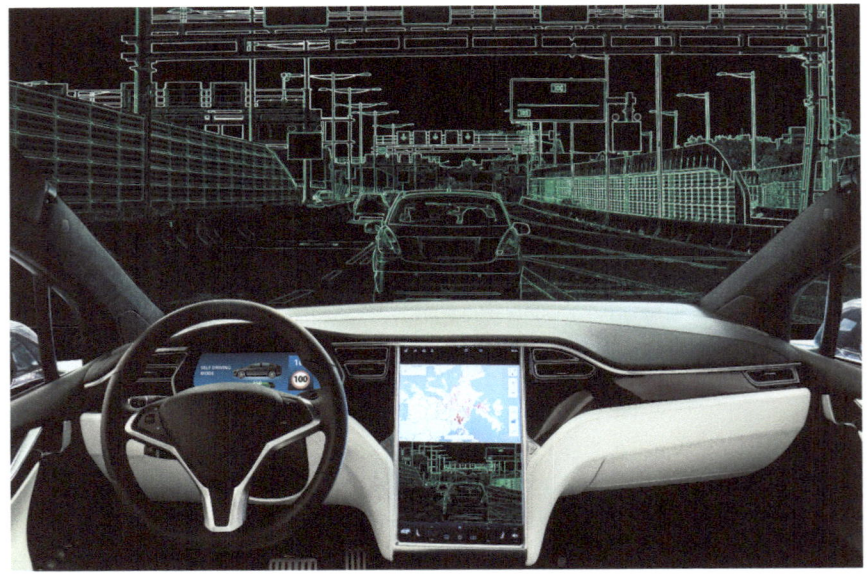

Scharfsinn/Shutterstock.com.

Key components of the system begin with IR sensors, which detect the heat emitted by objects, particularly those that are warmer than their surroundings, such as living beings. Thermal imaging cameras capture the thermal radiation emitted by objects and convert it into a visual image that can be displayed to the driver. This input is projected and presented onto a screen or HUD, allowing the driver to see beyond the range of traditional headlights.

The system identifies objects, pedestrians, or animals in the vehicle's path, especially in low-light or no-light conditions. The detected objects are highlighted on a screen to alert the driver to potential hazards. Some NVSs are integrated with alert mechanisms that warn the driver of imminent danger, such as an animal crossing the road.

NLP (Natural Language Processing)

NLP is a branch of AI that enables computers and systems to understand, interpret, and generate human language. In the context of ADAS, NLP facilitates human–machine interaction, enabling drivers and passengers to communicate with the vehicle through spoken or written language.

NLP in ADAS not only enhances safety but also offers unparalleled convenience. It enables voice-controlled commands, allowing drivers to interact with the vehicle's systems (e.g., navigation, infotainment, and climate control) without manual input. The natural language understanding (NLU) component of NLP processes spoken language to understand driver intentions, even if the commands are phrased in a conversational or nonstandard way. Text-to-speech (TTS) technology, part of NLP, converts text-based information, such as navigation instructions or alerts, into spoken language, providing audible feedback to the driver. The DMS portion of the NLP can analyze spoken language for signs of driver fatigue, distraction, or stress by detecting changes in speech patterns or tone, contributing to overall driver safety [6].

NVS (Night Vision System)

An NVS is an ADAS that enhances drivers' ability to see and identify objects in low-light or nighttime driving conditions. It typically uses IR sensors or thermal imaging cameras to detect heat emitted by objects, such as pedestrians, animals, and vehicles. It displays this information to the driver on a screen or HUD [7].

- IR or Thermal Imaging: NVSs capture IR radiation or heat emitted by objects, which is invisible to the human eye. These systems process the IR data to create a real-time visual representation, which is displayed to the driver.
- Object Detection and Recognition: Advanced NVS can differentiate between various types of objects (e.g., pedestrians and animals) and may highlight or alert the driver to potential hazards, enhancing situational awareness.
- Enhanced Visibility: NVS provides visibility beyond the range of the vehicle's headlights, making it easier to spot obstacles or hazards on dark roads or in conditions where visibility is compromised (e.g., fog and rain).

REFERENCES

1. Society of Automotive Engineers, "J2678_201609 Navigation and Route Guidance Function Accessibility While Driving Rationale," SAE Publishing, Warrendale, PA, 2016.

2. Society of Automotive Engineers, "J2364_201506 Navigation and Route Guidance Function Accessibility While Driving," SAE Publishing, Warrendale, PA, 2015.

3. Adalian, C., Fornells, A., and Parera, N., "Study of Differences in Overall Ratings in New Car Assessment Programs," SAE Technical Paper 2017-26-0017 (2017), doi:https://doi.org/10.4271/2017-26-0017.

4. Kusano, K.D. and Gabler, H.C., "Field Relevance of the New Car Assessment Program Lane Departure Warning Confirmation Test," *SAE Int. J. Passeng. Cars - Mech. Syst.* 5, no. 1 (2012): 253-264, doi:https://doi.org/10.4271/2012-01-0284.

5. Barakat, M. and Abdelaziz, M., "Neural Network Transmission Control," SAE Technical Paper 2016-01-0089 (2016), doi:https://doi.org/10.4271/2016-01-0089.

6. Lin, S., Zou, J., Zhang, C., Lai, X. et al., "Understanding User Requirements for Smart Cockpit of New Energy Vehicles: A Natural Language Process Approach," SAE Technical Paper 2022-01-7075 (2022), doi:https://doi.org/10.4271/2022-01-7075.

7. Ahire, A.S., "Night Vision System in BMW," *International Review of Applied Engineering Research* 4, no. 1 (2014): 1-10.

O

Object Fusion and Tracking

Object fusion and tracking are critical functions within the domain of ADAS and AVs. This process involves integrating and analyzing data from multiple vehicle sensors to identify, monitor, and track the positions and movements of various objects in the vehicle's environment. This is done by breaking the multi-object tracker (MOT) and multiple target trackers (MTTs) [1]. These objects can include other vehicles, pedestrians, animals, obstacles, and traffic signs (**Figure O.1**).

FIGURE O.1 Incorrect clustering of sensor data [1].

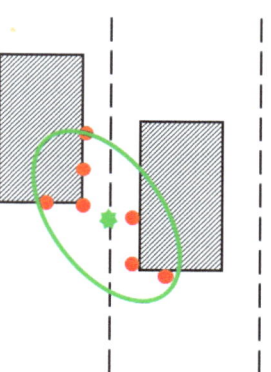

- Sensor Detection
- Clustered Results

© SAE International.

The primary goal of object fusion and tracking is to create a reliable and accurate representation of the surrounding environment, supporting decision-making processes in autonomous and semi-AVs. The process typically involves two main steps:
- **Sensor Fusion:** Data from diverse sensors such as cameras, radar, LiDAR, and ultrasonic sensors are combined to leverage their strengths and compensate for their weaknesses. For example, while cameras provide detailed visual information, radar offers reliable performance under poor visibility conditions. Sensor fusion algorithms integrate these data streams to provide a comprehensive understanding of the environment.
- **Object Tracking:** Once objects are detected, their positions, velocities, and trajectories are continuously tracked. This tracking relies on algorithms that predict the future states of objects based on their current and historical states, allowing the vehicle to anticipate potential environmental impacts. Object tracking helps enhance the ADAS's predictive capabilities, leading to more effective and timely decision-making.

Some of the vehicle functions that utilize object tracking and fusion are as follows:
- **Collision Avoidance:** By accurately tracking the position and velocity of nearby vehicles and obstacles, ADAS can predict potential collisions and automatically take corrective actions such as braking or steering adjustments.
- **ACC:** This feature uses object tracking to maintain a safe distance from the vehicle ahead, adjusting the vehicle's speed as necessary.
- **Pedestrian Detection:** Enhances safety in urban settings by identifying and tracking pedestrians, thereby preventing accidents in scenarios such as crosswalks, stop lights and parking lots.

Object Recognition

Object recognition in the context of ACVs refers to the ability of a vehicle's systems to identify and classify various objects within the environment through sensors and software algorithms. This includes other vehicles, pedestrians, animals, traffic signs, and any obstacles affecting the vehicle's path and driving decisions.

FIGURE O.2 Self-driving car systems [2].

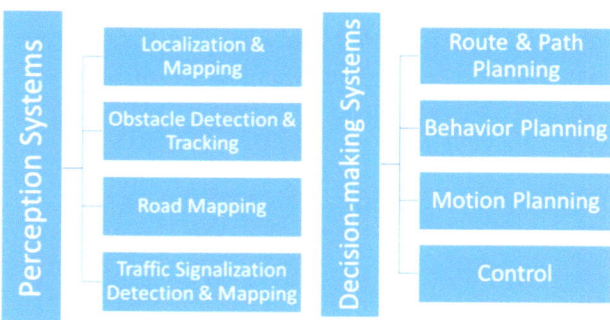

Object recognition is a critical component of ADAS and fully autonomous driving technologies (**Figure O.2**). It employs a combination of radar, LiDAR, cameras, and ultrasonic sensors to capture detailed images and measurements of the surrounding environment. These data are then processed using sophisticated ML models and computer vision techniques to accurately interpret and classify objects.

- **Collision Avoidance:** By recognizing objects around the vehicle, systems can predict potential hazards and act to avoid collisions.
- **Traffic Management:** Enhances navigation through complex traffic scenarios by recognizing traffic signals and adhering to traffic rules.
- **Pedestrian Safety:** Detects pedestrians, even in challenging conditions, to prevent accidents.
- **Parking Assistance:** Helps identify suitable parking spots and assists in the parking process without driver intervention.

Object recognition systems must contend with various challenges, including varying lighting conditions, weather-related impacts, sensor limitations, and unpredictable object movements. The reliability of object recognition is paramount to ensure safety and effectiveness in autonomous driving.

Obstacle Detection

Obstacle detection is a critical functionality within the broader ACV systems, specifically within ADAS. It involves the real-time identification and localization of objects in the vehicle's environment that may pose potential hazards.

This system utilizes a combination of sensors, cameras, radar, and, in some cases, LiDAR to continuously scan the vehicle's surroundings [3]. The data collected from these devices are processed using sophisticated algorithms to differentiate between various obstacles, such as other vehicles, pedestrians, animals, or inanimate objects like bollards and roadblocks [4].

The primary goal of obstacle detection is to enhance safety by providing the vehicle with the necessary information to make informed decisions. It enables functionalities such as automatic braking, steering adjustments, and ACC, which help prevent collisions and enhance overall driving dynamics.

OBU (Onboard Unit)

The OBU receives and broadcasts V2X messages to communicate with other OBU-equipped vehicles and infrastructure elements, such as RSU. Based on the transceiver, OBUs can exchange information in DSRC or C-V2X technologies or both. A vehicle that is actively utilized for public safety, such as emergency vehicles, is equipped with public safety OBUs.

An example of a message transmitted by OBU is BSM.

ODD (Operational Design Domain)

The ODD defines the specific conditions under which an AV is designed to function as intended. The conditions could be related to spatial, environmental, or temporal conditions.

Spatial conditions include geolocation, road type, road geometry, and type of lane. Temporal conditions include factors such as day or night, while environmental conditions encompass weather conditions, including snow, rain, and lighting conditions like low light and glare. ODD also takes into account the state of the vehicle, including speed, as well as the vehicle's overall condition. ODD is the key factor in an automated or driver assistance function. During feature operation, the ODD has to be monitored continuously to ensure the operating conditions are within the ODD. Most AVs have ODD management that tracks the operating conditions for which the vehicle is designed and predicts the ODD state for future time stamps. They use the data from radar, LiDAR, cameras, GNSS sensors, light, and rain sensors to determine if the vehicle is operating within the designed ODD. Suppose any of the ODD conditions are not met, the feature must be appropriately degraded or control has to be handed over to the driver sufficiently early to maintain safe operation.

OEDR (Object and Event Detection and Response)

OEDR refers to monitoring the driving environment, detecting objects and events, and taking actions to perform the DDT (**Figure O.3**). Depending on the level of automation, the OEDR is performed by the driver or vehicle [5]. In autonomous mode of operation, OEDR is entirely performed by the vehicle, utilizing sensors for perception and planning, and actuators to execute lateral and longitudinal motion [6]. In a partially AV, the OEDR is shared between the vehicle and the driver. If a handover of OEDR is warranted from the system to the driver, the feature needs to ensure sufficient time for the driver to comprehend the situation and respond safely.

FIGURE O.3 The hierarchical structure of the driving task [5].

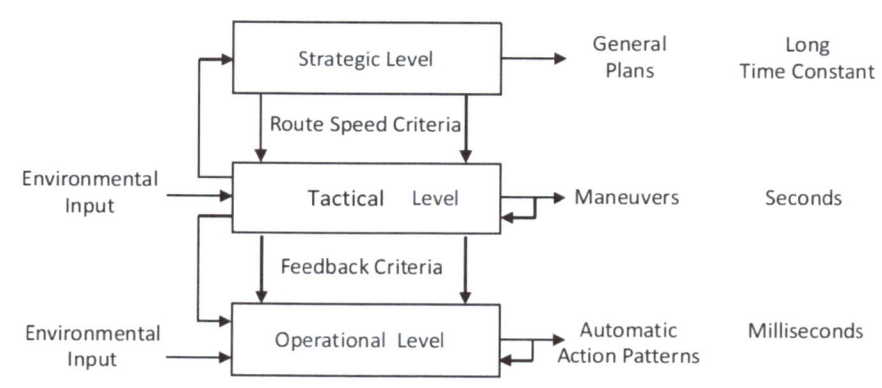

On-Demand Mobility

On-demand mobility refers to a system of transport services that can be accessed on an as-needed basis by consumers, typically facilitated by digital platforms and applications. This concept encompasses various modes of transportation including car sharing, bike-sharing, ride hailing, and AV services. The primary goal is to provide flexible, efficient, and personalized transportation options that reduce dependency on private vehicle ownership and improve urban mobility.

On-demand mobility takes a significant leap forward. ADAS-equipped and fully AVs can be integrated into on-demand mobility services, offering safer and more reliable transportation. These vehicles, connected through the Internet and capable of communicating with other vehicles and infrastructure (V2X), enhance the efficiency of on-demand services. They are equipped with automatic braking, ACC, and LKA, contributing to the safety and ease of mobility services.

On-demand mobility services have several implications:
- **Enhanced Safety:** Autonomous features reduce the risk of accidents caused by human error.
- **Increased Efficiency:** CVs can optimize routes in real time based on traffic conditions, weather, and road infrastructure, improving service speed and passenger experience.

- **Reduced Environmental Impact:** Efficient routing and reduced reliance on personal vehicles contribute to lower emissions.
- **Accessibility:** Improved access to transportation for a broader range of populations, including the elderly and those with disabilities.

OTA (Over-the-Air) Security

OTA security refers to the protective measures and protocols implemented to safeguard the wireless transmission of data, software, and firmware updates to connected and AVs (ADAS) [7]. This security is crucial for maintaining the integrity and functionality of vehicle systems, preventing unauthorized access, and ensuring compliance with privacy and safety regulations (**Figure O.4**).

FIGURE O.4 Overview of the data exchange between the entities involved in the OTA update development and approval process [7].

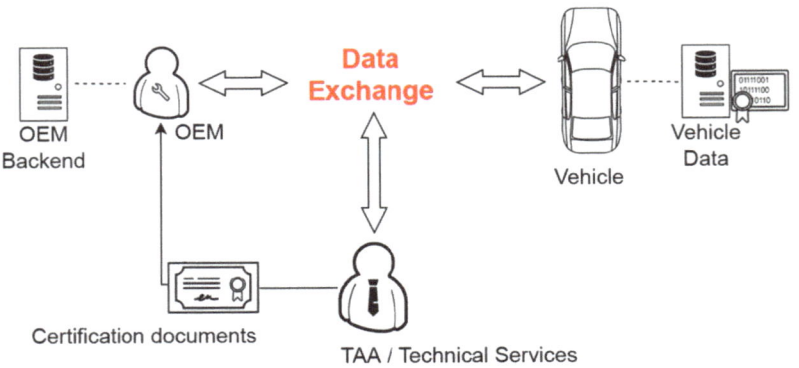

OTA security encompasses a range of technologies and strategies designed to protect vehicles from cyberthreats during wireless communications. It is an integral component of modern automotive systems, allowing manufacturers to remotely update software without requiring a physical connection to the vehicle. These updates can include improvements to vehicle performance, new features, bug fixes, and critical security patches.

Key Components:
- **Encryption:** Utilizes strong encryption protocols to secure data transmission between the vehicle and the manufacturer's servers, ensuring that data cannot be read or tampered with if intercepted.

- **Authentication:** Ensures that the data and updates sent and received are from legitimate, trusted sources. This often involves digital signatures and certificates.
- **Data Integrity:** Checks are performed to ensure that data have not been altered during transmission. Techniques such as cryptographic hash functions are commonly used.
- **Access Control:** This feature restricts access to vehicle systems and data to authorized entities only, preventing unauthorized access and manipulation.
- **Anomaly Detection:** Monitors network traffic for unusual patterns that may indicate a cyberattack, enabling proactive management of potential threats.

OTA (Over-the-Air) Updates

OTA updates, also referred to as download OTA (DOTA) updates, refer to the wireless transmission and installation of new software, firmware, or data to devices in ACVs [8]. This technology enables vehicle manufacturers and software providers to remotely update or patch vehicle systems, eliminating the need for the vehicle to visit a dealership or service center (**Figure O.5**).

FIGURE O.5 BMS parameter OTA update to optimize fleet health [8].

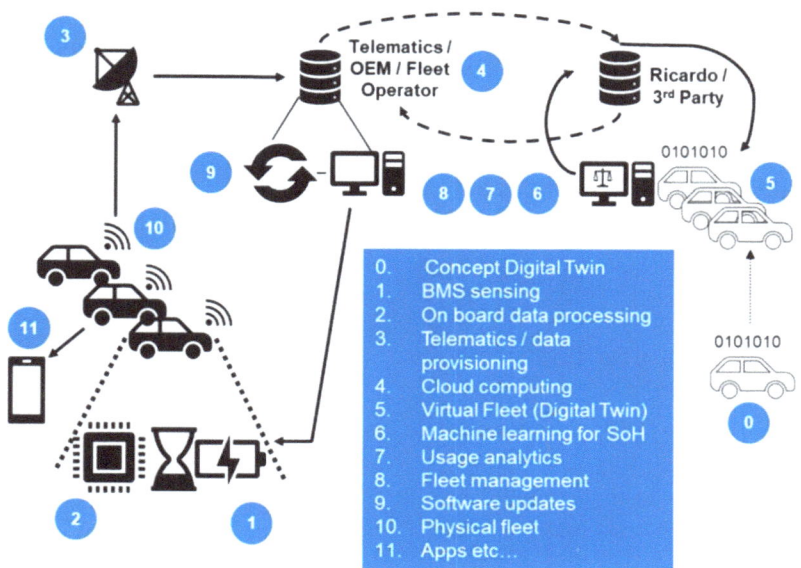

OTA updates are crucial for several reasons:
- **Enhancing Features and Capabilities:** OTA updates can introduce new functionalities to existing vehicle systems or improve their performance. This includes upgrades to navigation systems, entertainment interfaces, and ADAS functionalities.
- **Bug Fixes and Improvements:** Regular updates help address software bugs and enhance system stability and security, ensuring that the vehicle's systems operate efficiently and safely.
- **Responding to Regulations:** Changes in legal and safety requirements can necessitate updates to vehicle software to comply with new regulations or standards.
- **Security:** OTA updates can quickly roll out patches for vulnerabilities in in-vehicle software, helping protect against cybersecurity threats.

REFERENCES

1. Bassett, A., Cicotte, D., and Currier, P., "Object Tracking Comparison for Automated Vehicles Using MathWorks Toolsets," SAE Technical Paper 2021-01-0110 (2021), doi:https://doi.org/10.4271/2021-01-0110.

2. Pachhapurkar, N., Shah, R., Kale, J., Karle, M. et al., "Machine Learning Based Model Development with Annotated Database for Indian Specific Object Detection," *SAE Int. J. Adv. & Curr. Prac. in Mobility* 4, no. 3 (2021): 861-869, doi:https://doi.org/10.4271/2021-26-0127.

3. García, F., Escalera, A., and Armingol, J.M., "Enhanced Obstacle Detection Based on Data Fusion for ADAS Applications," in *16th International IEEE Conference on Intelligent Transportation Systems*, The Hague, The Netherlands, 2013.

4. Li, W., Li, W., Liu, J., and Chen, Y., "A Hybrid Method for Stereo Vision-Based Vehicle Detection in Urban Environment," SAE Technical Paper 2017-01-1975 (2017), doi:https://doi.org/10.4271/2017-01-1975.

5. Society of Automotive Engineers, "J3164_202301 Ontology and Lexicon for Automated Driving System (ADS)-Operated Vehicle Behaviors and Maneuvers in Routine/Normal Operating Scenarios," SAE Publishing, Warrendale, PA, 2023.

6. Ojha, G., Poudel, D., Khanal, J., Pokhrel, N. et al., "Design and Analysis of Computer Vision Techniques for Object Detection and Recognition in ADAS," *Journal of Innovations in Engineering Education* 5, no. 1 (2023): 47-54, doi:https://doi.org/10.4271/2022-01-1045.

7. Henle, J., Gierl, M., Guissouma, H., Müller, F. et al., "Concept for an Approval-Focused Over-the-Air Update Development Process," SAE Technical Paper 2023-01-1224 (2023), doi:https://doi.org/10.4271/2023-01-1224.

8. Azmin, F. and Mustafa, K., "Smart OTA Scheduling for Connected Vehicles Using Prescriptive Analytics and Deep Reinforcement Learning," *SAE Int. J. Adv. & Curr. Prac. in Mobility* 5, no. 3 (2023): 1120-1128, doi:https://doi.org/10.4271/2022-01-1045.

P

Parking Reservation System

The parking reservation system (parking information system) [1] is an automatic service that allows consumers to reserve a parking space in advance. The system utilizes the latest technology to inform specific applications about parking space availability and notify users accordingly to reserve the space. Sensors mounted in the parking infrastructure provide information on the availability of parking spaces based on the to-and-from traffic. Internet-based applications aggregate this information and inform their subscribers about vacant parking spaces in the vicinity of a selected POI.

Benefits:

- **Ease of Parking:** Securing a parking space, especially in major cities and events, can be a significant challenge. Securing a parking spot before arrival reduces time and provides comfort for the consumer.
- **Parking Space Optimization:** The parking reservation system achieves better management of parking facilities and space.
- **Reduced GHG Emissions:** Reducing traffic congestion and vehicle usage has a direct impact on GHG emissions, as less fuel is consumed, and CO_2 emissions are produced.

PAS (Parking Assistance System)

PAS [2] is a set of features designed to provide safety assistance to the driver so that the vehicle can be parked in the appropriate location. The system comprises various sensors, cameras, and ECUs that systematically communicate with each other to aid in vehicle parking (**Figure P.1**).

FIGURE P.1 Parking examples [2].

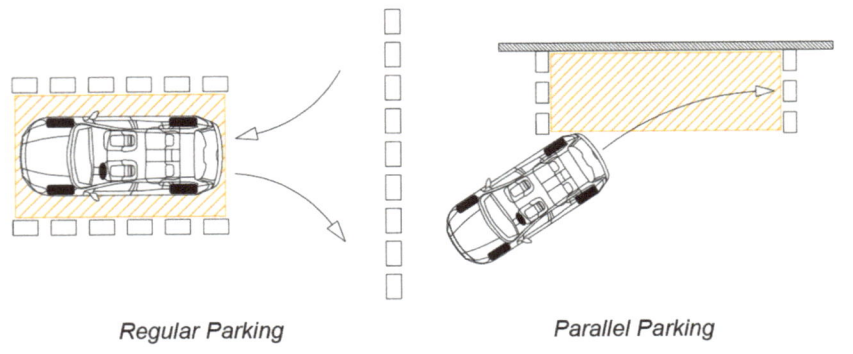

Regular Parking *Parallel Parking*

In addition, the system is designed to steer the vehicle semi-autonomously to the selected parking spot, ensuring the safety of the driver and the surrounding environment.

Key Features:
- **Sensor Fusion:** Various sensors, such as ultrasonic, radar, and camera, provide a clear view of the vehicle's obstacles, enabling decision-making to steer the vehicle in the appropriate parking area.
- **Parallel Park Assist:** This is the most common challenge in tight parking spots where the driver cannot maneuver the vehicle to park in the identified space. The vehicle can automatically steer into the parallel parking space using sensor fusion information.
- **Automatic Self-Parking:** With advancements in navigation and airborne technologies, vehicle computers can identify the available parking spot and use algorithms to perform self-parking without human intervention.

PCS (Pre-Collision System)

A PCS is an ADAS designed to detect an imminent collision with another vehicle, pedestrian, or obstacle and take preventive actions to avoid or mitigate the impact. The system uses sensors, including radar, cameras, and LiDAR, to monitor the vehicle's surroundings for potential hazards. When a potential collision is detected, the system can alert the driver with visual, auditory, or haptic warnings. In more advanced implementations, the PCS may automatically initiate emergency braking, reduce the vehicle's speed, or steer the vehicle to avoid or lessen the collision's severity. The primary objective of a PCS is to enhance safety by reducing the likelihood of accidents, particularly in situations where the driver may be distracted or unable to respond in time (**Figures P.2** and **P.3**).

FIGURE P.2 Sensor configuration [3].

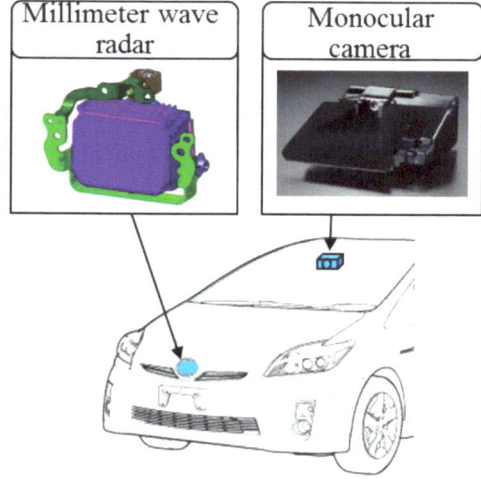

Benefits:
- Enhanced safety
- Reduced severity of accidents
- Assist in poor visibility conditions
- Increased confidence for drivers
- Insurance benefits

FIGURE P.3 PCS operation sequence [3].

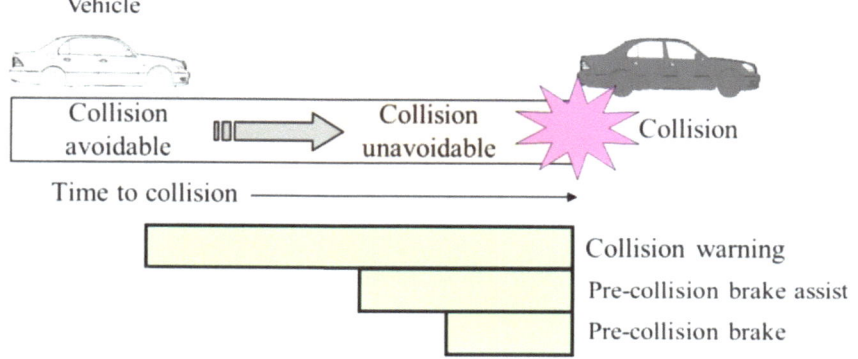

Pedestrian Collision Avoidance System

A pedestrian collision avoidance system is an advanced driver assistance feature designed to detect pedestrians in the vehicle's path and take preventive actions to avoid a collision. Utilizing sensors such as radar, cameras, and LiDAR, this system constantly monitors the environment around the vehicle for pedestrians, particularly in situations where the driver may not be able to react in time [4]. If a potential collision with a pedestrian is detected, the system may alert the driver, apply the brakes autonomously, deploy a pedestrian airbag, pop up the bonnet to reduce the impact of head injury, or take other corrective actions to prevent or mitigate an impact [5]. The primary goal of this system is to enhance pedestrian safety by reducing accidents, especially in urban environments or during low-speed driving.

Benefits:
- Improve road safety
- Reduces pedestrian injury
- Supports Level 4 autonomous capability
- Lower insurance premiums

Pedestrian Collision Warning

Pedestrian collision warning is a safety feature within an ADAS that alerts the driver to pedestrians in the vehicle's path, particularly when a collision is imminent. Using sensors such as cameras, radar, and LiDAR, the system continuously monitors the surroundings for pedestrians [5]. When a potential collision is detected, it activates visual, auditory, or haptic alerts to warn the driver in time to take evasive action. Sometimes, the system may work in conjunction with other features, such as automatic braking, to mitigate or avoid a collision. Pedestrian collision warning aims to enhance pedestrian safety and reduce the likelihood of accidents in urban or densely populated environments.

Benefits:
- Increased safety of pedestrians
- Prevention of accidents
- Lower insurance premiums
- Prominent application in urban city

Pedestrian Crosswalk Monitoring

Pedestrian crosswalk monitoring is an ADAS feature designed to detect pedestrians on or near crosswalks and alert drivers to potential hazards. The system uses sensors such as cameras, radar, and LiDAR to monitor crosswalks for pedestrians who may be crossing the vehicle's path. When a pedestrian is detected, especially in low-visibility or high-traffic areas, the system may activate visual, auditory, or haptic warnings to notify the driver [6]. Some systems can also provide automatic braking or other interventions if a potential collision is imminent. The primary goal of pedestrian crosswalk monitoring is to enhance pedestrian safety by reducing the risk of accidents at crosswalks, which are often high-risk areas for collisions [4, 5].

There are three different approaches to sensor fusion (**Figure P.4**) [6]:

- Early fusion method combines pixel-level information from images taken by an IR thermal camera and an red, blue, and green (RGB) camera to create a single image that is passed into an object detector.
- Mid-fusion strategy improves the object identification algorithm by extracting and fusing the characteristics from each sensor before feeding the images to the detector head.
- Late fusion methods make decisions after detections are made independently on RGB and thermal images. A final detection is made based on the confidence scores of the individual detections.

FIGURE P.4 Different fusion methods: (a) early fusion, (b) mid-fusion, and (c) late fusion [6].

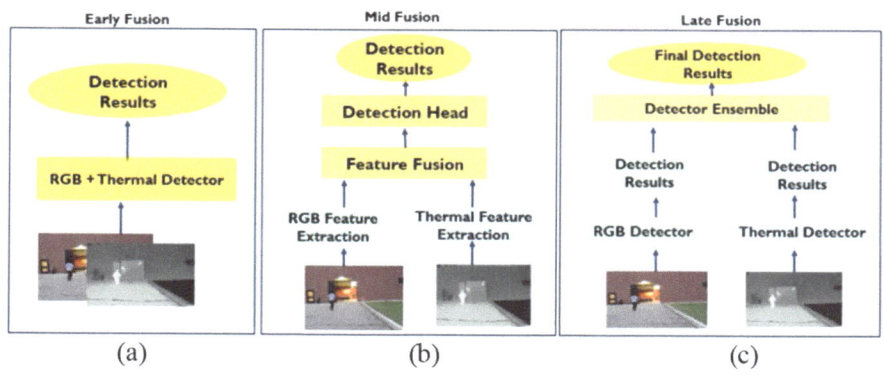

Pedestrian Detection

Pedestrian detection is an ADAS feature that utilizes sensors, including cameras, radar, and LiDAR, to identify and track pedestrians in or near the vehicle's path. The system continuously monitors the environment to detect pedestrians, particularly in low-light conditions or areas with heavy foot traffic. An example of an ML model that can perform pedestrian detection is a You Only Look Once—Pedestrian Detection ("YOLO-PD") [7]. When a pedestrian is detected, the system, equipped with a YOLO-PD model, can trigger visual, auditory, or haptic alerts to warn the driver of a potential collision. In some cases, pedestrian detection works in conjunction with AEB to prevent or mitigate the severity of a crash. The primary goal of pedestrian detection is to enhance safety by helping drivers avoid accidents involving pedestrians, particularly in urban environments or at crosswalks (**Figure P.5**).

FIGURE P.5 Examples of the Low-Light Visible Infrared and Paired (LLVIP) dataset: (a) RGB camera images and (b) corresponding IR thermal images [6].

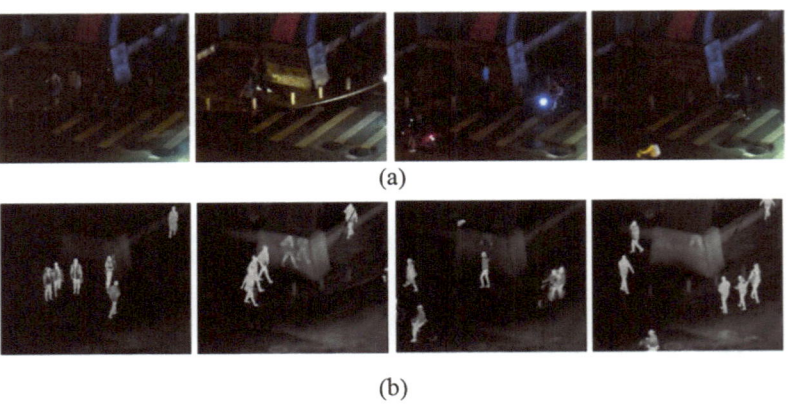

© SAE International.

Perception Algorithms

Perception algorithms are advanced computational techniques used in ADAS to process and interpret raw data from a vehicle's array of sensors, such as cameras, radar, LiDAR, ultrasonic sensors, and GPS (**Figure P.6**) [8]. These algorithms convert sensor data into a meaningful representation of the environment surrounding the vehicle, enabling the system to detect, classify, and track objects, as well as predict their motion [6].

FIGURE P.6 Data processing layers of active safety perception algorithms [8].

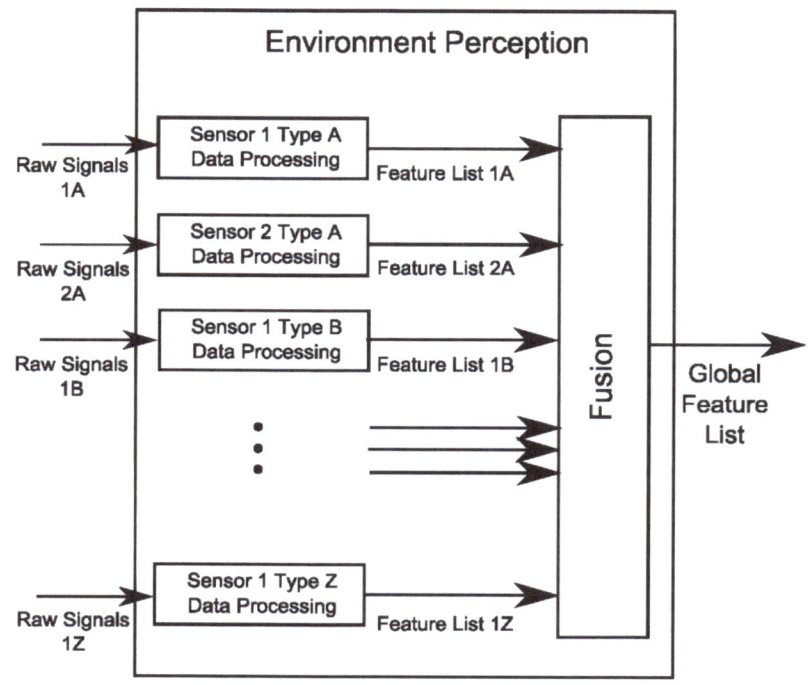

Typically, perception algorithms perform the following functions:

1. **Sensor Fusion:** Combining data from multiple sensor types to create a comprehensive and accurate environment model, accounting for individual sensors' strengths and weaknesses. For example, radar may provide long-range detection in poor visibility, while cameras offer high-resolution detail for identifying objects.

2. **Object Detection:** Identifying objects such as pedestrians, vehicles, road signs, lane markings, and traffic lights using ML and DL models. This is often done through techniques like CNNs applied to camera data for visual recognition.

3. **Object Tracking:** Continuously monitoring the position and movement of detected objects over time to assess potential risks and make dynamic decisions (e.g., for collision avoidance or ACC).

4. **Semantic Segmentation:** Dividing the environment into regions (e.g., roads, sidewalks, and buildings) and classifying each region. This enables the system to distinguish between various types of objects and areas, facilitating better navigation and informed decision-making.

5. **Sensor Calibration and Error Correction:** Ensuring that the sensors' data are aligned correctly in the vehicle's coordinate system and correcting for any sensor inaccuracies, such as misalignment or noise.
6. **Path Planning and Prediction:** Using object detection and tracking data to predict the future positions of moving objects (e.g., vehicles and pedestrians) and plan a safe path for the vehicle, including determining safe braking zones or adjusting speed.

The output of perception algorithms forms the foundation for various ADAS features, including emergency braking, LDWs, ACC, and pedestrian detection. These algorithms rely heavily on real-time data processing and AI techniques and intensive learning, to improve their accuracy and robustness under various driving conditions, such as adverse weather or complex traffic environments.

Perception Systems

Perception systems play a vital role in autonomous and semi-autonomous technologies, allowing them to comprehend and interpret their surroundings. These systems combine sensors, algorithms, and ML models to convert raw physical world data into actionable insights that guide decisions, operations, or interactions. They are utilized in diverse fields, including AVs, robotics, drones, smart city infrastructure, and many other applications.

Applications of Perception Systems:
- **AVs:** These systems are a cornerstone of self-driving cars, enabling them to detect and track pedestrians, vehicles, cyclists, and traffic signs. They help the vehicle comprehend road layouts, identify obstacles, and ensure safe navigation in dynamic conditions.
- **Smart Cities:** In urban infrastructure, perception systems enhance traffic monitoring, accident detection, streetlight control, and public safety.

PHEV (Plug-in Hybrid Electric Vehicle)

A PHEV is a hybrid vehicle that combines an ICE with an electric motor and a rechargeable battery. Unlike traditional hybrid vehicles, which rely solely on regenerative braking to charge the battery, a plug-in hybrid can be charged through an external electric power source (such as a home outlet or public charging station). This enables the vehicle to travel longer distances on electric power alone, thereby reducing fuel consumption and emissions, particularly in urban environments [9]. When the electric range is depleted, the ICE seamlessly takes over, providing continued range without the need for frequent recharging. Plug-in hybrids offer a flexible combination of electric and gasoline power, making them ideal for drivers who want the efficiency of an EV with the convenience of a gasoline engine for longer trips.

Benefits:
- **Improved Fuel Efficiency:** PHEVs consume less fuel than traditional vehicles by combining an electric motor with a gasoline engine. They can operate solely on electric power for shorter trips, maximizing efficiency.
- **Lower Emissions:** PHEVs produce fewer tailpipe emissions during electric-only operation, making them a greener alternative to standard gasoline or diesel-powered vehicles.
- **Versatility and Convenience:** Perfect for daily commutes on electric power and long journeys powered by gasoline, PHEVs offer the flexibility of an EV while retaining the extended range of a conventional car.
- **Financial Incentives:** Governments often provide tax breaks, rebates, and grants to encourage the adoption of PHEVs, making these vehicles more affordable upfront and supporting eco-friendly transportation.
- **Cost-Effective Operation:** Charging a PHEV is generally more economical than refueling with gasoline. For shorter drives, owners save significantly by relying on electricity instead of fuel.
- **Reduced Fossil Fuel Reliance:** Using electricity, PHEVs decrease dependence on fossil fuels. When paired with renewable energy sources, such as solar power, they become an even more sustainable choice.
- **Minimized Range Anxiety:** PHEVs can seamlessly switch to gasoline for longer distances, alleviating concerns about running out of charge, especially in areas with limited charging infrastructure.

Platooning

Platooning involves the formation of vehicle convoys, where the movements of the vehicles are synchronized using advanced technologies such as automation and communication systems (**Figure P.7**) [11]. This approach is commonly applied to autonomous or semi-AVs, allowing them to operate as a coordinated group to increase efficiency, lower fuel consumption, and enhance safety. For commercial vehicles, platooning is highly effective in improving the fuel efficiency of a fleet traveling in a specific route as a convoy. Many AV initiatives are focused on automated convoy trucking because of the efficiency improvements (**Figure P.8**).

FIGURE P.7 Potential vehicle operational situations (example) [10].

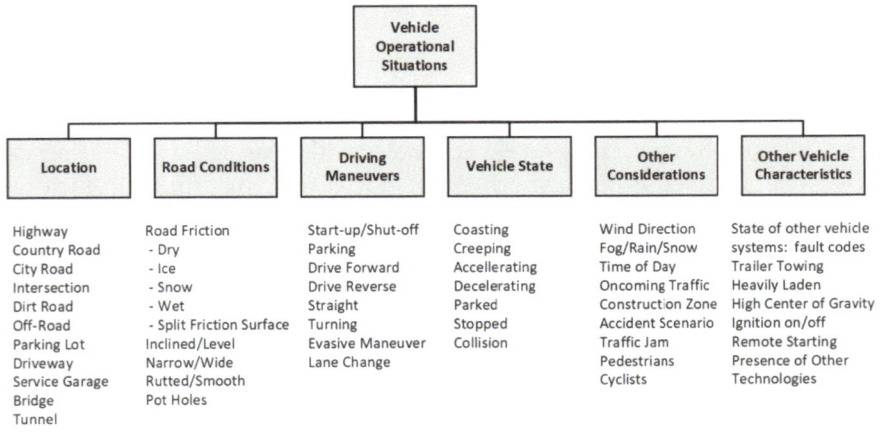

FIGURE P.8 CACC schematic overview [11].

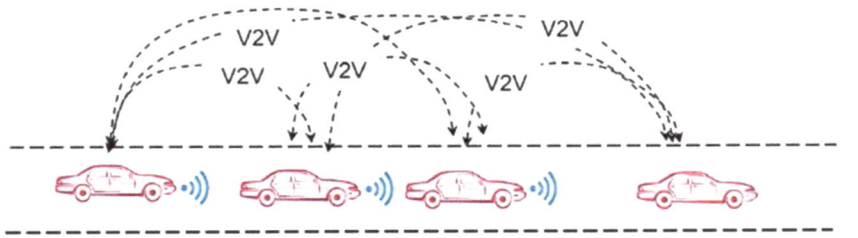

Benefits:

- **Improved Fuel Efficiency:** Platooning significantly reduces drag and optimizes the motion of vehicles in a convoy. The lead vehicle minimizes air resistance for trucks and other heavy-duty vehicles, allowing following vehicles to achieve noticeable fuel savings.
- **Enhanced Safety:** By leveraging communication between vehicles, platooning reduces the risk of human error. This enables quicker reaction to sudden stops, accidents, or obstacles. Synchronized speeds and braking also lower tailgating risks, further enhancing safety.
- **Reduced Traffic Congestion:** Efficiently managed platoons decrease congestion, particularly on highways and urban areas. With vehicles traveling closer together, fewer trips are needed to transport the same volume of people or goods, leading to better road space utilization.

- **Driver Productivity:** Semi-automated platooning minimizes the need for constant driver oversight. For instance, truck drivers can focus on route planning or rest breaks, enhancing productivity and reducing fatigue during long trips.
- **Lower Carbon Emissions:** By improving fuel efficiency and reducing emissions, platooning offers a greener alternative to traditional driving methods, helping to reduce overall carbon footprints.

Power Electronics

Power electronics is the technology that controls and converts electrical energy in vehicles, particularly within electric and hybrid powertrains. In the context of ADAS, power electronics play a crucial role in managing the flow of electrical power to various vehicle components, including motors, sensors, and other control systems. These systems include power converters, inverters, and controllers that regulate the voltage, current, and frequency of the power supplied to the electric motor or battery, ensuring efficient and reliable operation. Power electronics also play a vital role in energy recovery systems, such as regenerative braking, by converting kinetic energy into electrical energy to recharge the battery. The performance and efficiency of power electronics directly influence the responsiveness, energy efficiency, and overall functionality of ADAS features, particularly in electric and plug-in hybrid vehicles.

Predictive Analytics

Predictive analytics in ADAS uses statistical models, ML algorithms, and historical data to forecast potential risks, behaviors, and outcomes in a driving environment. By analyzing patterns from real-time data gathered from vehicle sensors (such as cameras, radar, and LiDAR), as well as external factors like traffic, weather, and road conditions, predictive analytics anticipate the likelihood of events such as collisions, lane departures, or sudden braking by other vehicles. This enables ADAS features to make proactive decisions, such as adjusting speed, initiating warnings, or activating safety interventions (e.g., AEB or ACC) [12]. Predictive analytics aims to enhance safety, optimize vehicle performance, and improve the driving experience by anticipating future events and enabling timely, preventive actions.

Benefits:

- Proactive decision-making
- Improved efficiency
- Risk management
- Cost savings
- Competitive advantage

Predictive Collision Alerts

Predictive collision alerts are an ADAS feature designed to warn the driver of an imminent collision based on real-time data analysis and predictive algorithms. Using sensors such as radar, cameras, and LiDAR, the system continuously monitors the vehicle's surroundings for potential hazards, including other vehicles, pedestrians, or obstacles. The system can predict the likelihood of a collision by analyzing the speed, trajectory, and distance of nearby objects. If the risk of a crash is detected, the system provides timely alerts, such as visual, auditory, or haptic warnings, to prompt the driver to take corrective action. In some cases, predictive collision alerts may work with other safety systems, like AEB or evasive steering assistance, to help mitigate the risk of an accident.

Benefits:
- Increased safety
- Reduced insurance costs
- Driver confidence
- Improved traffic flow
- Support for advanced driving features

Predictive Fuel Management

Predictive fuel management utilizes data analytics, advanced algorithms, and ML to enhance fuel efficiency and minimize waste across industries, particularly in transportation and logistics. By analyzing fuel usage patterns and predicting future needs, these systems enable businesses to cut costs, boost efficiency, and decrease their environmental footprint.

This technology is gaining significance as companies aim to optimize fleet operations, better manage fuel consumption, and align with sustainability objectives.

Benefits:
- Cost savings
- Enhanced efficiency
- Sustainability and regulatory compliance
- Improved fleet management
- Driver behavior optimization
- Accurate fuel forecasting

Predictive Maintenance
Predictive maintenance is a data-driven maintenance strategy in ADAS that uses real-time sensor data, analytics, and ML to monitor the health of system components and predict potential failures before they occur [13]. It enables proactive service interventions, improves vehicle safety, minimizes downtime, and extends the lifespan of critical ADAS elements such as sensors, cameras, ECUs, and actuators.

Predictive Maintenance Alerts
Advanced diagnostic systems generate notifications that inform vehicle operators or maintenance personnel of potential issues or required maintenance actions based on predictive analytics. These alerts are derived from real-time data collected from various vehicle sensors, OBD, historical performance data, and ML [13]. These are analyzed in the off-board infrastructure of the vehicle manufacturer, to identify patterns and anomalies indicative of impending failures or degradation of vehicle components. Predictive maintenance alerts facilitate timely interventions, allowing for proactive maintenance before a failure occurs, enhancing vehicle safety, reliability, and operational efficiency. Alerts can vary in urgency and detail, ranging from routine maintenance reminders to critical immediate warnings.

Predictive Road Maintenance
A proactive approach to road maintenance utilizes data analysis, ML, and predictive modeling techniques to forecast the future condition of road infrastructure. This method aims to identify potential issues before they become critical, allowing for timely repairs and maintenance interventions. Predictive road maintenance helps optimize resource allocation, reduce costs, and enhance road safety and longevity by analyzing factors such as traffic patterns, weather conditions, and historical maintenance data.

Predictive Roadside Assistance
An intelligent vehicle support system leverages real-time data, telematics, and predictive analytics to anticipate potential vehicle breakdowns or failures and proactively offer roadside services. Unlike traditional reactive models, predictive roadside assistance identifies issues, such as battery degradation, tire pressure anomalies, or engine faults, before they cause a breakdown. It enables timely alerts to drivers and automatically dispatches assistance when necessary. This enhances driver safety, reduces downtime, and improves the ownership experience.

Predictive Traffic Alerts

A real-time, data-driven notification system anticipates future traffic conditions by analyzing historical traffic patterns, current roadway data, and external factors such as weather or events. These alerts inform drivers or vehicle systems of expected congestion, incidents, or delays, enabling proactive route adjustments and improved travel efficiency.

Predictive Traffic Congestion

The projected occurrence of traffic slowdowns or bottlenecks is estimated using real-time sensor data, historical traffic patterns, GPS signals, weather conditions, and event data. Predictive traffic congestion enables smart mobility systems, navigation apps, and CVs to anticipate future traffic buildup and suggest alternative routes. This predictive insight helps reduce travel delays, optimize route planning, and improve fuel efficiency (**Figure P.9**).

- **Predictive Traffic Alerts**—Notifications generated based on predicted congestion or incidents.
- **Smart Navigation**—Route guidance systems that adapt in real time using predictive and live traffic data.
- **CVs**—Vehicles that share and receive data from infrastructure and other vehicles to improve mobility and safety.
- **TMS**—Platforms that monitor and control traffic flow, often integrating predictive analytics.

FIGURE P.9 Predictive traffic congestion.

- **Left panel** shows a traditional GPS route with live congestion (e.g., red zones on the route).
- **Right panel** shows a predictive model: green route ahead with a warning sign indicating **"Heavy Traffic Expected in 10 mins—Alternate Route Suggested."**

Predictive Traffic Modeling

A data-driven analytical process that uses historical traffic data, real-time sensor inputs, ML algorithms, and external factors such as weather, roadworks, and events to simulate and forecast future traffic conditions. Predictive traffic modeling supports urban planning, TMSs, and intelligent transportation solutions by estimating congestion levels, vehicle flow, and travel times ahead of actual events, enabling proactive decision-making.

- **Predictive Traffic Congestion**—The anticipated buildup of traffic based on model outputs.
- **Traffic Simulation**—The use of computational models to recreate and study traffic scenarios.
- **Intelligent Transportation Systems (ITSs)**—Advanced systems that apply technology for improved traffic flow and safety.
- **Digital Twin**—A virtual replica of transportation networks used in real-time simulation and prediction.

Predictive Traffic Patterns

Forecasted trends in traffic flow are derived from the analysis of historical travel data, real-time traffic inputs, and predictive algorithms. These patterns reveal anticipated vehicle movement behaviors, such as peak congestion times, high-traffic zones, and travel speed fluctuations, enabling navigation systems, city planners, and transportation networks to optimize traffic flow, reduce delays, and improve road safety.

- **Predictive Traffic Modeling**—The analytical framework used to forecast traffic conditions.
- **Traffic Flow Analysis**—The study of vehicle movement on road networks.
- **Smart Mobility**—Technology-enabled transportation systems using predictive insights.
- **Dynamic Routing**—Real-time route adjustments based on predicted and current traffic conditions.

Predictive Traffic Signal Timing

Analyzing and forecasting vehicular movement and congestion trends on roadways using historical data, real-time traffic information, and advanced algorithms, predictive traffic patterns aim to anticipate traffic flow, identify potential bottlenecks, and optimize route planning for drivers. This approach leverages data from various sources, including GPS devices, traffic cameras, and sensors, to provide insights that can enhance traffic management, reduce travel times, and improve overall transportation efficiency.

Privacy Protection

Privacy protection measures, practices, and legal frameworks are designed to safeguard an individual's personal information and ensure their right to control how their data are collected, used, and shared. It encompasses a range of strategies, including data encryption, secure storage, consent protocols, and compliance with regulations such as the GDPR. Privacy protection aims to prevent unauthorized access to personal data and maintain the confidentiality and integrity of sensitive information.

PSM (Personal Safety Message)

PSM is one of the messages defined by SAE J2735 V2X Communication Message Set Dictionary [14] that has data elements for the kinematic state of VRU. VRUs include pedestrians, cyclists, road construction workers, and scooterists. These messages play an important role in improving the safety of VRUs in situations such as a crosswalk near a signalized intersection. Examples of data elements in the PSM message are position, latitude, longitude, elevation, speed, and heading.

REFERENCES

1. Sadhana, M.V., Panday, A., and Chishti, F., "Reservation Based Smart Parking System," *International Journal of Creative Research Thoughts (IJCRT)* 6, no. 2 (2018): 972-975.

2. Lundkvist, A., Johnsson, R., Nykänen, A., and Stridfelt, J., "3D Auditory Displays for Parking Assistance Systems," *SAE Int. J. Passeng. Cars - Electron. Electr. Syst.* 10, no. 1 (2017): 17-23, doi:https://doi.org/10.4271/2017-01-9627.

3. Ono, R., Ike, W., and Fukaya, Y., "Pre-Collision System for Toyota Safety Sense," SAE Technical Paper 2016-01-1458 (2016), doi:https://doi.org/10.4271/2016-01-1458.

4. Society of Automotive Engineers, "J3116 Active Safety Pedestrian Test Mannequin Recommendation," SAE Publishing, Warrendale, PA, 2023.

5. Society of Automotive Engineers, "J3251 Cooperative Driving Automation (CDA) Feature: Perception Status Sharing for Occluded Pedestrian Collision Avoidance," SAE Publishing, Warrendale, PA, 2023.

6. Thota, B., Somashekar, K., and Park, J., "Sensor-Fused Low Light Pedestrian Detection System with Transfer Learning," SAE Technical Paper 2024-01-2043 (2024), doi:https://doi.org/10.4271/2024-01-2043.

7. Li, S., Wang, Q., Li, R., and Xiao, J., "Lightweight Neural Network Model and Algorithm for Pedestrian Detection," *SAE Intl. J. CAV* 8, no. 3 (2025): 1-13, doi:https://doi.org/10.4271/12-08-03-0027.

8. Cieslar, D., Kogut, K., Różewicz, M., and Orlowski, M., "Dynamic Input Generation for the Development of Active Safety Perception Algorithms," SAE Technical Paper 2016-01-0109 (2016), doi:https://doi.org/10.4271/2016-01-0109.

9. Quigley, J.M. and Starkey, F., *SAE International's Dictionary of Electric Vehicles* (Warrendale, PA: SAE International Publishing, 2024).

10. Society of Automotive Engineers, "J2980 (R) Considerations for ISO 26262 ASIL Hazard Classification," SAE Publishing, Warrendale, PA, 2023.

11. Society of Automotive Engineers, "J2945/6 Performance Requirements for Cooperative Adaptive Cruise Control (CACC) and Platooning," SAE Publishing, Warrendale, PA, 2023.

12. Aksan, N., Sager, L., Hacker, S., Marini, R. et al., "Forward Collision Warning: Clues to Optimal Timing of Advisory Warnings," *SAE Int. J. Trans. Safety* 4, no. 1 (2016): 107-112, doi:https://doi.org/10.4271/2016-01-1439.

13. Sun, Y., Xu, Z., and Zhang, T., "On-Board Predictive Maintenance with Machine Learning," SAE Technical Paper 2019-01-1048 (2019), doi:https://doi.org/10.4271/2019-01-1048.

14. Society of Automotive Engineers, "J3234 Active Safety Roadside Metal Guardrail Surrogate Recommendation," SAE Publishing, Warrendale, PA, 2022.

Radar

Radar is an acronym for radio detection and ranging. Radar sensors use electromagnetic waves to detect the position and velocity of surrounding objects such as barriers, pedestrians, and other vehicles [1]. They have a transmitter and receiver for electromagnetic waves. The time of flight of the wave is used to calculate the distance of the object. The shift in frequency of the emitted wave due to the Doppler effect is used to calculate the velocity of the object:

$$P_r = \frac{PtGt}{(4\pi r^2)} \sigma \frac{1}{(4\pi r^2)} * A_{eff},$$

where [2]
  Pt is the power transmitted by the radar (watts)
  Gt is the gain of the radar transmit antenna (dimensionless)
  r is the distance from the radar to the target (meters)
  σ is the radar cross section of the target (meters squared)
  A_{eff} is the effective area of the radar receiving antenna (meters squared)
  P_r is the power received back from the target by the radar (watts)

Automotive radars operate in the 77–79-GHz frequency range. Automated vehicles and ADASs have multiple radars of short, medium, and long range. A short- and medium-range radar with a higher FoV is used for blind spot monitoring systems and RCTA, while long-range radars with a narrower FoV are used for ACC and AEB (**Figure R.1**).

They have good performance in adverse weather conditions but are prone to noise caused by multiple reflections on metallic objects [3].

FIGURE R.1 Scattering mechanisms of different types of guardrail posts [3].

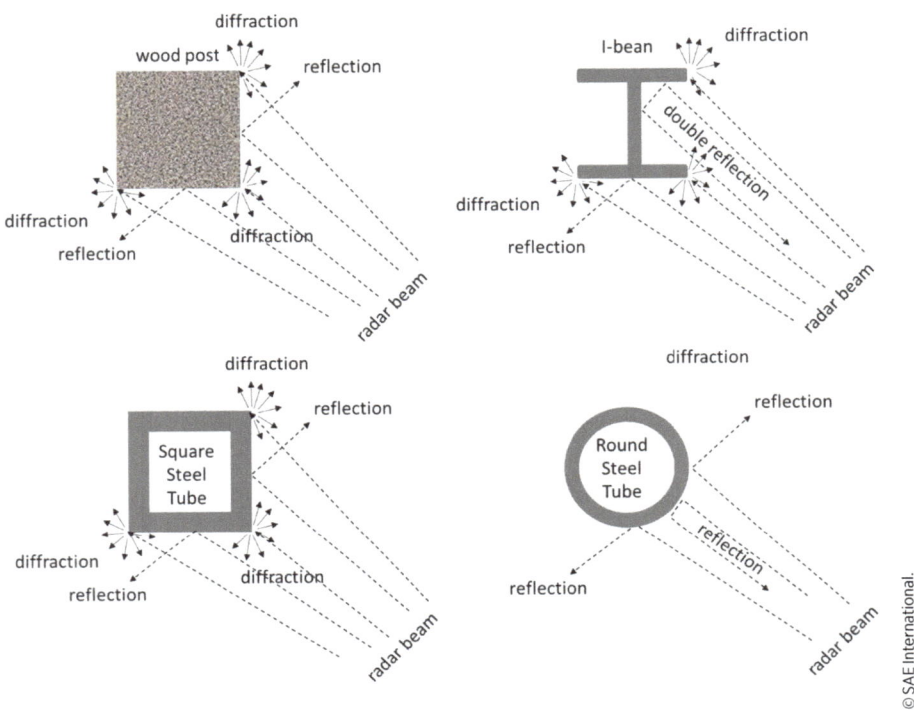

Range Anxiety

Range anxiety refers to the concern or fear experienced by drivers of EVs regarding the possibility of depleting the battery charge before reaching a charging station or their intended destination. This phenomenon significantly impacts user confidence and the adoption of EVs and CVs. This is also exacerbated by a lack of charging stations in certain geographies.

ADAS and CV technologies play a critical role in mitigating range anxiety. Features such as real-time route optimization, predictive energy consumption, and integration with charging infrastructure networks help drivers plan their journeys more effectively. Connected systems can also provide live updates on charging station availability, waiting times, and estimated charging duration, reducing uncertainty.

RCTA (Rear Cross Traffic Alert)

An RCTA system is an ADAS designed to enhance safety when a vehicle is reversing (**Figure R.2**). This feature uses radar sensors, cameras, or a combination of both to detect approaching vehicles, pedestrians, or objects that may cross the vehicle's rear path from the sides.

FIGURE R.2 Rear cross traffic visual warnings. Left depicts the rear vision system overlay, and the image on the right illustrates the outside mirror application using the side blind zone icon [4].

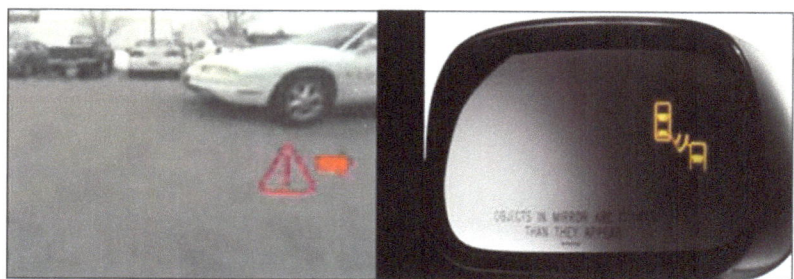

When the system identifies a potential hazard, it provides the driver with visual, auditory, or haptic warnings (e.g., steering wheel or seat vibrations) to prevent collisions during backing maneuvers. In some implementations, RCTA may also integrate with the braking system to automatically apply brakes if the driver does not respond to warnings.

This functionality is particularly beneficial in environments with limited visibility, such as crowded parking lots or driveways. It is a critical component of CV systems and contributes to overall traffic safety by addressing blind spot challenges and reducing the risk of backing accidents (**Figure R.3**).

FIGURE R.3 Staged event; dark-colored sedan in parking row (far left) is research vehicle, and white sedan in travel lane was used as the confederate and served as cross traffic [4].

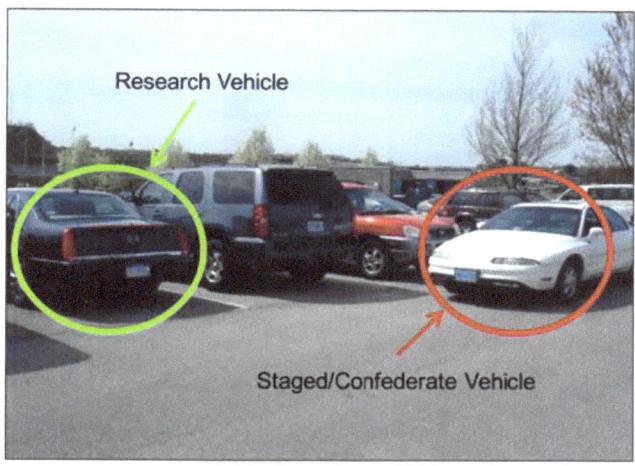

Real-Time Traffic Updates

Real-time traffic updates refer to the continuous and instantaneous transmission of traffic-related information to vehicles or devices equipped with ADAS and connected car technologies. This information includes current road conditions, congestion levels, accidents, construction zones, and other traffic-related events [5].

Purpose and Functionality:
Real-time traffic updates enhance situational awareness and improve decision-making for drivers and autonomous systems. They are transmitted through various communication channels, such as

- **GNSS:** Providing geolocation data to support real-time navigation.
- **Cellular Networks (e.g., 4G/5G):** Delivering high-speed data for real-time updates.
- **V2I Communication:** Sharing traffic signals, detours, and incident alerts from infrastructure systems.
- **Crowdsourcing Platforms:** Aggregating user-reported traffic data for accuracy and coverage.

Impact on ADAS and CVs:

1. **Enhanced Navigation:** Integration with predictive route planning to avoid congestion.
2. **Safety Improvements:** Early warnings about accidents or hazardous road conditions.
3. **Eco-Efficiency:** Optimizing routes to reduce fuel consumption and emissions.
4. **AV Decision-Making:** Supporting real-time adaptive driving strategies for Level 3+ autonomy.

Rear Collision Warning

Rear collision is an advanced driver assistance feature in which the driver is alerted of an impending collision while driving in the reverse direction (rear collision warning) [6]. The function uses sensors such as LiDAR, radar, ultrasound sensor, or camera mounted on the rear bumper or rear side of the vehicle to detect obstacles that are in collision course with the system vehicle. An audible, visual, or haptic warning is provided to alert the driver and under extreme circumstance application of brakes [6].

Rearview Camera

Rearview cameras or backup cameras are placed in the rear of vehicles to provide video of the reverse driving path to assist the driver in safely backing up the vehicle. They are connected to the vehicle's infotainment via a video link, such as low-voltage differential signaling (LVDS). The cameras receive the shifter state through the in-car network, as well as input from ultrasonic sensors and cameras at the rear of the vehicle [7]. Many regions, such as Europe, North America, and Japan, have regulations for installing backup cameras in vehicles [8].

Modern rear-view cameras also overlay the guiding lines while in reverse, based on the steering angle to improve the ability of the driver to navigate tight turns (**Figure R.4**).

FIGURE R.4 Example of parking assist via rearview camera.

Princess_Anmitsu/Shutterstock.com.

Redundancy

In ADAS and safety-critical engineering, redundancy refers to the deliberate inclusion of additional or duplicate components, subsystems, or processes within a system to ensure continued operation in the event of a failure [9]. This design principle is fundamental for risk management, system reliability, and safety assurance in automotive and other high-reliability domains (**Figure R.5**).

FIGURE R.5 Fail-operational safety architecture steps [9].

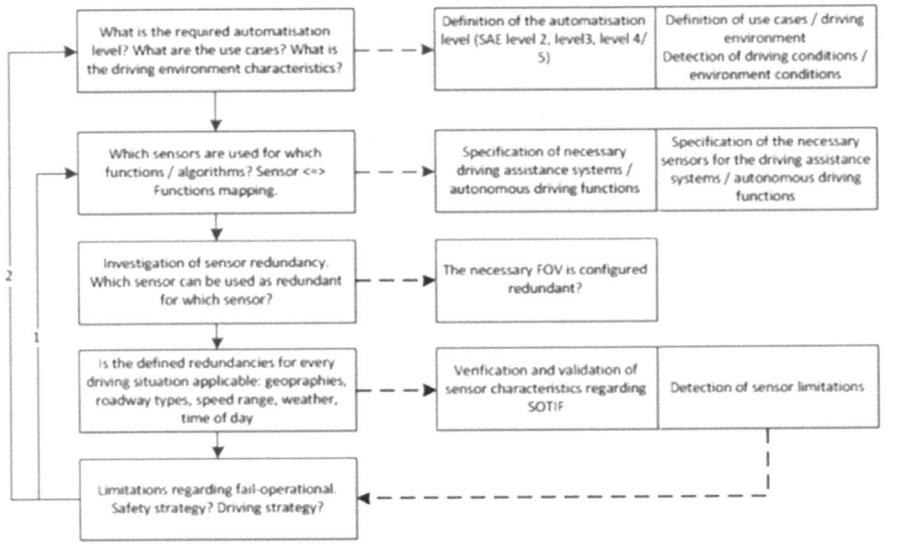

Redundancy is implemented to limit the impact of potential failures and other weaknesses in the system—such as hardware malfunctions, software errors, sensor insufficiencies or faults—by providing backup capabilities that maintain system functionality and safety.

Types of Redundancy [10]:
- **Hardware Redundancy:** Duplicating physical components (e.g., multiple sensors, processors, or actuators).
- **Software Redundancy:** Running parallel algorithms or processes to cross-check outputs.
- **Information Redundancy:** Repeating or cross-verifying data to detect and correct errors.
- **Time Redundancy:** Repeating operations or checks over time to ensure consistency.
- **Combined Redundancy:** Integrating several types for robust fail-safe mechanisms (e.g., both electronic and hydraulic braking systems).

In homogeneous redundancy, the function (hardware, software, communication, etc.) is duplicated exactly and this helps in addressing random hardware failures. This method does not help in preventing or mitigating common failure modes. In heterogeneous redundancy, the same function is implemented differently in terms of hardware, software or algorithms, which helps prevent common failure modes.

Regenerative Braking

Regenerative braking is an energy recovery mechanism used in electric and hybrid vehicles that converts kinetic energy during deceleration into stored electrical energy, enhancing efficiency and reducing reliance on traditional friction brakes [11].

The traction motors in vehicles receive energy from a RESS through an inverter to produce torque that is transferred to the wheels. When the brake pedal is pressed or the accelerator pedal is released, the motor acts as a generator, charging the battery [12, 13].

Regulatory Frameworks

Regulatory frameworks for ADAS are structured sets of rules, standards, and guidelines established by governmental and international bodies to ensure the safe design, testing, deployment, and operation of ADAS technologies in vehicles (Table R.1). These frameworks aim to strike a balance between safety, innovation, and market access by setting minimum requirements for performance, reliability, safety and compliance reporting.

TABLE R.1 Overview of major regulatory frameworks and standards governing ADASs worldwide.

Region/body	Framework/standard	Key focus areas
UNECE	WP.29 Regulations	International harmonization, ALKS, AEB
EU	Regulation (EU) 2019/2144	Mandatory ADAS features
US (NHTSA)	FMVSS, ADS guidance	Safety standards, voluntary assessments, incident reporting
US (States)	State-specific frameworks	Testing, permits, operational safety, incident reporting
Australia	Australian design rules (ADRs)	National standards, international harmonization
Draft/Other	DCAS (draft)	Hands-free driving regulations

© SAE International.

Remote Monitoring

Remote monitoring in the context of ADAS refers to the use of communication technologies and sensor networks to observe, collect, and analyze data from vehicles and drivers in real time from a location outside the vehicle itself [14]. This process enables continuous assessment of vehicle status, driver behavior, and environmental conditions, often supporting safety, maintenance, and autonomous driving functions.

Key characteristics of remote monitoring in ADAS include the following:

- **Data Collection:** Utilizes onboard sensors (such as cameras, radar, LiDAR, and tire pressure monitors) and external communication networks (e.g., wireless or mobile data) to gather information about the vehicle's operation, driver state, and surroundings [15].
- **Real-Time Analysis:** Processes and analyzes data remotely, allowing for immediate feedback, alerts, or interventions if abnormal or unsafe conditions are detected.
- **Integration with Autonomous Driving:** Supports autonomous and semi-AVs by enabling remote oversight and, in some cases, remote control, ensuring that vehicles can be monitored and managed without direct human presence inside the vehicle [15].
- **Applications:** Common uses include fleet management, predictive maintenance, driver health and performance monitoring, emission and fuel consumption tracking, and enhancing the safety and reliability of AV operations.

Remote Start

Remote start is a vehicle feature that enables the driver to start the engine remotely using a key fob or a smartphone app, without physically being inside the vehicle [14]. This system is designed primarily for convenience, enabling the vehicle to warm up or cool down the cabin before the driver enters by automatically activating the climate control system.

Key Features:

- **Convenience:** Start the car from inside a home, office, or across a parking lot, within the transmitter's range (typically 400–700 ft for key fobs; up to a mile or more for smartphone apps).
- **Climate Control:** Preconditions the interior temperature, warming or cooling the cabin as needed.
- **Security:** Encrypted signals and built-in safety features are used to prevent unauthorized use or theft.
- **Integration:** Some systems require a subscription service for smartphone connectivity.

Remote Vehicle Diagnostics

Remote vehicle diagnostics refers to detecting, identifying, and sometimes resolving issues within a vehicle's systems, including ADAS, from a distance, without requiring a technician to be physically present with the vehicle [16].

Remote diagnostics leverages connected vehicle technologies, such as OBD, telematics, and Internet connectivity, to transmit vehicle data to remote servers or expert technicians (**Figure R.6**). These data include DTC (diagnostic trouble codes), sensor readings, and system status information. The goal is to reduce the downtime and associated costs to the tune of 20 to 30% [18]. Using specialized software and interfaces, remote technicians can

- Access the vehicle's ECUs.
- Analyze fault codes and sensor data.
- Perform coding, programming, and some calibrations (including ADAS calibrations).
- Guide on-site personnel through repair steps or, in some cases, resolve issues remotely.

FIGURE R.6 Remote diagnostic system: Tester and vehicle are physically separated [17].

© SAE International.

Ride Sharing

Ride sharing refers to a transportation service model in which individuals use a digital platform, typically a mobile application, to arrange one-way, on-demand rides with drivers using their personal vehicles. The transaction is usually cashless, completed via credit card or digital payment. ADAS technologies are increasingly equipped in ride-sharing vehicles, which are designed to enhance safety, efficiency, and the UX for both drivers and passengers. There are driverless Robotaxi operations in selected cities in the US and China.

Technological Innovations:
- **AI:** Used for route optimization, dynamic pricing, and improving customer interactions [19].
- **IoT:** Enhances vehicle safety and maintenance, supporting features such as proximity verification to ensure accurate driver–passenger matching [19].
- **AVs:** The integration of self-driving cars is an emerging trend, promising 24/7 service, reduced labor costs, and improved safety, though full deployment faces regulatory and technological hurdles [20].

RL (Reinforcement Learning)

RL is an ML paradigm where an agent learns optimal decision-making strategies by interacting with a dynamic environment, receiving feedback through rewards or penalties, and adjusting its actions to maximize cumulative rewards. In ADAS, RL enables systems to improve safety and efficiency through real-world experience, adapting adaptively, thereby transcending rigid rule-based programming [21, 22].

Key Components
- **Agent:** The ADAS that learns and makes decisions (e.g., controlling acceleration or steering) [21].
- **Environment:** The driving context, including road conditions, traffic, and vehicle dynamics [21].
- **State:** A representation of the current situation (e.g., vehicle speed, proximity to obstacles) [23].

- **Action:** Decisions like braking, accelerating, or lane changes [21].
- **Reward Function:** Feedback guiding the agent (e.g., +1 for maintaining safe distance and −10 for near collision) [23].

ADAS Applications

- **ACC:** Automatically adjusts speed to match the traffic flow [21].
- **Collision Avoidance:** Learns optimal emergency braking strategies through simulated scenarios [21].
- **Lane Keeping:** Refines steering adjustments by balancing smooth navigation and safety rewards [24].
- **TSR:** Improves detection accuracy via reward-based learning from driver feedback.

Road Departure Mitigation
See **Lane Departure**.

Road Edge Detection
Road edge detection refers to the technological capability of ADAS to identify the physical boundaries of a road, such as curbs, shoulders, or unpaved edges, using sensors and computer vision algorithms (**Figure R.7**). This function enables vehicles to maintain safe positioning and avoid collisions with roadside obstacles or off-road terrain [25].

FIGURE R.7 Overall architecture of unstructured road area detection and classification algorithm [25].

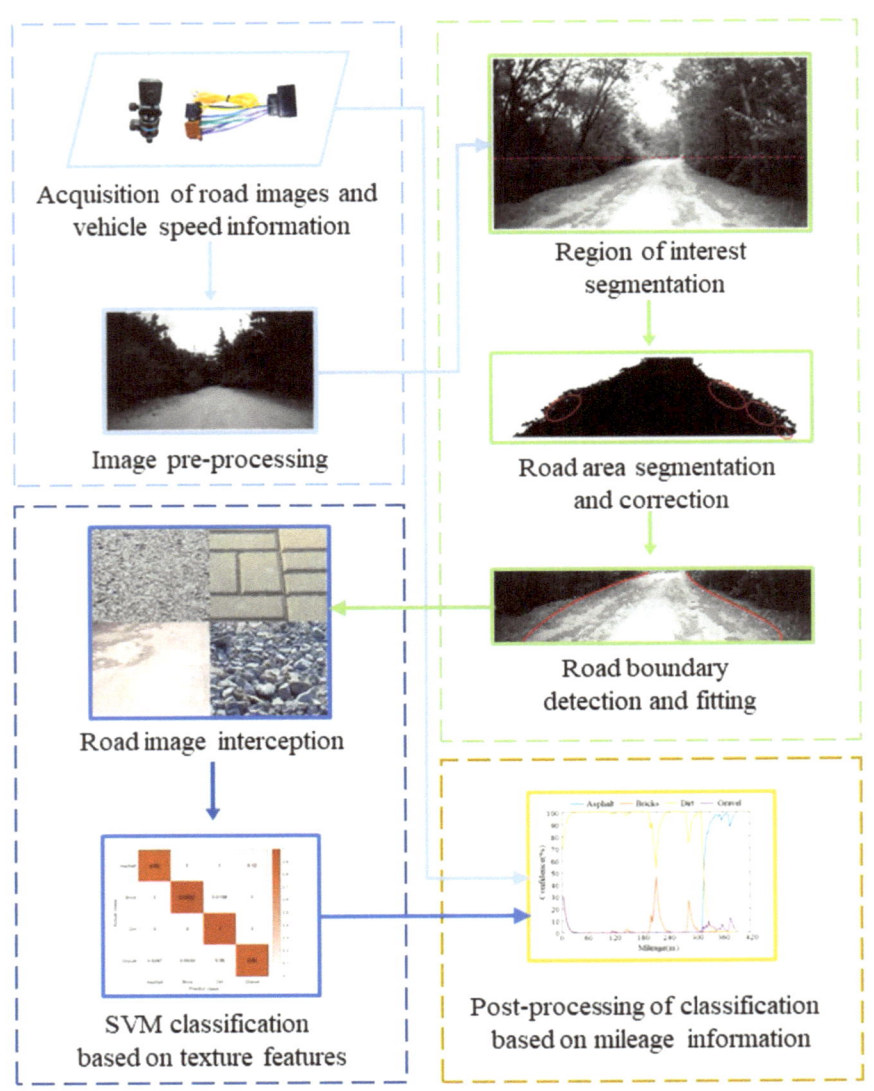

Key Features [25]
- **Real-Time Environmental Analysis:** Processes data from cameras, LiDAR, or radar to detect road edges in diverse conditions, such as low-light conditions and uneven terrain.
- **Edge Detection Algorithms:** Employs techniques like Canny edge detection and Gaussian blur to isolate road boundaries from background noise (Figure R.8).
- **Integration with Safety Systems:** Works in conjunction with LKA and adaptive steering to prevent unintended lane departures.

Technological Components [25]
- **Sensors:** Cameras capture road images, while LiDAR/radar maps surroundings to distinguish edges from obstacles.
- **Image Processing:** Converts raw data into gray scale, reduces noise, and applies geometric analysis to highlight edges.
- **ML Models:** Improves accuracy by adapting to varying road types (e.g., highways and rural roads) and markings.

FIGURE R.8 (a) The effect of road edge point detection. (b) Road boundary fitting [25].

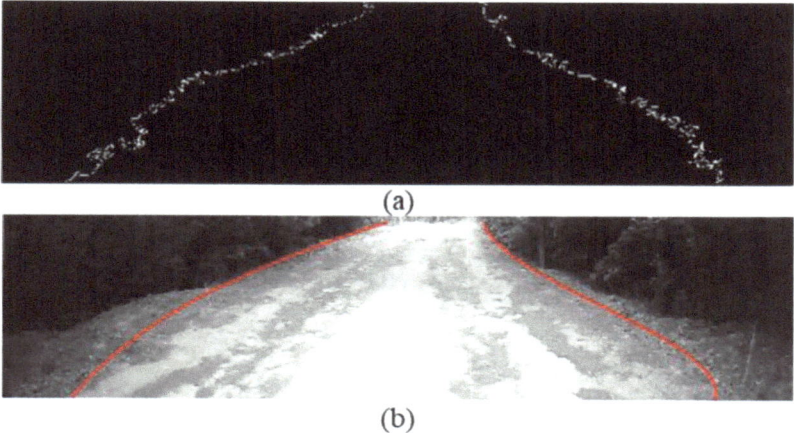

Road Infrastructure

Road infrastructure refers to the physical and digital elements of the transportation network—including roads, intersections, traffic signals, signage, and communication systems—that enable and support the safe, efficient movement of vehicles and users (**Figure R.9**). In the context of ADAS and CVs, road infrastructure also encompasses the smart technologies and data interfaces that allow vehicles to interact with their environment and with TMSs, sensors, cameras, communication modules (e.g., V2I) and data processing units embedded in or alongside roadways to collect and disseminate information.

See also **Connected Vehicle Ecosystem**.

FIGURE R.9 Possible mounting locations of chip and reader [26].

© SAE International.

Digital/smart infrastructure includes sensors, cameras, communication modules (e.g., V2I), and data processing units embedded in or alongside roadways to collect and disseminate information. CVs share and receive real-time information about road conditions, hazards, and traffic status via infrastructure-based communication networks (V2I and V2X), enhancing situational awareness and enabling cooperative driving features. Road infrastructure forms the backbone of ITS, integrating big data analytics, AI, and IoT to monitor, predict, and manage traffic, reduce congestion, and support eco-friendly driving.

Road Safety

Road safety in the context of ADAS and connected cars refers to the comprehensive set of technologies and systems designed to enhance the safety of vehicle occupants, pedestrians, and other road users by preventing accidents, mitigating collision impacts, and improving overall traffic conditions through real-time sensing, communication, and automated intervention.

Road safety in ADAS and connected car technology encompasses the use of sensors, cameras, radar, LiDAR, V2X communication, and AI to detect hazards, assist drivers, and enable vehicles to respond proactively to dynamic road environments. These technologies aim to reduce human error—the leading cause of most traffic accidents—and improve decision-making, situational awareness, and vehicle control to prevent crashes and save lives [27].

See also **Connected Vehicle Ecosystem**.

Road Side Equipment

See **RSU (Road Side Unit)**.

Roadside Assistance

Roadside assistance refers to the suite of services and capabilities offered by CAVs to aid and support drivers and vehicles in case of emergencies, breakdowns, or accidents occurring on the road. CAVs are equipped with eCall systems that can automatically or manually notify emergency services in the event of an accident or emergency. These systems may include automatic crash notification (ACN) functionalities.

CAVs can conduct real-time diagnostics of vehicle systems and components, detecting engine malfunctions, tire pressure abnormalities, or battery problems. This information is relayed to the driver and may also be transmitted to service providers for remote diagnostics via cellular telemetry systems.

CAVs can establish remote communication with service centers or roadside assistance providers. This enables drivers to receive guidance, troubleshooting assistance, or remote vehicle control (such as unlocking doors) in case of roadside issues. Roadside assistance functionalities are integrated with navigation systems, allowing CAVs to provide accurate location information to emergency responders or roadside assistance teams.

Roadside Assistance App

A roadside assistance app is a digital application, typically available on smartphones, smart watches, or integrated vehicle infotainment systems, that allows drivers to request emergency services and support for common vehicle issues while on the road. These apps are a key component of connected car ecosystems and often integrate with ADAS to enhance driver safety and convenience.

Route Planning

Route planning in the context of CAVs refers to the process of autonomously selecting and optimizing the most efficient and safe route from a starting point to a destination, considering various factors such as traffic conditions, road infrastructure, vehicle capabilities, and user preferences.

CAVs are equipped with advanced navigation systems that use GNSS data (e.g., GPS) along with mapping databases to determine the vehicle's location and plan routes. CAVs receive real-time traffic data through V2I communication, connected TMSs, and other vehicles (V2V). These data include information on traffic congestion, accidents, road closures, and construction zones, enabling CAVs to dynamically adjust their routes.

Route planning considers user preferences such as preferred routes (e.g., fastest route, shortest route, lower number of turns, avoid highways, and scenic route), charging station availability for EVs, and accessibility options for passengers with specific needs. CAVs prioritize safety by considering factors such as road conditions (e.g., weather and terrain), speed limits, traffic regulations, and compliance with safety standards outlined in automotive regulations.

RSA (Road Sign Assist)

RSA is an ADAS and connected car feature that uses vehicle-mounted cameras and image processing algorithms to detect, recognize, and interpret road signs as the vehicle travels (**Figure R.10**). The system displays key road sign information—such as speed limits, stop signs, yield signs, and warning signs—directly on the vehicle's dashboard or multi-information display, providing real-time alerts and reminders to the driver (**Figure R.11**) [29].

FIGURE R.10 Training set gallery and corresponding feature image library [28].

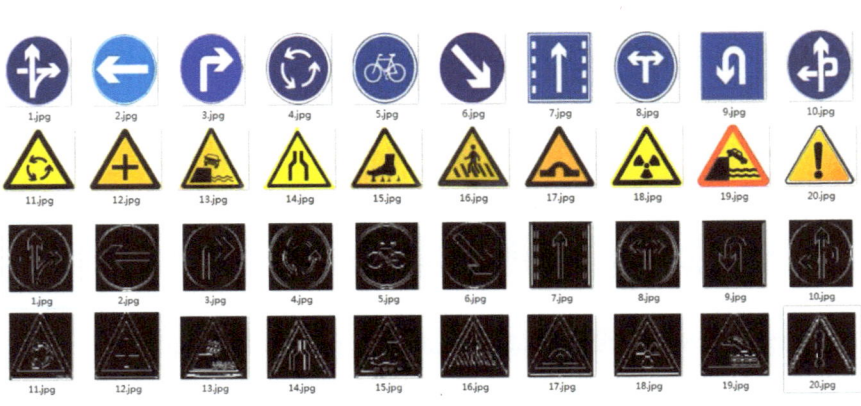

© SAE International.

FIGURE R.11 How automated driving technology can behave like human driving processes [29].

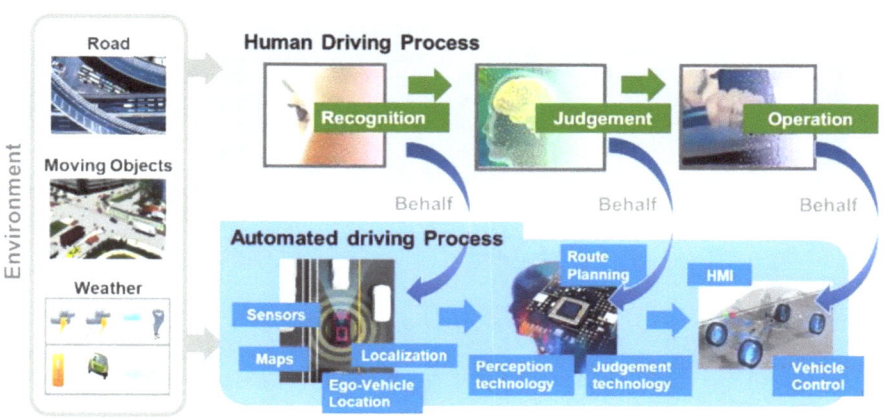

How It Works

- A forward-facing camera mounted on the vehicle continuously scans the environment for road signs.
- Image recognition algorithms, often powered by machine learning, identify and classify detected signs, even under varying lighting and weather conditions.
- Recognized signs are transmitted to the driver via visual alerts on the instrument panel, HUD, or connected car interface.
- Some systems may provide additional alerts, such as a flashing icon if the driver exceeds the posted speed limit (**Figure R.12**).

FIGURE R.12 System configuration [29].

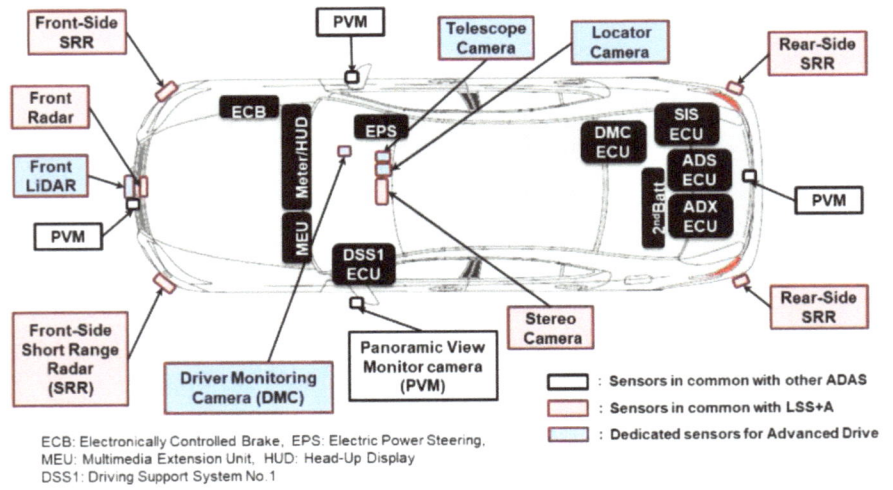

ECB: Electronically Controlled Brake, EPS: Electric Power Steering,
MEU: Multimedia Extension Unit, HUD: Head-Up Display
DSS1: Driving Support System No.1

RSU (Roadside Unit)

An RSU is an electronic communications device installed along roadways as part of ITS and CV environments (**Figure R.13**). RSUs enable wireless data exchange via DSRC or C-V2X communication based on SAE J2735 message set and GNSS data [30], between vehicles and transportation infrastructure, supporting ADAS and connected car technologies.

RSUs act as gateways, facilitating two-way communication between vehicles (via OBUs) and infrastructure. This includes real-time exchange of safety messages, traffic management data, and alerts. RSUs are central to V2I and V2X communications, allowing vehicles, traffic signals, road signs, and other infrastructure elements to share information. This enhances situational awareness for both human-driven and AVs.

FIGURE R.13 Figure V2P system [30].

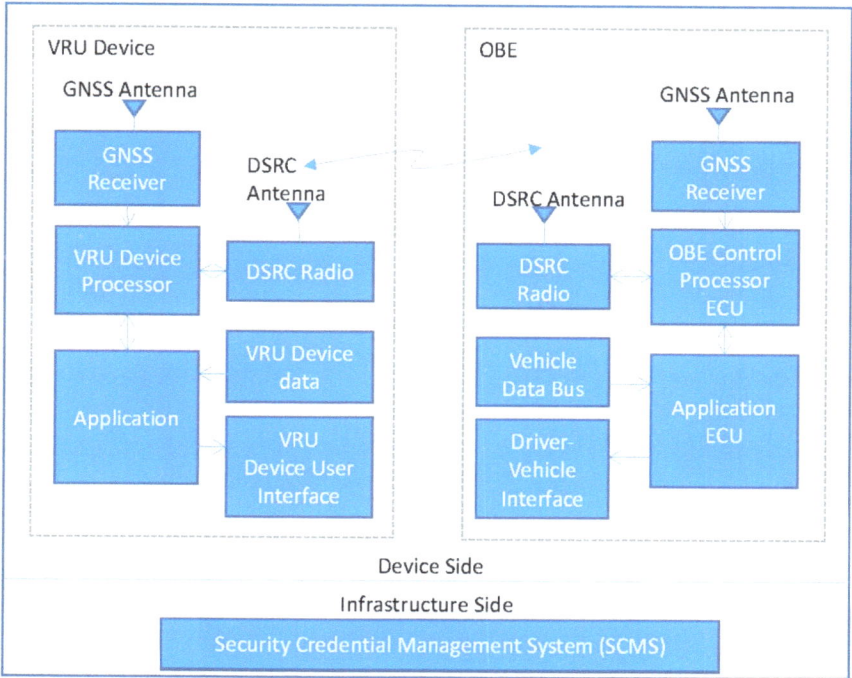

RTOS (Real-Time Operating Systems)

RTOS is a specialized operating system designed to process data and execute tasks within a guaranteed time, ensuring deterministic behavior, which is critical for time-sensitive applications. Safety-critical systems such as propulsion systems, braking systems, and steering systems in vehicles use RTOS in their software architecture. In ADAS and CVs, RTOS platforms provide the foundation for managing hardware resources and executing complex algorithms in real time [31].

Key Features:

- **Deterministic Performance:** Ensures predictable task execution with minimal jitter, essential for safety-critical operations. They meet strict timing requirements often of the order of microseconds.
- **High Reliability and Safety:** Robust and fault-tolerant architecture to maintain consistent operation under varying conditions and fault condition, which improves safety.
- **Concurrency Management:** Efficiently handles multiple simultaneous tasks, such as sensor data processing, actuator control, and communication protocols.

- **Scalability:** Supports diverse hardware configurations, from MCUs to high-performance processors.
- **Memory Management:** RTOS ensures that memory allocated to task is not utilized by a different task thereby ensuring freedom from interference of memory.
- **Prioritization:** RTOS executes safety-critical and high-priority tasks before low-priority tasks.

Applications in ADAS and CVs:

1. **Sensor Fusion:** Processes real-time data from cameras, LiDAR, radar, and ultrasonic sensors to create an accurate representation of the vehicle's environment.
2. **Control Systems:** Manages low-latency execution of braking, steering, and throttle functions for systems such as ACC and AEB.
3. **Communication Protocols:** Ensures timely data exchange over automotive networks, such as CAN, LIN, Ethernet, and V2X communication.
4. **Autonomous Driving:** Powers decision-making algorithms for navigation, obstacle avoidance, and path planning in real time.
5. **Functional Safety:** Complies with ISO 26262 standards to mitigate risks associated with software failures in safety-critical applications.

Rural Mobility

Rural mobility refers to the ability of CAVs to navigate and operate effectively in rural areas, including remote regions with limited infrastructure and challenging terrain conditions.

CAVs in rural mobility are equipped with advanced navigation systems that utilize GPS, GNSS, and high-definition maps to accurately determine their location and plan optimal routes, even in areas with limited or no cellular network coverage. These vehicles have various sensors such as LiDAR, radar, cameras, and ultrasonic sensors. These sensors enable the CAVs to perceive their surroundings, detect obstacles, identify road signs, and adapt to varying road conditions in rural environments.

While rural areas may have limited communication infrastructure, CAVs can still leverage satellite communication and offline data storage to exchange critical information with other vehicles and infrastructure when connectivity is available.

RWIS (Road Weather Information System)

The RWIS in CAVs refers to a network of sensors and communication technologies that gather real-time data about weather conditions on roadways, enabling vehicles to adapt their operations accordingly for enhanced safety and performance (**Figure R.14**). This information is articulated to vehicles through data distribution services (DDSs) (**Figure R.15**) [32].

FIGURE R.14 Reference physical architecture [security credential management system (SCMS)] [32].

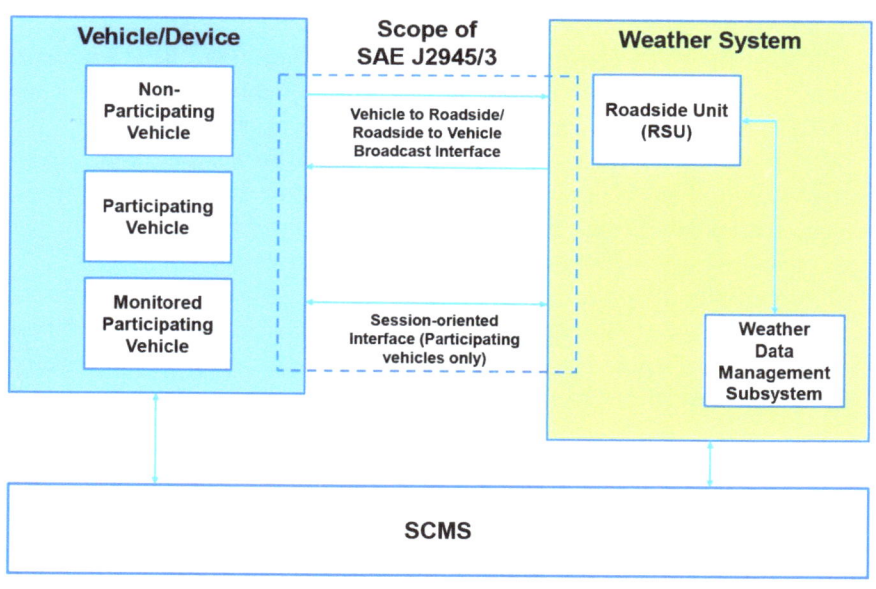

FIGURE R.15 Main DDS data flow [32].

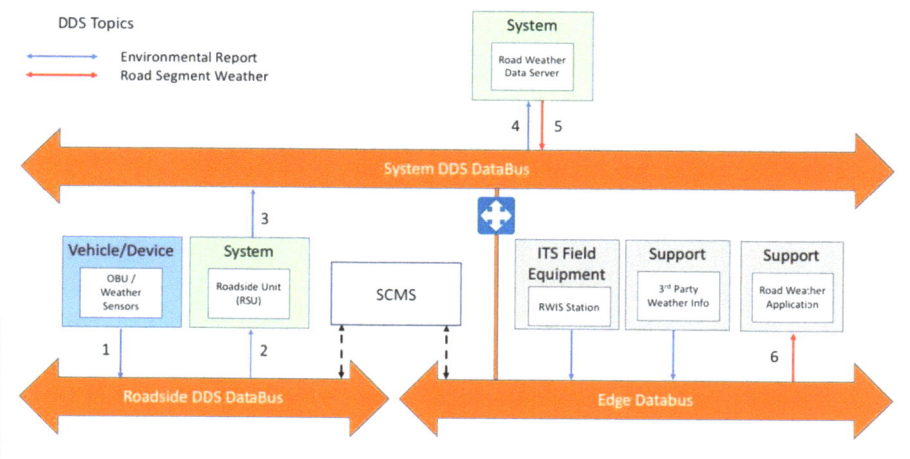

CAVs are equipped with weather sensors such as temperature sensors, humidity sensors, precipitation sensors (rain and snow), wind sensors, and visibility sensors. These sensors continuously monitor environmental conditions on roadways.

CAVs utilize V2I communication to access data from RWIS networks deployed along roadsides. This communication enables vehicles to receive up-to-date weather information specific to their location and route.

Advanced algorithms and AI systems process the weather data collected from sensors and RWIS networks. This data integration allows CAVs to analyze current weather conditions, predict weather changes, and make informed decisions regarding driving strategies and safety measures.

CAVs employ adaptive control systems that adjust vehicle parameters such as traction control, stability control, braking systems, and acceleration based on real-time weather data received from RWIS. These systems enhance vehicle stability and control in adverse weather conditions.

REFERENCES

1. Society of Automotive Engineers, "J3116 Active Safety Pedestrian Test Mannequin Recommendation," SAE Publishing, Warrendale, PA, 2023.

2. Society of Automotive Engineers, "J3122 Test Target Correlation - Radar Characteristics," SAE Publishing, Warrendale, PA, 2020.

3. Society of Automotive Engineers, "J3234 Active Safety Roadside Metal Guardrail Surrogate Recommendation," SAE Publishing, Warrendale, PA, 2022.

4. Neurauter, M., Llaneras, R., Li, B., and Green, C., "Issues Related to the Use and Design of a Backing Rear Cross Traffic Alert System," *SAE Int. J. Passeng. Cars - Mech. Syst.* 4, no. 1 (2011): 462-466, doi:https://doi.org/10.4271/2011-01-0578.

5. Darbari, J.C., Chavan, U.V., Bharti, K.G., Kavedia, M. et al., "Real-Time Traffic Management System," *International Journal for Research in Applied Science & Engineering Technology (IJRASET)* 10, no. IV (2022): 2343-2348.

6. Rictor, A. and Chandrasekar, C., "Model-Based Systems Engineering of the Aft Collision Assist Advanced Driver Assistance System," *SAE Int. J. Passeng. Veh. Syst.* 16, no. 3 (2023): 183-209, doi:https://doi.org/10.4271/15-16-03-0012.

7. Ramagopal, B., "Comfort Backup Assist Function," SAE Technical Paper 2022-28-0395 (2022), doi:https://doi.org/10.4271/2022-28-0395.

8. Department of Transportation, "FMVSS 111 Rear Visibility," Washington, DC, 2019.

9. Sari, B. and Reuss, H., "Fail-Operational Safety Architecture for ADAS Systems Considering Domain ECUs," SAE Technical Paper 2018-01-1069 (2018), doi:https://doi.org/10.4271/2018-01-1069.

10. Visual Knowledge Share, "Redundancy," 2024, accessed April 28, 2025, https://vksapp.com/dictionary/redundancy.

11. Quigley, J.M. and Starkey, F., *SAE International's Dictionary of Electric Vehicles* (Warrendale, PA: SAE International Publishing, 2024).

12. Yao, Y., Zhao, Y., and Yamazaki, M., "Integrated Regenerative Braking System and Anti-Lock Braking System for Hybrid Electric Vehicles & Battery Electric Vehicles," *SAE Int. J. Adv. & Curr. Prac. in Mobility* 2, no. 3 (2020): 1592-1601, doi:https://doi.org/10.4271/2020-01-0846.

13. Heydari, S., Fajri, P., Lotfi, N., and Rasheduzzaman, M., "Maximum-Current Curve Operation of Electric Vehicles for Improved Energy Recuperation during Regenerative Braking," *SAE Int. J. Elec. Veh.* 12, no. 2 (2023): 145-155, doi:https://doi.org/10.4271/14-12-02-0007.

14. Society of Automotive Engineers, "J3114 Human Factors Definitions for Automated Driving and Related Research Topics," SAE Publishing, Warrendale, PA, 2016.

15. Anritsu, "AD/ADAS," accessed April 28, 2025, https://www.anritsu.com/en-us/test-measurement/solutions/automotive/ad-adas?tm_navigation=solution.

16. SAE International, "J3067 Candidate Improvements to Dedicated Short Range Communications (DSRC) Message Set Dictionary [SAE J2735] Using Systems Engineering Methods," SAE Publishing, Warrendale, PA, 2020.

17. Subke, P., Moshref, M., Vach, A., and Steffelbauer, M., "Measures to Prevent Unauthorized Access to the In-Vehicle E/E System, Due to the Security Vulnerability of a Remote Diagnostic Tester," *SAE Int. J. Passeng. Cars - Electron. Electr. Syst.* 10, no. 2 (2017): 422-429, doi:https://doi.org/10.4271/2017-01-1689.

18. Biehl, M., Prater, E., and McIntyre, J.R., "Remote Repair, Diagnostics, and Maintenance," *Communications of the ACM* 47, no. 11 (2004): 101-106.

19. LISNR, "Innovative Ridesharing Tech Trends," April 12, 2024, accessed April 29, 2025, https://lisnr.com/resources/blog/innovative-ridesharing-tech-trends-how-9-ridesharing-companies-use-ai-iot-and-ultrasonic-proximity/.

20. Admin, "The Future of Ride-Sharing: Innovations to Watch," Appicial Applications - Grepix Infotech, November 12, 2024, accessed April 29, 2025, https://www.appicial.com/blog/the-future-of-ride-sharing-innovations-to-watch.html.

21. Malik, A., "Driving into the Future: The Transformative Role of Reinforcement Learning in ADAS," *International Journal for Multidisciplinary Research* 7, no. 1 (2025): 1-11.

22. Jiang, Y., Deng, W., Wang, J., and Zhu, B., "Studies on Drivers' Driving Styles Based on Inverse Reinforcement Learning," SAE Technical Paper 2018-01-0612 (2018), doi:https://doi.org/10.4271/2018-01-0612.

23. NVIDIA, "What Is Reinforcement Learning?," accessed April 28, 2025, https://www.nvidia.com/en-us/glossary/reinforcement-learning/.

24. The AllBusiness Team, "Reinforcement Learning," TIME, April 3, 2025, accessed April 28, 2025, https://time.com/collections/the-ai-dictionary-from-allbusiness-com/7274566/reinforcement-learning/.

25. Xie, F., Zhang, J., Wang, C., Liu, Q. et al., "Unstructured Road Region Detection and Road Classification Algorithm Based on Machine Vision," SAE Technical Paper 2023-01-0061 (2023), doi:https://doi.org/10.4271/2023-01-0061.

26. Jain, P., "A Smart Transportation System for Existing Vehicles and Roads Infrastructure to Ease Traffic and Toll Problems in India," SAE Technical Paper 2024-26-0181 (2024), doi:https://doi.org/10.4271/2024-26-0181.

27. Department of Transportation, "FMVSS 126: Electronic Stability Control System," Washington, DC, 2007.

28. Wang, L., "Road Sign Recognition System Based on Wavelet Transform and OPSA Point Set Distance," SAE Technical Paper 2018-01-1609 (2018), doi:https://doi.org/10.4271/2018-01-1609.

29. Kawasaki, T., Caveney, D., Katoh, M., Akaho, D. et al., "Teammate Advanced Drive System Using Automated Driving Technology," *SAE Int. J. Adv. & Curr. Prac. in Mobility* 3, no. 6 (2021): 2985-3000, doi:https://doi.org/10.4271/2021-01-0068.

30. Society of Automotive Engineers, "J2945/9 Vulnerable Road User Safety Message Minimum Performance Requirements," SAE Publishing, Warrendale, PA, 2017.

31. Kim, S., Kim, J., Nugraha, D., Wan, V. et al., "Automotive ADAS Camera System Configuration Using Multi-Core Microcontroller," SAE Technical Paper 2015-01-0023 (2015), doi:https://doi.org/10.4271/2015-01-0023.

32. Society of Automotive Engineers, "J2945/3 Requirements for Road Weather Applications," SAE Publishing, Warrendale, PA, 2022.

Safety Standards

Safety standards for CAVs encompass voluntary and regulatory standards established to ensure their safe operation, performance, and interaction within the transportation ecosystem. These standards cover both **active safety** (accident avoidance) and **passive safety** (injury mitigation during accidents), as well as software, cybersecurity [1], and functional integrity of complex electronic systems [2].

Standards created by bodies such as ISO, SAE, American National Standards Institute (ANSI), and Underwriters Laboratories (UL) are typically not legally binding, but they are widely adopted or at least provide a base in product development. These standards are established by technical committees that include industry experts and academia and are state of the art in the respective domains. Regulatory standards are mandatory and are published by regulatory bodies. Examples of regulatory standards include FMVSS and UNECE. They have longer approval times because they involve governmental organizations, which lag significantly behind technology.

Safety standards such as ISO 26262 and **SAE J3061** focus on functional safety and cybersecurity respectively, including hazard analysis, threat analysis, risk assessment, and developing secure and safety-critical systems within CAVs. These standards aim to mitigate risks associated with autonomous driving functions and ensure fail-safe mechanisms.

SAE J2945 standards define communication protocols and message sets for V2V communication, enabling CAVs to exchange safety-critical information with other vehicles in real time. This facilitates cooperative driving strategies, collision avoidance, and improved situational awareness [3].

Standards like **SAE J2735** and IEEE 802.11p govern V2I communication, allowing CAVs to interact with roadside infrastructure, TMSs, and smart city components. This communication enhances safety by providing real-time traffic updates, road condition alerts, and emergency notifications.

FMVSS standards, such as FMVSS 208 for occupant crash protection and FMVSS 214 for side-impact protection, set requirements for vehicle design, crash testing, and safety features such as airbags, seat belts, and structural integrity. These standards ensure that CAVs prioritize occupant safety in the event of a crash [4].

CMVSS standards, such as CMVSS 108 for lighting systems and CMVSS 135 for brake systems, address pedestrian and cyclist safety by defining requirements for visibility, lighting, and braking performance. These standards help CAVs detect and respond to VRUs effectively [5].

Scenario-Based Testing

Scenario-based testing is a method used in the development and validation of CAVs to evaluate their performance under various simulated real-world scenarios [6]. These scenarios encompass a wide range of driving conditions, weather challenges, and edge cases to ensure the safety, reliability, and robustness of CAV systems (**Figure S.1**).

FIGURE S.1 Scenario-based testing is integral to iterative development and test process of ADS.

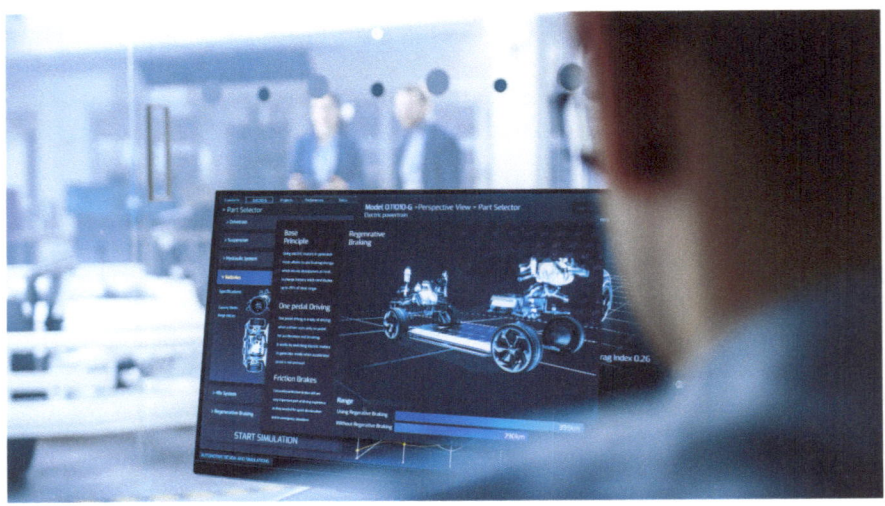

Engineers create diverse scenarios based on real-world driving data, traffic patterns, environmental conditions, and potential hazards. Scenarios may include complex intersections, a range of adverse weather conditions, pedestrian crossings, emergency maneuvers, and interactions with other vehicles. Scenarios are also created from real-world test vehicle data using data mining tools. Near misses, disengagements, infractions, etc., are converted into scenarios to be executed in a simulation platform.

Advanced simulation tools and platforms are used to replicate scenarios in a virtual environment. This includes simulating sensor inputs, vehicle dynamics, traffic interactions, scene and road characteristics, and communication protocols to mimic real-world driving scenarios accurately. The simulation environment for connected and AVs consists of multiple simulation tools operating in co-simulation mode. This interoperability is essential as each of the tools has a specific focus area, such as scene characteristics or vehicle dynamics, and multiple such tools form a complete simulation platform.

During scenario-based testing, data are collected from simulations, test vehicles, and controlled experiments. These data are then analyzed to evaluate CAV performance metrics such as decision-making accuracy, response time, adherence to traffic laws, and system robustness under challenging conditions.

Scenario-based testing adheres to industry standards and regulations to ensure the safety and compliance of CAVs. Relevant standards include guidelines from the SAE, FMVSS, and CMVSS (**Figure S.2**).

FIGURE S.2 Scenario-based testing.

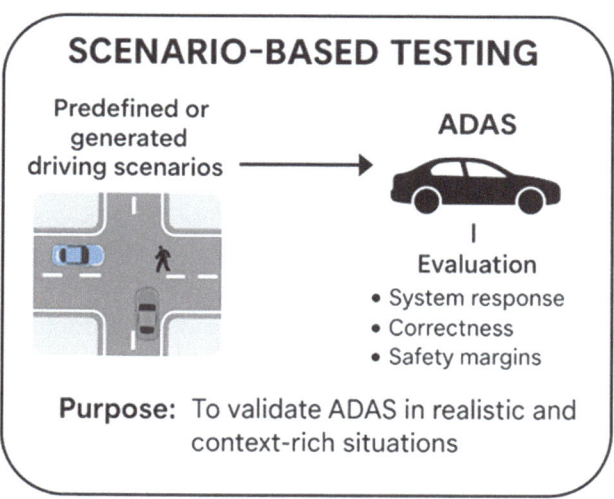

SDC (Software-Defined Chassis)

An SDC refers to a vehicle chassis system that utilizes software to control and optimize its functions dynamically. This approach integrates various vehicle subsystems such as steering, braking, suspension, and powertrain into a cohesive, programmable unit that can be adjusted in real time based on driving conditions and preferences.

These systems incorporate advanced sensors, actuators, and controllers that communicate through a vehicle's electronic and software architecture. Technologies such as ECUs and VCUs are fundamental to the software-defined chassis. Software algorithms manage the integration and operation of different chassis systems to enhance vehicle handling, stability, and comfort. The chassis settings can be altered on the fly to suit different driving scenarios, such as changing road conditions, vehicle load dynamics, and driver preferences. Telemetry system offers the capability to update and customize the vehicle's driving characteristics via software updates, similar to updating the operating system or apps on a smartphone.

SDC (Software-Defined Cockpit)

An SDC in CVs refers to the integration of software solutions that control the primary and secondary interfaces of the vehicle's cockpit. This technology emphasizes flexibility, customization, and integration of the cockpit's features through software, moving away from traditional hardware-based controls. The transition to an SDC is also accelerated by the use of large display screens for infotainment and the instrument panel cluster [7]. These displays support the integration of many applications into the display. The software architecture for SDCs is complex, as they handle data for safety-critical displays and infotainment data.

SDCs feature advanced HMIs that allow drivers and passengers to interact with vehicle systems through touchscreens, voice commands, and gesture controls. Integration of vehicle systems centralizes the control of various systems such as navigation, media, climate control, and vehicle diagnostics into a single interface, enhancing the UX [7]. Customization and personalization offer users the ability to personalize settings and profiles, adapting the vehicle environment according to individual preferences.

SDCs represent a significant shift in vehicle design philosophy, emphasizing software over hardware and promoting a more flexible, upgradeable, and customizable cockpit environment. This approach aligns with the broader trends toward fully integrated and AVs, highlighting the importance of software in the future of automotive design.

SDC (Software-Defined Connectivity)

SDC refers to the control and management of vehicle communication systems through software configurations rather than traditional hardware. This approach allows for more dynamic networking capabilities within and outside the vehicle, including adjustments and optimizations that can be made via software updates without altering physical components.

V2X communications: Facilitates the dynamic management of vehicle communications with other vehicles, infrastructure, devices, and networks. See **V2X (Vehicle-to-Everything)** entry for additional details. OTA updates enable remote software updates that can introduce new functionalities or improve existing ones without the need for physical modifications to the vehicle [8]. Dynamic network configuration allows for the real-time adaptation of network parameters to optimize connectivity based on current needs and conditions.

SDC represents a paradigm shift in how vehicles communicate and interact with their surroundings. It offers enhanced flexibility, scalability, and the potential for continuous improvement and integration of new technologies, which are essential as the automotive industry evolves toward increased automation and connectivity.

SDCAV (Software-Defined Connected and Autonomous Vehicle)

An SDCAV refers to a vehicle that leverages software for core functionalities across connectivity, automation, vehicle functions, and user interaction, fundamentally relying on software to define and control vehicle operations, characteristics, and features.

Software updates can introduce new features or improvements without changes to physical components, enhancing vehicle longevity and adaptability. Connectivity is maintained via integration of a variety of communication technologies (e.g., V2X and cellular) to improve interaction with other vehicles, infrastructure, and cloud services. Complex algorithms and sensor systems are used to enable autonomous driving functions, which are governed through software configurations.

OTA updates allow for remote software updates that can improve vehicle performance, add new features, or address security vulnerabilities. User settings enable customization that can be adapted or enhanced according to user preferences or driving conditions. Data-driven insights from data collection and software algorithms provide valuable assessments of vehicle performance and usage patterns, facilitating predictive maintenance by monitoring vehicle components and enhancing UXs.

The development of SDCAVs is expected to accelerate with advancements in AI, ML, and communications technology. The industry aims for higher levels of vehicle autonomy and enhanced connectivity, leading to more intelligent, efficient, and user-centric transportation solutions.

SDS (Software-Defined Suspension)

SDS refers to an advanced vehicle suspension system that utilizes software algorithms to adjust and control the suspension characteristics dynamically. This system is designed to improve vehicle handling, comfort, and safety by adapting to changing road conditions, vehicle dynamics, user preferences during ride hailing, or driver preferences in real time.

SDS systems typically incorporate sensors, ECUs, and actuators. The sensors monitor various parameters such as vehicle speed, wheel load, road surface characteristics and irregularities, and driver inputs. These data are processed by ECUs that use sophisticated algorithms to control actuators, which adjust the stiffness and damping characteristics of the suspension. The adjustments can be made continuously and almost instantaneously, allowing for optimal performance across varying conditions.

By continuously adjusting to road conditions, SDS can reduce the impact of bumps and potholes, enhancing passenger comfort. The system allows for stiffer suspension settings during aggressive driving maneuvers, such as cornering, thereby reducing body roll and improving vehicle stability. These systems can adapt to different driving scenarios, such as highway cruising or off-road driving, optimizing suspension settings for each situation. SDS is integrated with the other vehicle systems to provide cohesive and enhanced driving experience.

SDT (Software-Defined Telemetry)

SDT in the context of CVs refers to the advanced telemetry systems where both data collection and processing functionalities are driven by software. This allows for high flexibility, dynamic reconfigurability, and scalability in the management and analysis of vehicular data streams such as vehicle performance, maintenance needs, and usage patterns.

SDT leverages the principles of software-defined networking (SDN) and software-defined anything (SDx) to separate the data plane from the control plane in telemetry systems. This abstraction enables vehicle manufacturers and operators to deploy customized telemetry applications that can be updated or modified without altering the underlying hardware. SDT systems typically use cloud-based services for data aggregation and analytics, supporting real-time decision-making and predictive maintenance.

SDTM (Software-Defined Traffic Management)

SDTM refers to the application of SDN principles to the management and optimization of traffic flows in urban environments. This approach leverages CV technologies and real-time data analytics to control traffic signals dynamically, manage traffic congestion, and optimize route selections based on current traffic conditions.

SDTM systems use real-time data and data processing from various sources, including CVs, traffic sensors, and surveillance cameras, to monitor traffic conditions continuously. By analyzing real-time data, SDTM systems can adjust traffic signals, modify speed limits, and reroute traffic to alleviate congestion and enhance safety. V2I communication technologies allow vehicles to communicate with TMSs to share data on traffic conditions, vehicle speeds, and other relevant information. Traffic lights can adapt their timing based on actual traffic demands, reducing unnecessary waiting times at intersections and improving overall traffic flow.

SDTM is increasingly relevant in smart city initiatives aimed at reducing traffic congestion, minimizing environmental impact, and improving urban mobility. Its application helps in creating more efficient urban transport networks, reducing commute times, and improving air quality.

SDVs (Software-Defined Vehicles)

An SDV refers to a vehicle where core functionalities, traditionally characterized and managed by hardware, are controlled by software (**Figure S.3**). This transformation allows for dynamic updates and upgrades through software, enabling continuous improvements and the addition of new features after the vehicle has left the production line.

FIGURE S.3 SDV block diagram.

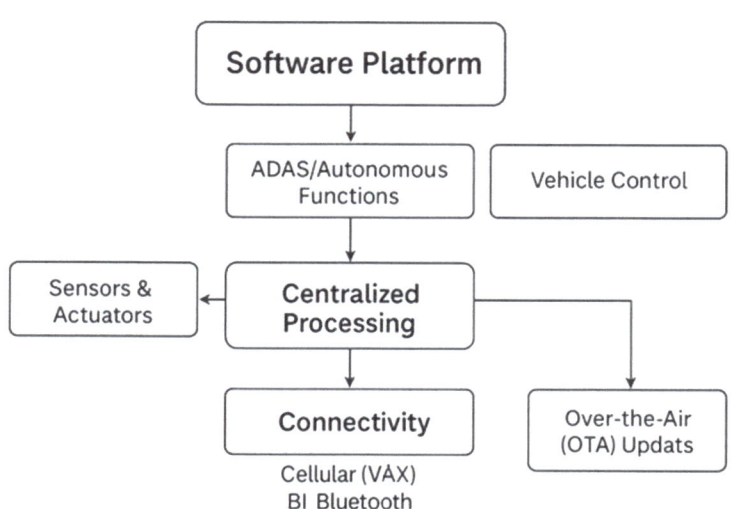

Traditional vehicle hardware becomes standardized and modular, serving as shared resources across platforms, while the software delivers the primary value and differentiation. In this way, the vehicle hardware (ECU) is decoupled from the software. By consolidating functionalities that used to be spread across multiple ECUs into a single high-performance central computing system, SDVs reduce complexity and cost, while enhancing performance and security [9].

Service-oriented architecture (SOA) facilitates flexible and scalable integration of in-vehicle and cloud-based services. Data-driven and AI-centric analysis supports real-time analytics, personalization, and ML models for perception and decision-making.

The system consists of elements. The infrastructure layer involves not only the vehicle but also the associated telecommunications equipment, and back end systems of original equipment manufacturers (OEMs), which support continuous development and operational services through data insights. Hybrid cloud and AI platforms are integral to the infrastructure, to facilitate flexible software deployment and an AI data platform for managing and optimizing vehicle functionalities and security. AI plays a crucial role in the development and operation phases, enhancing everything from cybersecurity to UX through adaptive and predictive technologies.

SDVD (Software-Defined Vehicle Diagnostics)

SDVD refers to the methodologies and technologies used to assess, monitor, and maintain the operational integrity of vehicle systems through software-based processes [10]. This approach leverages the integration of advanced software, sensors, and network communications to provide real-time diagnostics, predictive maintenance, and remote troubleshooting capabilities for vehicles (**Figure S.4**).

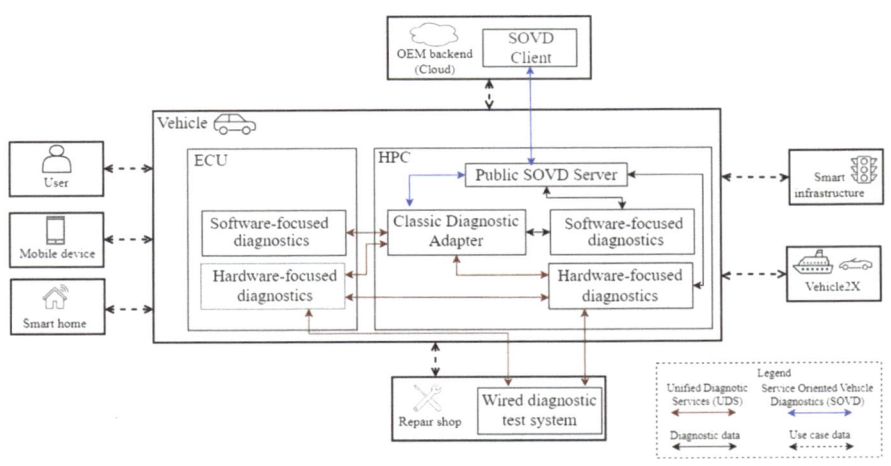

FIGURE S.4 The concept for future vehicle diagnostics with influencing stakeholders [10].

SDVD utilizes data from various onboard sensors and external sources to facilitate immediate diagnostic processes and predictive maintenance. This ensures optimal vehicle performance and safety through continuous monitoring and updates, which can be conducted OTA. The integration of SDVD with CV technologies also enhances UX by allowing vehicle owners to receive updates and maintenance alerts directly on their personal devices.

OBD with advanced software algorithms processes data from vehicle sensors to detect anomalies and potential system failures. Telematics systems use wireless communication systems to transmit vehicle data to external servers for further analysis and support. Predictive maintenance software algorithms analyze historical and real-time data to predict when maintenance should be performed to prevent unexpected failures. Remote firmware OTA (FOTA) updates to software can be wirelessly downloaded and installed on vehicles, improving diagnostic capabilities and resolving identified issues without the need for physical service center visits.

Secure Boot

Secure boot is a security mechanism used in electronic vehicle systems to verify the integrity of software being executed on ECUs (**Figure S.5**). It ensures that only verified and trusted software can boot and run on the system, protecting against unauthorized modifications and malware [11].

FIGURE S.5 Secure boot.

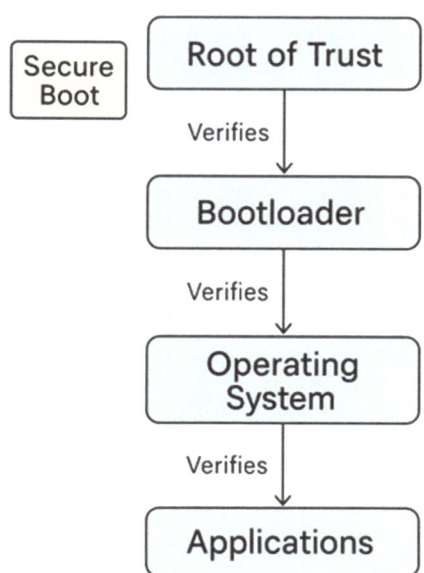

Secure boot involves cryptographic techniques such as digital signatures and hashing to validate the authenticity and integrity of software before it is allowed to execute. During the boot process, the system's firmware checks whether the software's cryptographic signature matches a prestored valid signature. If validation fails, the system prevents the software from executing, thereby blocking potential threats [12].

Applications in ADAS:
- Ensures safe booting of perception, fusion, and actuation software.
- Protects ECU software for ACC, emergency braking, and lane assist.
- Critical in vehicles with OTA update capability.

Self-Driving Vehicle

A self-driving vehicle, also known as an AV or driverless car, is a vehicle equipped with advanced technologies that enable it to navigate and operate without direct human intervention. These vehicles use a combination of sensors, cameras, radar, LiDAR, GPS, and AI algorithms to perceive their environment, make driving decisions, and execute maneuvers safely and efficiently.

SAE J3016_202104 is a globally recognized standard published by SAE International that classifies six levels of driving automation (Levels 0 through 5) (**Figure S.6**) [13].

- Level 0: No automation: The driver is in full control of all aspects of driving.
- Level 1: Driver assistance: The vehicle can assist with specific tasks, such as steering or braking, but the driver remains primarily responsible.
- Level 2: Partial automation: The vehicle can control both steering and acceleration/deceleration simultaneously under certain conditions, but the driver must remain engaged and monitor the environment.
- Level 3: Conditional automation: The vehicle can handle most driving tasks under certain conditions, allowing the driver to disengage temporarily, but must be ready to take over when prompted.
- Level 4: High automation: The vehicle can perform all driving tasks within specific ODDs without human intervention but may still require human control in certain situations.
- Level 5: Full automation: The vehicle can operate autonomously under all conditions and environments without human intervention or oversight.

FIGURE S.6 SAE J3016 Levels of driving automation [14].

Sensor

CAVs are replete with devices that detect and measure physical phenomena or environmental conditions. Sensors play a critical role in enabling CAVs to perceive their surroundings, gather data, and make informed decisions autonomously.

Key types of sensors in CAVs:

- LiDAR sensors use laser pulses to measure distances to objects, creating high-resolution 3D maps of the vehicle's surroundings. They are crucial for object detection, obstacle avoidance, and precise mapping in autonomous driving systems (**Figure S.7**).
- Radar sensors emit radio waves to detect objects, determine their distance, relative speed, and size. They are effective in various weather conditions and are used for ACC, collision avoidance, and blind spot monitoring.

FIGURE S.7 Vehicles have a multitude of sensors detecting the state of the ambient environment.

Cameras capture visual information, including images and videos, to identify lane markings, traffic signs, pedestrians, and other vehicles. They are essential for object recognition, traffic light detection, and LKA.

Ultrasonic sensors use high-frequency sound waves to detect nearby objects and measure distances. They are commonly used for parking assistance, object detection at close range, and low-speed maneuvering. IR detects heat signatures for low-light object or pedestrian detection, which are commonly used for night vision driving and DMSs. GNSS provides precise location data that are used for geofencing and autonomous planning of routes for ADAS applications.

Sensor Fusion

Sensor fusion refers to the integration and synchronization of data from multiple sensors within CAVs, enhancing perception and decision-making capabilities (**Figure S.8**).

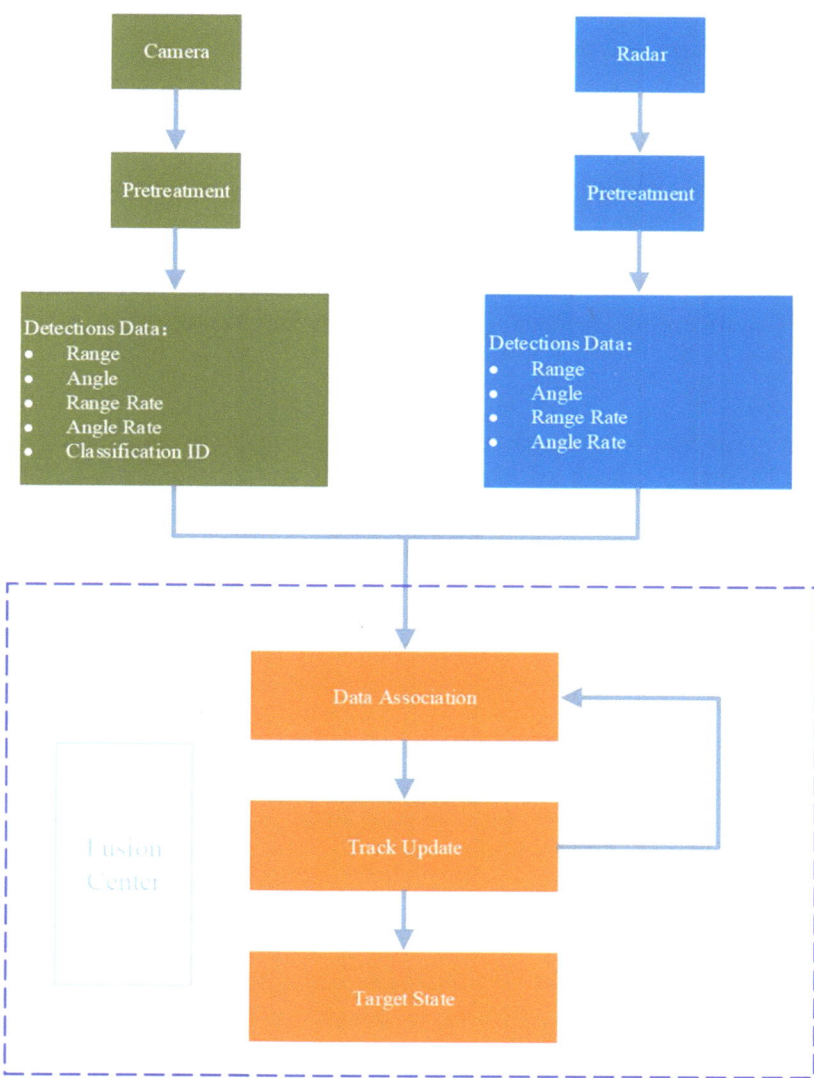

FIGURE S.8 Centralized fusion structure [15].

Sensor fusion algorithms combine data from different sensors to create a comprehensive and accurate representation of the vehicle's environment. This integration enables CAVs to detect and track objects, identify potential hazards, and make informed decisions in real time [15].

Sensor fusion enables CAVs to detect and track objects such as vehicles, pedestrians, cyclists, and road signs. By fusing data from multiple sensors, CAVs can achieve robust object detection capabilities even in challenging environmental conditions.

Sensor fusion enhances environmental perception by providing a 360° view of the vehicle's surroundings, including object classification, localization, and trajectory prediction. This perception is essential for safe navigation, collision avoidance, and lane keeping in various driving scenarios.

Sensor Range

Sensor range refers to the maximum and effective operational distance within which a vehicle's sensors (such as radar, LiDAR, ultrasonic, and cameras) can accurately detect and respond to external objects, conditions, or events (Table S.1). This range is crucial for the functionality of various driver assistance and ADSs. In the context of autonomous and ADAS applications, sensor range is a critical performance metric that determines how early a vehicle can perceive environmental elements and respond appropriately.

TABLE S.1 Typical sensor range.

Sensor type	Typical range	Characteristics/notes
Radar	100–250 m	Reliable in poor weather, long-range detection of vehicles and large objects
LiDAR	50–200 m	High-resolution mapping and object contouring; 3D object representation, weather-sensitive
Camera	Up to 150 m	Object detection and classification and lane detection; performance depends on lighting and visibility
Ultrasonic	0.1–5 m	Used for close-range detection, parking, and low-speed maneuvers
IR/thermal	Up to 100 m	Used for pedestrian and animal detection in low light or nighttime
GNSS	Global (satellite-based)	Provides position data but not obstacle range

© SAE International.

Sensor Resolution

Sensor resolution refers to the slightest change that it can measure. Another term that is often discussed together with resolution is sensor accuracy. Sensor accuracy is the closeness of the estimated value to the actual value. For sensors used in CAVs, sensor resolution represents the level of detail with which sensors in CAVs can perceive and interpret their surroundings, including objects, obstacles, and environmental conditions. High-resolution sensors enable vehicles to detect smaller objects, precise distances, and fine features in the driving environment. This is critical for object classification, pedestrian detection, lane tracking, and safe maneuvering at both low and high speeds.

Cameras in CAVs capture images and videos of the vehicle's surroundings. Sensor resolution for cameras is measured in terms of pixels. Pixel is the short form of "picture element." Images are represented as 3-D arrays of numbers, with integers between [0, 255]. Higher resolutions provide sharper and more detailed imagery, enhancing object recognition and scene analysis capabilities.

LiDAR sensors emit laser beams to create 3D maps of the environment. LiDAR sensor resolution refers to the density of points or data points per square meter, impacting the accuracy of object detection, distance measurement, and terrain mapping.

Radar sensors use radio waves to detect objects and measure their relative speed and distance. Radar resolution is determined by factors such as frequency bandwidth and pulse repetition frequency, affecting the sensor's ability to distinguish between closely spaced objects and detect small targets.

Ultrasonic sensors emit high-frequency sound waves to detect objects and measure distances. Sensor resolution for ultrasonic sensors relates to the precision of distance measurements, influencing collision avoidance and PSAs at shorter distances (**Table S.2**).

TABLE S.2 Sensor-specific resolution characteristics.

Sensor type	Resolution characteristics	Impact on ADAS/autonomy
Camera	High spatial resolution (up to 4K or greater); color and texture detail	Enables object recognition, lane detection, traffic sign classification
LiDAR	Angular and range resolution (e.g., 0.1–0.5° at 200 m)	Enables 3D object contouring, curb/lane mapping, obstacle avoidance
Radar	Lower spatial resolution; improving with imaging radar	Good for motion detection, less accurate for shape or exact position
Ultrasonic	Very low resolution; binary presence/absence detection	Useful for near-field proximity (e.g., parking assist)
IR/thermal	Low spatial resolution; detects heat contrast	Effective in nighttime or low-visibility human/animal detection
GNSS	Position resolution varies; sub-meter with RTK	Accurate vehicle localization in combination with correction data broadcasted from reference stations and map data

© SAE International.

Side Collision Warning

Side collision warning is a safety feature integrated into CAVs designed to alert drivers or take autonomous action to prevent or mitigate side-impact collisions with other vehicles, objects, or pedestrians, to detect and warn vehicles, objects, or road users approaching from the side, especially in blind spots or adjacent lanes, and to prevent unsafe lateral maneuvers. This enhances situational awareness and improves vehicle safety during lane changes, turns, and merging scenarios.

CAVs employ a combination of sensors such as radars, cameras, and ultrasonic sensors to detect vehicles approaching from the sides, as well as objects or pedestrians near the vehicle's path. When a potential side collision threat is detected, the vehicle's warning system alerts the driver through visual, auditory, or haptic feedback. The warning alerts provide drivers with additional situational awareness, enabling them to take evasive action or adjust their driving behavior to avoid potential collisions. In AVs, side collision warning systems can autonomously initiate braking or steering maneuvers to prevent collisions, further enhancing safety.

Side-Impact Airbags

Side-impact airbags are safety features integrated into CAVs to protect occupants during side-impact collisions by deploying airbags from the vehicle's interior panels (**Figure S.9**). In the context of autonomous and ADAS, side airbags complement active safety features by providing critical protection in crash scenarios where avoidance is not possible.

SAE International's Dictionary of ADAS and Connected Vehicles   **249**

FIGURE S.9 Example of deployed side-impact airbags from a collision.

Ronald Rampsch/Shutterstock.com.

Side-impact airbag systems consist of airbag modules installed within the vehicle's doors, seats, and side panels. These modules deploy rapidly upon detecting a side-impact collision, providing a cushioning barrier between occupants and the vehicle's interior structure.

CAVs utilize advanced crash sensors, such as accelerometers and gyroscopes, to detect the severity and direction of a side-impact collision. These data trigger the deployment of side-impact airbags to mitigate potential injuries. ECUs manage the operation of side-impact airbag systems. They receive input from crash sensors and make split-second decisions to deploy airbags at the optimal timing and force level to protect occupants effectively.

Some advanced CAVs incorporate occupant position sensors that adjust the deployment strategy of side-impact airbags based on the occupant's seating position, height, weight, and proximity to the impact zone.

Integration with ADAS and autonomous systems:
- Paired with side collision warning and lateral collision mitigation systems to optimize timing of deployment.
- Activation based on inputs from accelerometers, crash sensors, and occupancy classification systems.
- In AVs, linked to precrash sensing systems for early airbag staging and occupant repositioning.

Governing regulations and standards:
- FMVSS 214 (NHTSA): US regulation defining side-impact protection standards [16].
- UN R135 (UNECE): Global regulation for pole side-impact tests [17].

- NCAP [Euro, US, Association of Southeast Asian Nations (ASEAN)]: Consumer-facing test programs for safety ratings.
- Insurance Institute for Highway Safety (IIHS) Side-Impact Test: Includes updated, more severe crash scenarios for advanced ratings [18].

Signalized Intersection

A signalized intersection is an intersection managed by traffic control signals that regulate the flow of vehicles and pedestrians (**Figure S.10**). In the context of CAVs, these intersections are equipped with communication infrastructure that enables real-time data exchange between the infrastructure and vehicles. This communication supports advanced functionalities like intersection collision avoidance, optimized traffic flow, and enhanced situational awareness for both human-operated and AVs [20].

FIGURE S.10 Right-turning hazard vehicle and approaching through vehicle. The through vehicle is facing a green light [19].

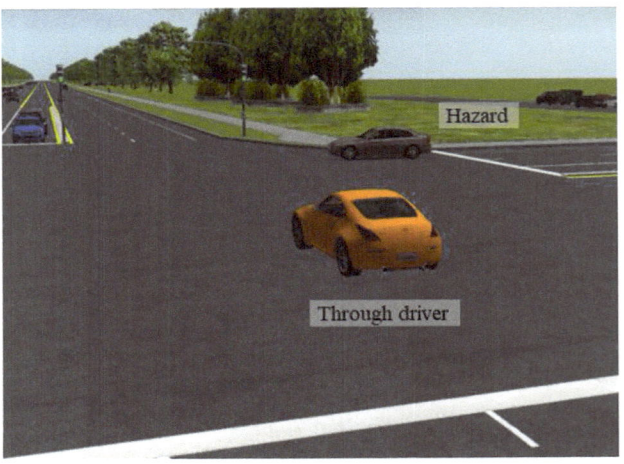

Signalized intersections are critical nodes within the urban traffic network where CAVs utilize V2I communications to interact with traffic signals and other connected entities. These intersections contribute to the operational safety and efficiency of CAVs by providing timely and precise SPaT [20] information and geographic intersection descriptions (MAP) through DSRC or cellular networks (**Figure S.11**).

FIGURE S.11 Vehicles communicating with a signalized intersection [20].

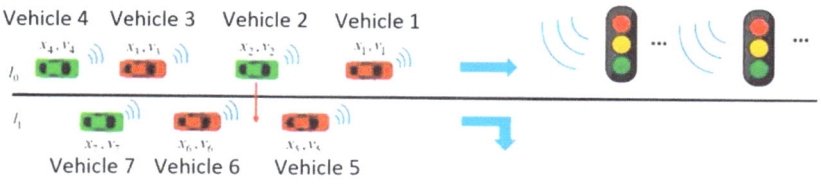

With the advancement of smart cities and the integration of IoT technologies, signalized intersections are expected to evolve into more dynamic and responsive systems. These improvements will likely enhance the interaction between CAVs and infrastructure, leading to better traffic management solutions and reduced congestion.

Situational Awareness

Situational awareness refers to the capability of CAVs to comprehensively perceive, analyze, and understand their surrounding environment, including road conditions, traffic, obstacles, pedestrians, and other vehicles. This awareness enables CAVs to make informed decisions and take appropriate actions to ensure safe and efficient operation.

CAVs are equipped with a sophisticated sensor suite comprising LiDAR, radar, cameras, ultrasonic sensors, and other technologies. These sensors provide real-time data about the vehicle's surroundings, including object detection, lane markings, traffic signs, and pedestrians.

Computation platform and in-vehicle communication channels: The data collected by the sensor suite are sent to a computational platform through a communication channel appropriate for the specific sensor data. These computation platforms consist of high-performance computing hardware, their interfaces to sensor data channels, and interfaces to in-vehicle systems. The software for perception, planning, and control runs on these platforms.

Advanced sensor data fusion algorithms integrate information from multiple sensors to create an accurate representation of the vehicle's environment. This fusion enhances the accuracy and reliability of situational awareness by reducing sensor errors and improving object recognition and tracking.

CAVs leverage V2V and V2I communication technologies to exchange data with other vehicles, roadside infrastructure, and centralized TMSs. This connectivity enhances situational awareness by providing access to real-time traffic updates, road condition information, and potential hazards.

AI-powered perception systems analyze sensor data to identify and classify objects, predict their behavior, and assess potential risks. Perception function includes detecting moving vehicles, stationary obstacles, pedestrians, cyclists, and road hazards such as potholes or debris, determining their position, velocities in the x, y, and z directions, direction of movement, and acceleration.

CAVs employ sophisticated decision-making algorithms that consider the information gathered from sensors, communication systems, and environmental models. These algorithms enable CAVs to make decisions in real time, such as adjusting speed, changing lanes, executing evasive maneuvers, and interacting safely with other road users.

SLA (Speed Limit Assist)

SLA refers to the vehicle technology that helps drivers adhere to the speed limits by either notifying the driver when the vehicle exceeds the speed limit or automatically adjusting the vehicle's speed to stay within legal limits. SLA systems typically use a combination of GPS data and TSR cameras to detect speed limits. These systems may also integrate with a vehicle's cruise control system to automatically adjust the vehicle's speed. Drivers are often given the choice to activate or deactivate the assist feature and can override it manually in most vehicles.

SLAM (Simultaneous Localization and Mapping)

SLAM is a critical technique used in the field of robotics and AVs for creating or updating a map of an unknown environment while simultaneously keeping track of the vehicle's location within that environment. SLAM integrates data from various sensors (such as LiDAR, cameras, and GPS) to compute the vehicle's current position and orientation, while also constructing a detailed spatial map. This process is essential for AVs to navigate safely and efficiently in their surroundings without prior knowledge of the environment.

In the context of CAVs, SLAM enables the vehicle to navigate complex environments by recognizing and avoiding obstacles, planning paths, and performing tasks like docking and maneuvering in dynamic scenarios. SLAM technology supports various levels of vehicle automation, contributing to the enhancement of situational awareness and real-time decision-making capabilities.

As shown in **Figure S.12**, SLAM enables vehicles to navigate safely and autonomously without relying solely on preexisting maps or external localization systems (e.g., GPS). It allows for real-time environment perception, path planning, and obstacle avoidance, especially in dynamic or GPS-denied environments such as tunnels, parking garages, or urban canyons.

FIGURE S.12 SLAM block diagram.

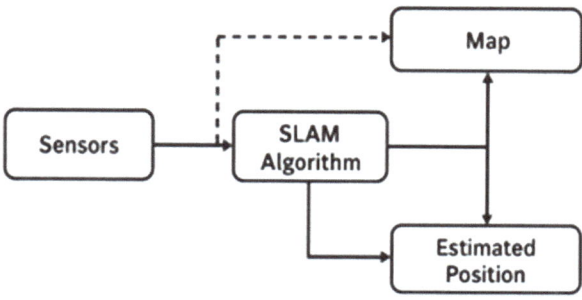

Smart City

In the context of a smart city, CAVs are advanced vehicles equipped with technologies that allow them to communicate with each other and with infrastructure elements (such as mass transit systems, traffic lights, and sensors). These vehicles use a combination of sensors, cameras, radar, and AI to navigate roads without human intervention. Integration into smart cities aims to enhance traffic efficiency, reduce accidents, and lower emissions. CAVs operate under specific regulatory standards to ensure safety and interoperability in urban environments.

Smart cities support connected and cooperative mobility through V2X communication, enabling autonomous and ADAS-equipped vehicles to make more informed, predictive, and safe driving decisions based on infrastructure data, real-time traffic conditions, and dynamic regulations (**Figure S.13**).

FIGURE S.13 Smarty city for ADAS and AV.

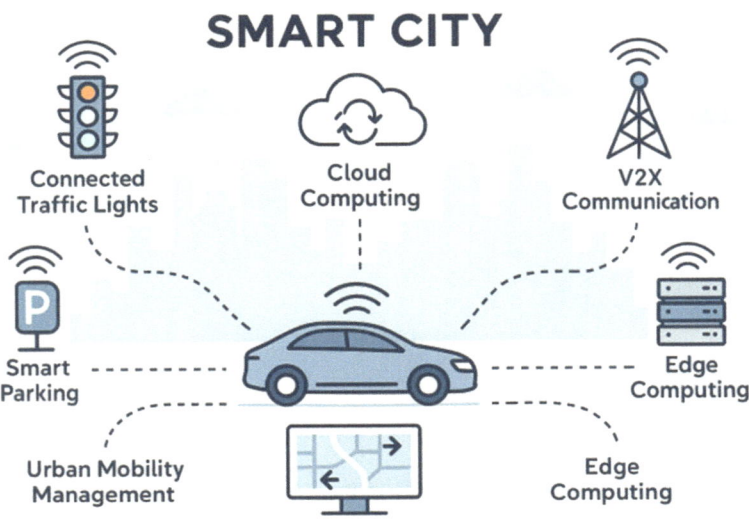

Smart Grid

Smart grid refers to an electricity supply network that uses digital communication technology to detect and react to local changes in usage. In the context of CAVs, the smart grid plays a critical role in managing the energy demands of EVs, including their charging requirements, and integrating V2G technologies. This integration allows for more efficient energy management, where vehicles not only consume power but can also return energy to the grid during peak load times, helping to stabilize the grid.

To support energy-aware mobility by enabling bidirectional communication between EVs and the power grid, smart grids ensure that CAVs can charge efficiently, support grid stability, and participate in energy-balancing strategies.

CAVs rely on the smart grid to optimize their energy consumption and contribute to energy distribution, necessitating robust communication and operational standards. Although specific SAE International, FMVSS, and CMVSS that directly reference smart grids may not exist, related standards include the following:

- **SAE J2954**—This standard deals with wireless power transfer for EVs, directly impacting how vehicles interact with the smart grid during charging processes [21].

- **SAE J2847/1**—Specifies communication between plug-in EVs and the utility grid, facilitating the effective integration of vehicles with smart grid technologies [22].
- **FMVSS and CMVSS**—While these primarily focus on vehicle safety, the integration of smart grid technologies touches aspects of vehicle electronics and cybersecurity, influenced by broader standards that ensure the safe operation of electric and AVs within the grid infrastructure.

Smart Intersections

Smart intersections refer to technologically enhanced road junctions designed to improve traffic flow, safety, and communication between vehicles and infrastructure. These intersections utilize various technologies, including sensors, cameras, and connectivity systems, to manage and optimize the movement of vehicles, pedestrians, and cyclists.

Smart intersections, guided by these standards, enhance traffic management by reducing congestion and improving the speed and accuracy of vehicular and pedestrian movement decisions. They are vital in urban planning to accommodate the increasing number of CAVs, thereby supporting more efficient urban mobility. Smart intersections are connected to TMSs, map servers, and security back ends.

As technology evolves, the integration of advanced communication systems between vehicles and traffic management infrastructure will play a critical role. This may lead to updates in existing standards or the introduction of new ones to govern the interactions at smart intersections better.

Smart Parking Solutions

Smart parking solutions refer to technologies and systems that utilize automation and connectivity to enhance parking efficiency and UX. These solutions involve the use of sensors, cameras, and software algorithms to manage the allocation of parking spaces, guide vehicles to available spots, and optimize parking resources. Smart parking is a component of the broader smart mobility ecosystem, aiming to reduce congestion, improve traffic flow, and decrease carbon emissions by minimizing the time spent searching for parking.

Smart parking solutions use real-time data to show available parking spots, often accessible via mobile apps or vehicle infotainment systems. Automated parking guidance guides drivers directly to vacant parking spaces using dynamic signage or direct vehicle communication. In advanced incarnations of the vehicle systems, the vehicles park themselves without human intervention. Payment integration automates parking payments and ticketing through connected applications, reducing the need for physical payment methods.

Smart parking solutions, particularly those integrated with vehicle systems and AV technologies, are subject to various regulatory standards to ensure safety and interoperability. As urbanization increases and AVs become more prevalent, smart parking solutions are expected to become more integrated into urban planning and vehicle design, representing a critical component of smart city infrastructures and connected transportation systems.

Software-Defined Battery Management

Software-defined battery management refers to the approach of using software-based systems to control and optimize the functions of a BMS in vehicles. This approach leverages advanced software algorithms to dynamically manage the charging, discharging, health monitoring, and overall efficiency of battery cells in EVs and hybrid systems.

The system architecture typically involves a centralized or distributed BMS architecture where software algorithms provide real-time data analysis, predictive maintenance, and adaptive control strategies based on the state of charge and health of the battery. These systems are often integrated with vehicle telematics and other onboard systems for enhanced diagnostics and user feedback. System performance is optimized to maximize battery life and efficiency through adaptive charging techniques and load management. Real-time health monitoring of battery and conditions to preemptively identify and address issues such as cell imbalance or degradation. The system optimally manages the energy stored in the battery to support vehicle functions and energy regeneration techniques [23].

Software-defined battery management represents a pivotal shift in how vehicle power systems are controlled and maintained, promising enhanced reliability, efficiency, and adaptability of battery technologies in modern vehicles.

Software-Defined Braking System

A software-defined braking system refers to the advanced braking technology where software algorithms predominantly control the braking functions in a vehicle. Unlike traditional braking systems that rely on mechanical and hydraulic components, software-defined braking systems utilize electronic signals to operate brakes, often in conjunction with electric motors that drive the brake calipers to provide the brake torque (**Figure S.14**).

FIGURE S.14 BBW system structure [24].

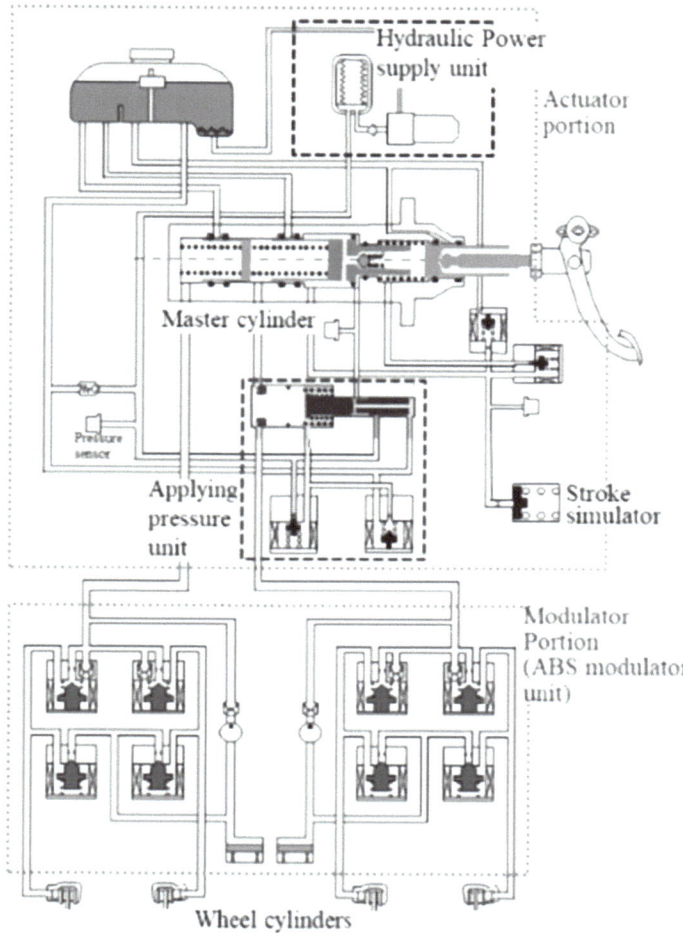

Software-defined braking systems are typically implemented as part of a broader "by-wire" technology, which may also include SbW and throttle-by-wire systems. These systems are often integrated with ADAS to enhance vehicle safety features such as AEB (**Figure S.15**).

FIGURE S.15 Variables S (brake pedal stroke), G (deceleration), and controller control logic [24].

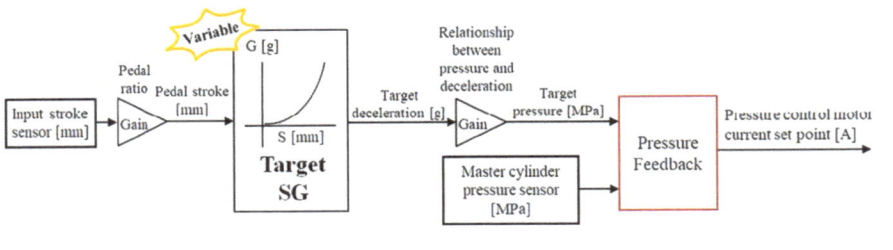

Software-defined braking systems represent a significant shift from traditional mechanical systems to more integrated and intelligent vehicle systems. They are a key component of modern vehicles' transition toward higher levels of automation and enhanced safety functionalities. The challenge associated with software-defined full-by-wire braking systems is that they lack mechanical backup, necessitating additional safety mechanisms and measures to ensure safety. Adherence to safety standards such as ISO 26262 [2] is critical for full-by-wire brake system development.

Software-Defined Climate Control

Software-defined climate control refers to the management of a vehicle's HVAC systems through software, enabling dynamic adjustments based on vehicle conditions, environmental factors, and user preferences. This approach allows for greater flexibility, efficiency, and personalization in climate control. These systems typically include a combination of sensors (temperature, humidity, occupancy), control units, and UI elements, all governed by sophisticated software algorithms, that can incorporate predictive algorithms to pre-adjust cabin conditions based on expected weather changes or user habits.

Software-Defined IC (Instrument Cluster)

A software-defined IC is an advanced type of vehicle dashboard that uses software to display vehicle information digitally rather than traditional analog or mechanical gauges. This allows for greater flexibility and customization in presenting data such as speed, fuel level, engine stats, and navigation details [7].

Users can personalize the display according to their preferences and the specific driving context. The system is capable of integrating information from various vehicle systems and external sources, such as ADAS and V2X communications. Software updates can introduce new functionalities or improve existing ones without the need for hardware changes remotely via telematics systems.

The trend toward more connected and AVs will likely increase the prevalence and sophistication of software-defined ICs. These systems are expected to become more interactive and integrated with other in-vehicle and external data systems, potentially incorporating AR elements to blend physical and digital driving cues seamlessly.

Software-Defined Sensors

Software-defined sensors in vehicles refer to sensor technology that can be dynamically configured or adjusted via software updates to adapt to new functions or enhance existing capabilities without the need for physical modifications. This flexibility allows for continual improvement and customization of sensor operations as part of a broader SDV architecture.

Adaptive functionality enables vehicles to adjust their sensor functions according to specific needs, thanks to the implementation of new features that leverage available sensors or environmental conditions, thereby enhancing vehicle safety and performance. Maintenance and updates enable remote software updates to fix bugs, improve performance, or add new features, significantly reducing the need for physical repairs or replacements.

Software-defined sensors represent a significant shift in vehicle sensor technology, offering enhanced adaptability and upgradability, which are crucial for the future of connected and AVs. As vehicles become more integrated with software, the role of software-defined sensors will become increasingly critical in enabling flexible, responsive, and efficient vehicle systems.

Software-Defined V2X (Vehicle-to-Everything) Architecture

Software-defined V2X architecture refers to a flexible and programmable framework within CAVs that enables seamless communication and interaction with various entities, including other vehicles (V2V), infrastructure (V2I), pedestrians (V2P), and the cloud (V2C), through SDN and protocols.

SDN uses V2X architecture, which utilizes SDN principles, allowing for dynamic and efficient management of communication resources. SDN enables the allocation of bandwidth, prioritization of data traffic, and the implementation of customized communication policies based on real-time requirements. V2X architecture employs standardized protocols such as IEEE 802.11p (wireless access in vehicular environments [WAVE]) and cellular–V2X (C-V2X) technologies (based on LTE/5G), ensuring interoperability and compatibility between CAVs and external entities. Robust security mechanisms, compliant with standards such as ISO/SAE 21434, SAE J3061, and NHTSA's Cybersecurity Best Practices for Modern Vehicles [25], are integrated into the V2X architecture to protect data integrity, privacy, and mitigate cyber threats. V2X architecture includes data management solutions that handle large volumes of heterogeneous data from diverse sources, applying techniques such as edge computing, cloud integration, and distributed processing to optimize data processing and decision-making.

SOTIF (Safety of the Intended Functionality)

SOTIF refers to a concept within the realm of CAVs that focuses on ensuring the safe operation of ADSs beyond the traditional scope of functional safety measures. The focus of SOTIF is to ensure hazards are not caused when the vehicle is operating without any functional safety-related failure modes. Functional safety-related failure modes include random hardware failure and systematic failures in hardware and software. SOTIF-related hazards are caused by limitations in the performance of the system, triggering conditions, or reasonably foreseeable direct and indirect misuses. ISO 21448 [26] is the ISO standard that defines an SOTIF life cycle for vehicles of autonomy Levels 1 through 5. SOTIF as a philosophy can be extended to all applications, not just autonomy.

While functional safety (addressed by standards like ISO 26262) deals with mitigating the risks associated with system malfunctions and failures, SOTIF extends beyond this by considering potential hazards arising from the intended functionality of automated systems, including scenarios where the system operates correctly but leads to unsafe outcomes due to unforeseen circumstances [2].

SOTIF involves a comprehensive risk assessment process that identifies potential scenarios where the automated system's intended functionality could lead to unsafe situations, such as ambiguous sensor inputs, environmental uncertainties such as bad weather, or human–machine interaction complexities. Conditions that are typically external to the vehicles that trigger the weaknesses in the autonomy system are called triggering conditions, for example, glare that blinds the camera, resulting in a loss of object detection. Here, glare is a triggering condition.

SOTIF requires rigorous validation and testing procedures to evaluate the system's response to various real-world scenarios, including edge cases and corner cases that traditional functional safety assessments may not cover. This validation process often involves simulation, testing on public roads, and scenario-based testing.

SPaT (Signal Phase and Timing)

SPaT is a message defined by the **SAE 2735a** standard that contains data elements related to signal phases and their changes in a connected intersection [27]. Signal phases are red, yellow, and green and their variants such as blinking red and red arrows. The SPaT messages are used in conjunction with the MAP messages for the intersection to define features for a connected intersection.

Steering Assist

Steering assist refers to systems in vehicles that support the driver by making it easier to steer the vehicle (**Figure S.16**). These systems typically assist by reducing the physical effort required to turn the steering wheel, enhancing comfort and control, particularly during low-speed maneuvers or parking. This may include autonomous evasive steering assist (ESA) solutions [28].

FIGURE S.16 Simplified architecture of ESA [28].

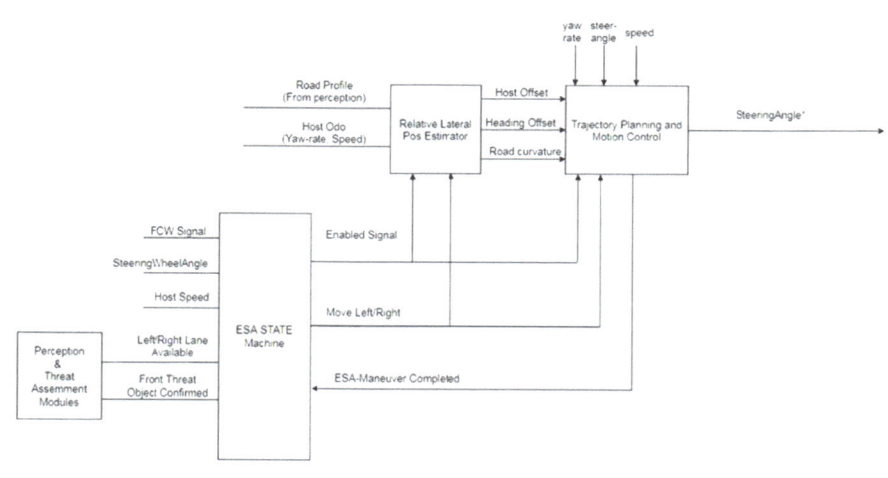

Suburban Mobility

Suburban mobility in the context of CAVs refers to the ability of these vehicles to navigate and operate effectively in suburban environments, which typically include mixed-use areas, residential zones, commercial districts, and connecting roads between urban and rural areas.

CAVs are equipped with advanced navigation systems that utilize GPS, satellite data, and mapping algorithms to accurately navigate suburban road networks, including complex intersections, roundabouts, and diverse traffic conditions. These vehicles employ a range of sensors such as LiDAR, radar, cameras, and ultrasonic sensors to perceive their surroundings. This includes detecting pedestrians, cyclists, vehicles, and obstacles, ensuring safe and efficient navigation in suburban settings.

CAVs integrate with TMSs and adhere to relevant traffic regulations and standards. This includes recognizing traffic signs, obeying traffic signals, and interacting safely with other vehicles, pedestrians, and cyclists on suburban roads. CAVs use V2X communication technologies to exchange data with other vehicles, infrastructure, and centralized systems. This enables real-time information sharing, such as traffic updates, road conditions, and safety warnings, enhancing suburban mobility and traffic management.

Supervised Learning

Supervised learning in the context of CAVs refers to a type of ML that involves training algorithms on a labeled dataset. Here, "labeled" means that each piece of training data is paired with an output value (or label) that the model should aim to predict. This method is fundamental for developing features such as perception systems in AVs, where the algorithms learn to recognize and respond to complex environments based on examples that humans have previously annotated.

In autonomous driving, supervised learning algorithms are used to train perception models to accurately identify and classify objects, such as vehicles, pedestrians, and road signs, from sensory data. The training involves feeding large amounts of data into the algorithms, allowing them to learn from the labeled examples about how to respond under various driving conditions. This learning process can involve techniques like federated learning, where data from multiple sources are utilized to enhance the learning process without compromising data privacy.

Sustainability

Sustainability refers to the integration of environmentally friendly practices, energy efficiency, and reduced emissions in the design, operation, and life cycle of autonomous transportation systems. CAVs are often equipped with electric or hybrid propulsion systems that reduce reliance on fossil fuels and minimize GHG emissions. These systems align with sustainability goals by promoting cleaner transportation alternatives (**Figure S.17**).

FIGURE S.17 Achieving sustainability goals through the integration of renewable energy and low-emission mobility systems.

Aliaksei Kaponia/Shutterstock.com.

Advanced energy management systems in CAVs optimize energy consumption by regulating power usage based on driving conditions, traffic patterns, and battery performance. This enhances overall efficiency and reduces the environmental impact of vehicle operations. CAVs may feature regenerative braking systems that capture and store energy during braking maneuvers, converting kinetic energy into electrical energy. This technology improves energy efficiency and contributes to sustainable driving practices.

Vehicles incorporating lightweight materials and efficient designs, in compliance with relevant standards such as **SAE J2580**, help reduce weight, leading to lower energy consumption and improved fuel efficiency in both traditional and AVs. Adherence to emission standards set by regulatory bodies such as FMVSS and CMVSS ensures that CAVs meet stringent requirements for pollutant emissions, contributing to environmental sustainability and air quality improvement.

SVS (Surround View System)

SVS in CAVs refers to an ADAS designed to provide a 360° panoramic view around the vehicle. This system typically utilizes four to six wide-angle cameras positioned around the vehicle, which helps in eliminating blind spots and assists in parking and navigating tight spaces. The integration of SVS enhances situational awareness and safety by giving drivers and autonomous systems a comprehensive view of the vehicle's surroundings. The SVS is highly useful in navigating tight parking spaces, connecting pickup trucks and trailers, and accurately augmenting the vehicle in parking spaces.

REFERENCES

1. Society of Automotive Engineers, "J3061 Cybersecurity Guidebook for Cyber-Physical Vehicle Systems," SAE Publishing, Warrendale, PA, 2021.

2. ISO, "ISO 26262 Road Vehicles - Functional Safety Part 1-12, Second Edition," Geneva, 2018.

3. National Highway Traffic Safety Administration, "Vehicle-to-Vehicle Communications," accessed August 6, 2025, https://www.nhtsa.gov/technology-innovation/technology-innovation/vehicle-vehicle-communications.

4. Department of Transportation, "FMVSS 208 Occupant Crash Protection," Washington, DC, 2019.

5. Society of Automotive Engineers, "J3116 Active Safety Pedestrian Test Mannequin Recommendation," SAE Publishing, Warrendale, PA, 2023.

6. Zhu, Z., Philipp, R., and Howar, F., "Leveraging Triggering Conditions for Efficient Scenario-Based Testing of Automated Vehicles," *SAE Int. J. CAV* 8, no. 4 (2025), doi:https://doi.org/10.4271/12-08-04-0035.

7. Taylor, K., "Digital Cockpit in the Era of the Software-Defined Vehicle," SAE Technical Paper 2024-01-2391 (2024), doi:https://doi.org/10.4271/2024-01-2391.

8. Fuchs, A., *Automotive Telematics: An Introduction into the Technical Aspects of Automotive Telematics with Reference to Business Model and User Needs* (Warrendale, PA: Society of Automotive Engineers, 2002).

9. Jiang, S., "Vehicle E/E Architecture and Key Technologies Enabling Software-Defined Vehicle," SAE Technical Paper 2024-01-2035 (2024), doi:https://doi.org/10.4271/2024-01-2035.

10. Bickelhaupt, S., Hahn, M., Morozov, A., and Weyrich, M., "Towards Future Vehicle Diagnostics in Software-Defined Vehicles," SAE Technical Paper 2024-01-2981 (2024), doi:https://doi.org/10.4271/2024-01-2981.

11. Society of Automotive Engineers, "J2640 General Automotive Embedded Software Design Requirements," SAE Publishing, Warrendale, PA, 2008.

12. Silva, J., "Advanced Firmware Device Manager for Automotive: A Case Study," *SAE Int. J. Passeng. Cars - Electron. Electr. Syst.* 5, no. 1 (2012): 34-45, doi:https://doi.org/10.4271/2012-01-0013.

13. Society of Automotive Engineers, "J3016 Taxonomy and Definitions for Terms Related to Driving Automation Systems for On-Road Motor Vehicles," SAE Publishing, Warrendale, PA, 2021.
14. Society of Automotive Engineers, "SAE Standards News: J3016 Automated-Driving Graphic Update," January 7, 2019, accessed July 23, 2024, https://www.sae.org/news/2019/01/sae-updates-j3016-automated-driving-graphic.
15. Wu, X., Ren, J., Wu, Y., and Shao, J., "Study on Target Tracking Based on Vision and Radar Sensor Fusion," SAE Technical Paper 2018-01-0613 (2018), doi:https://doi.org/10.4271/2018-01-0613.
16. Department of Transportation, "FMVSS 214S (STATIC) Side Impact Protection," Washington, DC, 2006.
17. United Nations, "UN R135 (UNECE): Global Regulation for Pole Side Impact Tests," United Nations Publishing, Geneva, 2015.
18. IIHS-HLDI Vehicle Research Center, "Insurance Institute for Highway Safety Highway Loss Data Institute," 2025, accessed August 6, 2025, https://www.iihs.org/ratings/about-our-tests/side.
19. Ziraldo, E., Attalla, S., Kodsi, S., and Oliver, M., "Driver Response to Right Turning Path Intrusions at Signalized Intersections," *SAE Int. J. Adv. & Curr. Prac. in Mobility* 2, no. 3 (2020): 1623-1632, doi:https://doi.org/10.4271/2020-01-0884.
20. Du, Z., Xu, B., and Pisu, P., "Cooperative Mandatory Lane Change for Connected Vehicles on Signalized Intersection Roads," SAE Technical Paper 2020-01-0889 (2020), doi:https://doi.org/10.4271/2020-01-0889.
21. Society of Automotive Engineers, "J2954 Wireless Power Transfer for Light-Duty Plug-In/Electric Vehicles and Alignment Methodology," SAE Publishing, Warrendale, PA, 2022.
22. Society of Automotive Engineers, "J2847-1 Communication between Plug-In Vehicles and Off-Board DC Chargers," SAE Publishing, Warrendale, PA, 2023.
23. Dange, D., Ballal, R., Murali, M., Sonawane, D. et al., "Design and Development of Automotive Battery Management System," *SAE Int. J. Adv. & Curr. Prac. in Mobility* 2, no. 2 (2020): 1116-1127, doi:https://doi.org/10.4271/2019-28-2498.
24. Kakizoe, K. and Bull, M., "Real-World Application of Variable Pedal Feeling Using an Electric Brake Booster with Two Motors," *SAE Int. J. Adv. & Curr. Prac. in Mobility* 3, no. 2 (2021): 1020-1029, doi:https://doi.org/10.4271/2020-01-1645.

25. Society of Automotive Engineers, "J3061 Cybersecurity Guidebook for Cyber-Physical Vehicle Systems," SAE Publishing, Warrendale, PA, 2021.

26. ISO, "ISO 21448:2022 Road Vehicles - Safety of the Intended Functionality," Geneva, 2022.

27. Society of Automotive Engineers, "J2735 V2X Communications Message Set Dictionary," SAE Publishing, Warrendale, PA, 2023.

28. Zhang, G., Wang, Q., Mujumdar, T., and Sugiarto, T., "Path Planning and Motion Control in Evasive Steering Assist," SAE Technical Paper 2022-01-0088 (2022), doi:https://doi.org/10.4271/2022-01-0088.

TBP (Traffic Behavior Prediction)

TBP is a crucial aspect of CAVs, referring to the capability of these vehicles to anticipate and predict the movements and behaviors of surrounding traffic participants, including vehicles, pedestrians, and cyclists. TBP relies on advanced sensor systems, ML algorithms, and real-time data processing to make informed predictions about the actions of these entities, enhancing safety and efficiency in dynamic traffic environments.

CAVs utilize sensor fusion techniques, integrating data from various sensors such as LiDAR, radar, cameras, and V2X communication systems. This multi-source data fusion improves the accuracy and reliability of predictions by providing a comprehensive view of the surrounding environment.

Advanced ML algorithms, including DL models such as CNNs and recurrent neural networks (RNNs), analyze sensor data and historical patterns to learn and predict traffic behaviors. These algorithms continuously adapt and improve their predictions based on real-world feedback. TBP algorithms predict the future trajectories of vehicles, pedestrians, and other entities based on their current positions, velocities, accelerations, and environmental factors. Predicted trajectories help CAVs make proactive decisions, such as adjusting speed, changing lanes, or planning alternative routes to avoid potential collisions or conflicts.

TBP incorporates behavioral modeling techniques to understand and predict complex human behaviors on the road, such as lane changes, merging maneuvers, braking patterns, and pedestrian crossing intentions. By modeling diverse behaviors, CAVs can anticipate and react appropriately to diverse scenarios. TBP systems continuously receive and process real-time updates from surrounding vehicles, infrastructure, and TMCs. This dynamic data exchange ensures that predictions remain accurate and up to date, even in rapidly changing traffic conditions.

TCA (Traffic Congestion Assistance)

TCA refers to a set of advanced features and technologies integrated into CAVs aimed at mitigating traffic congestion, improving traffic flow, and enhancing overall transportation efficiency [1]. These features leverage real-time data, communication technologies, and intelligent algorithms to optimize vehicle movements in congested areas and reduce the impact of traffic jams on road users.

TCA systems utilize data from various sources, including V2V and V2I communication, GPS, traffic sensors, and historical traffic patterns. By analyzing these data, the system can identify areas of congestion and dynamically adjust routes or recommend alternative paths to avoid traffic bottlenecks.

In TCA-enabled CAVs, vehicles can cooperate with each other through V2V communication to optimize their driving behavior in congested conditions. This includes cooperative merging, CACC, platooning (driving in close proximity to lead vehicle to reduce aerodynamic drag), and synchronized acceleration and deceleration to maintain smooth traffic flow.

TCA systems incorporate predictive analytics algorithms that anticipate traffic congestion based on real-time data and historical trends. This enables proactive measures such as rerouting vehicles before congestion occurs, adjusting traffic signal timings, and optimizing lane usage to prevent gridlock.

CAVs equipped with TCA capabilities can receive real-time traffic updates and dynamic routing recommendations based on current traffic conditions. These recommendations consider factors such as congestion levels, road closures, accidents, and construction zones to guide vehicles along the most efficient paths.

TCS (Traction Control System)

A TCS is an electronic system designed to prevent loss of traction of the driven road wheels (**Figure T.1**). TCS detects wheel slip and intervenes by controlling throttle, brake, and engine power outputs to maintain adequate traction during acceleration, particularly on slippery surfaces.

FIGURE T.1 TCS in action.

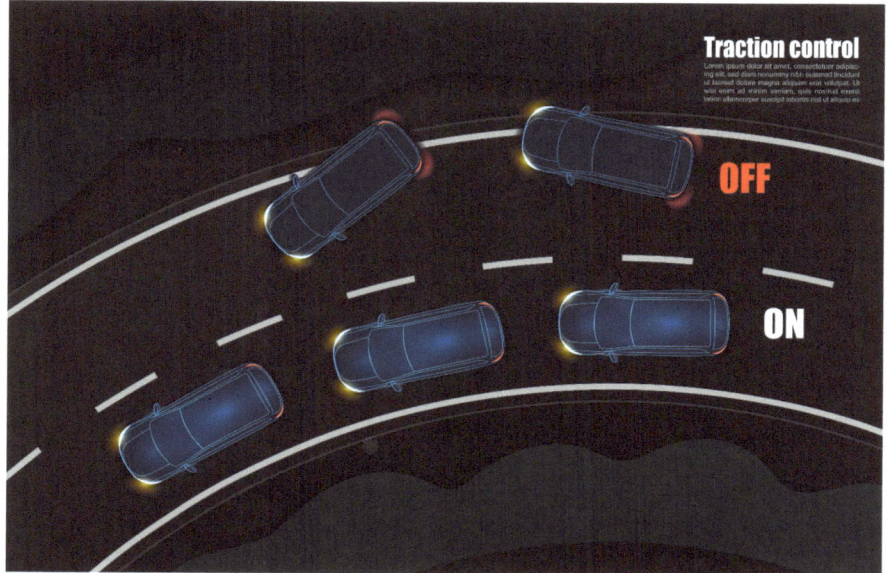

TCS helps to stabilize the vehicle's traction by adjusting the brake force on individual wheels during acceleration to prevent wheelspin. It is closely integrated with the vehicle's ABS and, in many vehicles, is a component of the ESC system [2]. The system uses wheel speed sensors to detect when one or more wheels begin to lose traction and spin faster than the wheel(s) with traction.

Advances in vehicle sensor technology and integration of TCS with more sophisticated vehicle dynamics control systems are expected. This includes enhancements in AI and ML to predict and effectively manage traction control in real time.

Telematics

Telematics is the integrated use of telecommunications and informatics to collect, transmit, and analyze data from vehicles (**Figure T.2**). It involves hardware and software systems that gather information such as location, speed, engine diagnostics, driver behavior, and sensor data and then send this information to a central server or cloud platform for processing and actionable insights [3].

FIGURE T.2 Telematics end-to-end system framework [3].

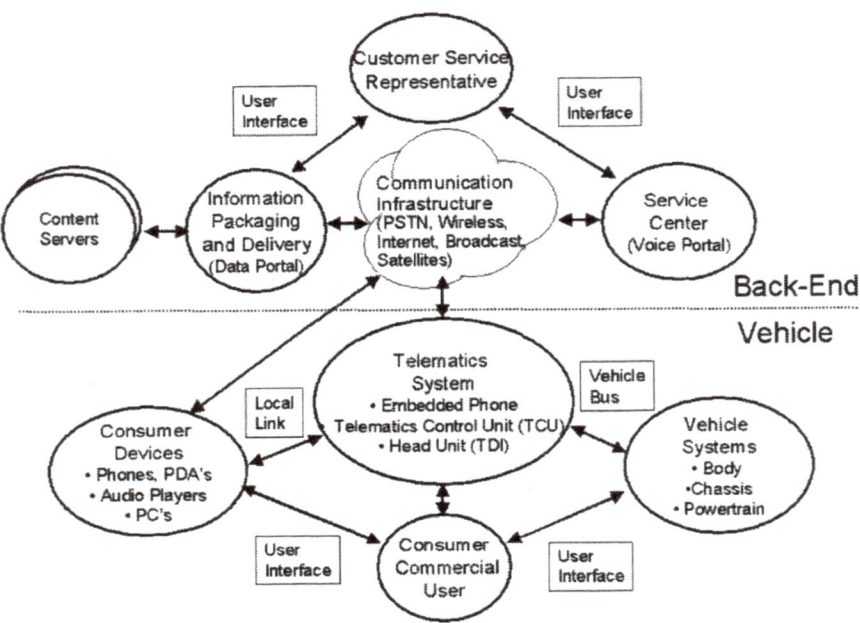

A telematics system typically includes a telematics control unit (TCU), an ECU device (or "black box") installed in the vehicle, often connected via the OBD-II or CAN-BUS port (**Figure T.3**). This device collects data from GPS modules, OBD, accelerometers (for collision detection), and various vehicle sensors. The collected data are then transmitted—usually over cellular networks—to cloud servers or fleet management platforms, where they can be accessed and analyzed in real time.

FIGURE T.3 Fully integrated in-vehicle system [3].

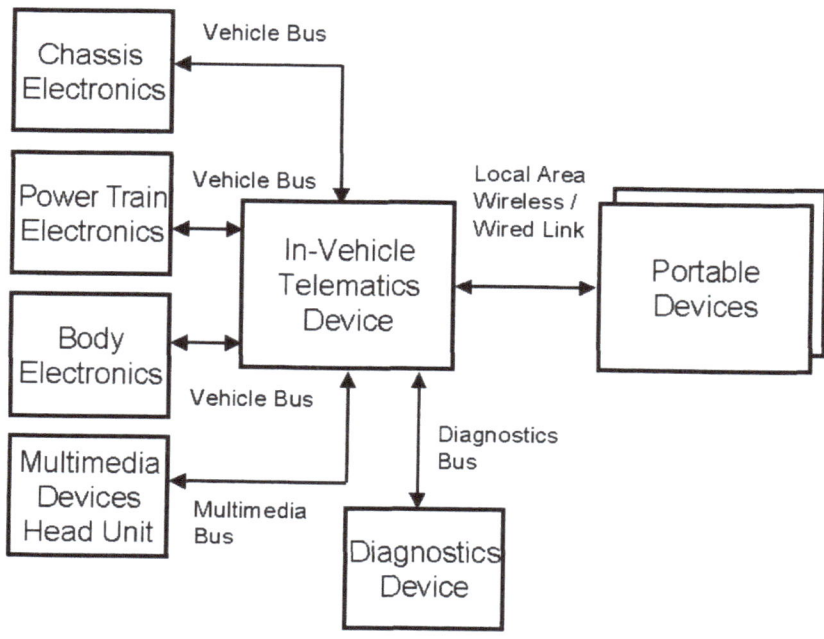

Telematics works in conjunction with ADAS by transmitting data from ADAS sensors, such as cameras, radar, and LiDAR, to enable features like collision avoidance, LDW, and ACC. This real-time data exchange enhances vehicle safety and paves the way for higher levels of automation.

In connected cars, telematics acts as a central hub for data, supporting remote diagnostics, predictive maintenance, OTA updates, and integration with driver apps. It enables communication between vehicles (V2V), infrastructure (V2I), and cloud services, forming the backbone of modern CV platforms.

Telematics Insurance

Telematics insurance is a dynamic automotive insurance model that utilizes telematics technology embedded within vehicles to collect and transmit real-time data on vehicle usage, driver behavior, and environmental conditions. This data-centric approach allows insurance companies to offer personalized insurance policies based on actual vehicle usage and driving behavior rather than historical data and generalized risk assessments.

Telematics devices integrated into vehicles under these standards collect a wide range of data points, including speed, location, time of day, and driving patterns such as hard braking or rapid acceleration. This information is crucial for insurers to assess risk and adjust premiums accordingly. More precise data allow for more tailored insurance policies, potentially lowering costs for safer drivers while providing incentives for riskier drivers to improve their driving habits.

Teleoperated Assistance

Teleoperated assistance refers to the capability of CAVs to be remotely controlled or assisted by human operators through wireless communication, enhancing safety, navigation, and operational efficiency.

CAVs are equipped with interfaces that allow human operators to control or assist the vehicle remotely. This can include interfaces such as joysticks, touchscreens, and specialized control panels. Teleoperated assistance relies on high-speed and reliable wireless communication technologies, such as 5G networks, to transmit real-time data between the vehicle and the remote operator. This communication facilitates commands, feedback, and video/audio streams for remote monitoring.

The vehicle is equipped with sensors and cameras that provide the remote operator with a comprehensive view of the vehicle's surroundings, including traffic conditions, obstacles, and navigation cues. Teleoperated assistance systems implement safety protocols and fail-safe mechanisms to ensure safe operation, such as emergency stop capabilities, redundant communication channels, and sensor-based collision avoidance.

For autonomous operation, latency is a critical factor as there is a transmission delay in the wireless technologies that enable communication to the remote operation command center. Emergency intervention may not be highly effective through remote takeover, as vehicle data need to be received at the remote command center, analyzed, and command is given and actuated. Milliseconds matter for emergency operations.

Teleoperated assistance plays a crucial role in scenarios where human intervention or oversight is needed, such as complex driving environments, emergencies, or testing/validation phases of AVs. Compliance with relevant standards ensures the safety, reliability, and interoperability of teleoperated systems in CAVs.

Teleoperation

Teleoperation refers to the ability of CAVs to be remotely controlled or monitored by human operators, typically in situations where autonomous capabilities may be insufficient or when intervention is required (**Figure T.4**).

FIGURE T.4 An example of a teleoperated vehicle.

Teleoperation systems include interfaces that allow human operators to remotely control essential vehicle functions such as steering, acceleration, braking, and navigation. CAVs use advanced communication technologies such as 5G, LTE, and satellite communication to establish reliable and low-latency connections between the vehicle and the remote command center. Teleoperation systems provide real-time feedback to operators through live video feeds, sensor data (e.g., LiDAR and cameras), and vehicle telemetry, enabling operators to make informed decisions and control the vehicle effectively. Teleoperation systems adhere to safety standards such as those outlined by SAE, FMVSS, and CMVSS to ensure secure and reliable remote control operations. Redundant communication channels and fail-safe mechanisms are often implemented to mitigate risks.

Teleoperation plays a crucial role in enhancing the versatility and safety of CAVs, allowing for remote assistance, intervention in complex situations, and operational flexibility in diverse environments. Adherence to established standards ensures that teleoperation systems meet stringent safety, security, and performance requirements.

Thermal Imaging

Thermal imaging in the context of CVs involves the use of IR cameras to detect heat signatures from the vehicle's surroundings. This technology enhances visibility and detection capabilities under various environmental conditions, such as darkness, fog, smoke, or glare, where traditional optical cameras might fail. This technology can reduce pedestrian and vehicle accidents caused by crossing animals, which are among the major contributors to traffic injuries and fatalities.

Enhanced night vision assists drivers by improving visibility beyond the range of standard headlights, detecting the heat signatures of vehicles, pedestrians, animals, and other obstacles. Pedestrian detection helps in identifying and alerting drivers about pedestrians in or near the roadway, which is crucial for preventing accidents, especially in poor visibility conditions. These components support autonomous driving systems by providing additional data input for obstacle detection and navigation, complementing other sensors like radar and LiDAR.

Thermal imaging represents a critical advancement in vehicle safety technology, providing significant benefits in visibility and obstacle detection. Its integration into CV systems continues to evolve, driven by developments in sensor technology and vehicle automation.

Threat Detection

Threat detection in the context of CVs refers to the processes and technologies used to identify, assess, and mitigate potential cybersecurity threats and vulnerabilities within vehicle systems. This includes the monitoring and analysis of vehicle networks and communication systems to address security risks preemptively.

TJP/TJA (Traffic Jam Pilot/Traffic Jam Assist)

TJP/TJA is a connected autonomous function that supports the smooth operation of the vehicle during traffic congestion or slowdowns. Traffic congestion causes more GHG emissions, air pollution, and increased driver fatigue, resulting in more accidents and loss of productivity.

Vehicles equipped with TJP/TJA are equipped with radars, LiDARs, and cameras and are connected to V2X infrastructure. They are **SAE J3016** Level 3 systems that can relieve the driver of the driving task. For the TJP mode to be active, vehicles have a speed threshold (e.g., 45 mph), and the system operates within a pre-mapped, access-controlled highway system, such as an interstate in the US or an autobahn in Germany. When the traffic jam conditions are present in an access-controlled highway and the preconditions are met, the system requests the driver to take over, and the driver can release control to the system. The system provides the longitudinal and lateral control commands and follows the vehicle in front, maintaining a safe separation distance. Some systems are also capable of lane change maneuvers. The driver has to be ready for a takeover when requested by the vehicle. This can help drivers from getting exhausted while driving in traffic jam conditions.

TMS (Traffic Management System)

A TMS is a combination of technologies and techniques used to monitor and manage traffic flow on roads and highways (**Figure T.5**). TMS employs a range of applications from basic TSC to more complex responses to traffic conditions and incidents, improving road safety and reducing congestion.

FIGURE T.5 TMSs are entangled with the myriad of attributes of the driving environment.

Real-time traffic monitoring and control use cameras, sensors, and other data sources to monitor traffic conditions and adjust signal timings and road usage dynamically. Incident management coordinates responses to traffic incidents to minimize their impact on traffic flow and enhance safety. Information dissemination provides real-time traffic information to motorists via variable message signs, mobile apps, and other communication channels.

TMSs play a vital role in modern transportation infrastructure, aiming to make travel safer, faster, and more efficient. As part of the broader smart city framework, TMS solutions are increasingly integrated with other urban systems to support sustainable mobility initiatives.

TPMS (Tire Pressure Monitoring System)

A TPMS is an electronic system designed to monitor the air pressure inside the pneumatic tires of various types of vehicles (**Figure T.6**). TPMS reports real-time tire pressure information to the driver of the vehicle, either via a gauge, a pictogram display, or a simple low-pressure warning light [4].

FIGURE T.6 Low tire indicator [4].

© SAE International.

Direct TPMS measures tire pressure directly from within the tire via sensors placed in each tire that report pressure data to the vehicle's computer system (**Figure T.7**). Indirect TPMS uses the wheel speed sensors connected to the vehicle ABS to estimate air pressure by monitoring the rotational speeds of the tires. It is less accurate than direct TPMS but also less expensive [5].

FIGURE T.7 Low tire by location [4].

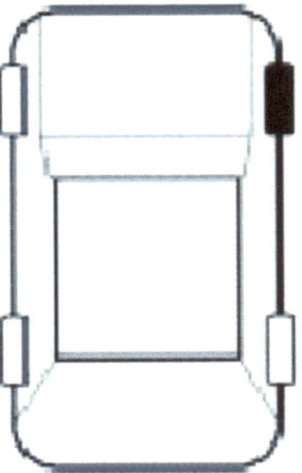

© SAE International.

TPMSs are a critical safety feature in modern vehicles, helping to ensure that tires are properly inflated, which enhances the overall safety, efficiency, and cost-effectiveness of vehicle operation. During CAV operation, the actuation commands for steering, propulsion, and braking may be effective if the tire pressure is outside of the thresholds. As technology progresses, TPMS is expected to become more sophisticated and integrated within the broader ecosystem of vehicle monitoring and maintenance technologies [6].

Traffic Analysis

Traffic analysis in the context of CAVs refers to the process of gathering, analyzing, and interpreting traffic-related data to optimize vehicle operations, improve safety, and enhance traffic flow within a transportation network.

CAVs utilize various sensors, including LiDAR, radar, cameras, and V2X communication systems, to collect real-time data about surrounding traffic conditions, road infrastructure, and vehicle movements [7]. Advanced algorithms and computing systems process the collected data to identify traffic patterns, detect potential hazards, predict traffic congestion, and assess the overall state of the transportation environment (**Figure T.8**).

FIGURE T.8 Application of maneuver in a traffic simulation model [7].

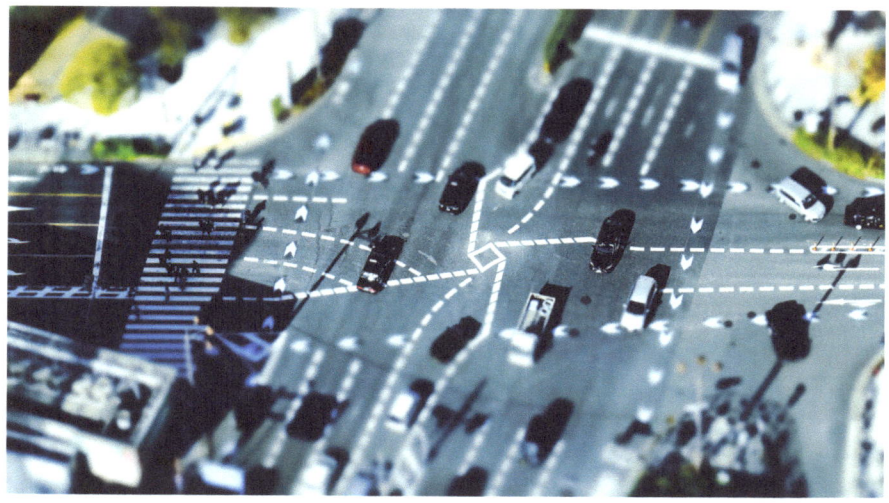

By analyzing traffic data, CAVs can optimize their routes, speeds, and driving behavior to minimize delays, reduce fuel consumption, and improve overall traffic efficiency. Traffic analysis enables CAVs to proactively identify and respond to safety-critical situations, such as abrupt lane changes, pedestrian crossings, and potential collisions, by applying autonomous driving algorithms and safety protocols. Traffic analysis algorithms and systems adhere to relevant standards and regulations, such as those defined by the SAE, FMVSS, and CMVSS, ensuring that CAVs operate safely and legally within traffic environments.

Traffic Analytics

Traffic analytics in the context of CAVs refers to the use of advanced data collection, processing, and analysis techniques to understand and optimize traffic patterns, vehicle movements, and overall transportation efficiency. These analytics are crucial for enhancing safety, reducing congestion, improving infrastructure planning, and facilitating the development of ITSs.

Traffic Behavior Analysis

Traffic behavior analysis refers to the systematic study and evaluation of the interactions and behaviors exhibited by CAVs within the traffic environment. This analysis encompasses a range of factors, including V2V and V2I communications, sensor data processing, decision-making algorithms, and compliance with regulatory standards [8].

CAVs utilize standardized communication protocols such as **SAE J2735** to exchange safety messages, environmental information, and cooperative maneuvers with other vehicles and infrastructure elements. These communication protocols enable real-time data sharing for enhanced traffic behavior analysis [8].

Traffic behavior analysis includes the assessment of systems such as collision avoidance systems as per FMVSS [9]. These systems utilize sensor data, including radar, LiDAR, and cameras, to detect and mitigate potential collisions through automated braking or steering interventions. Connected and AVs equipped with LKA and ACC systems adhere to standards such as CMVSS 500 [10]. Traffic behavior analysis evaluates the effectiveness of these systems in maintaining safe distances from other vehicles, changing lanes, and navigating complex traffic scenarios.

In urban environments, CAVs must analyze the behavior of pedestrians and cyclists to ensure safe interactions. Standards like **SAE J2945** outline protocols for detecting and responding to VRUs, contributing to comprehensive traffic behavior analysis. Traffic behavior analysis also considers ethical and regulatory aspects, including compliance with local traffic laws, prioritization of safety over convenience, and transparent communication of CAV capabilities to other road users.

Traffic Congestion

Traffic congestion refers to the condition in which the flow of vehicles on roads or highways slows down significantly or comes to a standstill due to a high volume of vehicles exceeding the road's capacity. This phenomenon is a significant concern in urban areas and can lead to increased travel times, fuel consumption, air pollution, and driver frustration [1].

CAVs play a significant role in addressing and mitigating traffic congestion through various technologies and strategies

V2V communication: CAVs utilize V2V communication to exchange real-time traffic information with nearby vehicles. This includes data on traffic flow, congestion levels, and alternative routes. By sharing this information, CAVs can collectively optimize their routes to avoid congested areas, improving overall traffic efficiency [11].

Through traffic flow optimization, AVs can adapt their speed and driving behavior based on traffic conditions and congestion levels (**Figure T.9**). Advanced algorithms and sensors enable CAVs to maintain optimal spacing between vehicles, reduce unnecessary acceleration and braking, and smooth out traffic flow patterns [10].

FIGURE T.9 Traffic flow, or lack thereof, is a source of frustration and accidents.

Coordinated intersection management: This is the mechanism by which CAVs can communicate with traffic signals and infrastructure through V2I communication. This allows them to receive signal timing information (SPaT and MAP messages) and optimize their approach to intersections, reducing congestion at busy junctions [12].

Dynamic routing and navigation are facilitated via CVs' access to real-time traffic data and predictive analytics. They can suggest or autonomously choose alternative routes to avoid congested areas, construction zones, or accidents, thereby reducing congestion on primary routes [13].

Traffic Pattern Recognition

Traffic pattern recognition refers to the technological process by which data about vehicular movement and behaviors are analyzed to identify recurring trends and patterns. This technology is used in intelligent transportation systems to optimize traffic flow, predict traffic conditions, and enhance road safety.

Traffic management assists in the dynamic control of traffic lights and signs based on recognized traffic patterns, improving overall traffic efficiency. Urban planning provides valuable insights for urban planners to design better road networks and manage traffic more effectively. This helps urban planners identify additional infrastructure needs and build the required turning lanes, roundabouts, etc. These support autonomous driving systems with real-time traffic pattern data, enhancing navigational decisions and safety protocols.

Traffic Prediction

Traffic prediction involves the use of methodologies and technologies to forecast traffic conditions, flow, and patterns over a short or long term. It leverages data analytics, ML models, and historical traffic data to anticipate future traffic states, helping in efficient traffic management and planning.

Traffic management enhances the capability of traffic control systems to manage congestion and optimize traffic flow. Route planning assists navigation applications in providing more accurate travel times and optimal routing suggestions based on predicted traffic conditions.

Traffic prediction is a pivotal component of modern TMSs, helping to alleviate congestion, enhance safety, and improve the overall efficiency of transportation networks. As technology evolves, so does the potential for more accurate and impactful traffic forecasting.

Traffic Queue Assistance

Traffic queue assistance refers to ADAS that aid drivers in navigating and managing slow-moving or stationary traffic conditions. This technology is designed to reduce driver fatigue and enhance safety during traffic congestion by automatically adjusting the vehicle's speed, maintaining a safe distance from other vehicles, and providing guidance to pass through traffic optimally.

Congestion management assists in maintaining consistent vehicle flow by adjusting speeds based on the surrounding traffic conditions detected through sensors and communication systems. This reduction in traffic congestion helps in reducing fuel consumption and emissions by avoiding unnecessary idling, stops, and starts in traffic and maintaining optimal speed.

These systems typically integrate radar, cameras, and ultrasonic sensors to monitor traffic conditions around the vehicle, and ML algorithms are used to predict traffic patterns and make real-time decisions.

Traffic queue assistance is a pivotal component of modern ADAS technologies, aimed at making driving in congested conditions safer and more bearable for drivers. By reducing the workload on drivers and smoothing the flow of traffic, these systems contribute significantly to the overall efficiency and safety of road transportation.

Traffic Simulation

Traffic simulation encompasses the use of computer-based models and software tools to replicate and study the behavior and interactions of elements within traffic systems. This includes vehicles, road users, and infrastructure under various conditions and scenarios. Traffic simulation serves as a critical tool in the evaluation of traffic management strategies, road network design, and vehicle system performance.

Road network design and traffic management simulations help in planning and optimizing road layouts, signal timings, and traffic flow patterns. Traffic simulation is used to test and refine ADAS and automated driving technologies in a controlled yet realistic virtual environment. Environmental impact studies are used to predict the effects of traffic on pollution and to develop strategies to mitigate adverse environmental impacts.

Traffic simulation plays a pivotal role in the development and evaluation of transportation systems, offering insights that help reduce congestion, enhance safety, and improve the overall efficiency of traffic networks. As technology evolves, so does the fidelity and application scope of traffic simulation tools, making them indispensable in modern traffic management and vehicle system development.

Traffic Violation Detection

Traffic violation detection refers to the automated identification and reporting of traffic law violations by vehicles, typically detected using various types of sensors and video surveillance technology integrated with AI and data analysis systems.

Traffic violation detection commonly involves the use of high-resolution cameras, radar interceptors, LiDAR, and ultrasonic sensors combined with sophisticated algorithms that can recognize patterns indicative of violations such as speeding, running red lights, illegal turns, and other traffic infractions. Advanced data processing techniques, including ML and pattern recognition, are employed to analyze the sensor data and accurately identify violations without human intervention. It involves automating the detection of traffic violations to support law enforcement efforts, enhancing the efficiency and accuracy of traffic monitoring, and integrating violation data with TMSs to improve road safety and reduce congestion based on trends and recurrent issues identified through data analysis.

Traffic violation detection systems represent a significant advancement in ITSs, offering the potential to enhance road safety and efficiency significantly. As technology progresses, these systems are expected to become more integrated within the CV ecosystem, supported by evolving standards and practices.

Training ML (Machine Learning) Model

The training of ML models for CAVs refers to the process of developing and refining ML algorithms that enable vehicles to perceive, interpret, and respond to their environment autonomously. This training involves collecting and labeling large datasets and separating them into a training dataset, a validation dataset, and a test dataset. The training dataset is used to train the model; the validation dataset is used for hyperparameter tuning, and the test dataset is used for testing the model's performance. The trained model is evaluated in virtual environments and adhering to relevant standards for safety, performance, and regulatory compliance.

CAVs gather data from various sensors, including LiDAR, cameras, radar, and GPS. These data include information about road conditions, traffic, pedestrian behavior, signage, and environmental variables. Annotated (labeled) data are crucial for training ML models. This process involves labeling collected data to classify objects, recognize patterns, and identify relevant features for autonomous decision-making.

Simulations play a vital role in training ML models for CAVs. Virtual environments replicate real-world scenarios, allowing for extensive testing and validation of algorithms before deployment on actual vehicles. ML algorithms, such as deep neural networks, are trained using labeled data. Training involves optimizing model parameters, adjusting hyperparameters, and validating performance metrics.

Trained ML models undergo rigorous validation and testing to ensure accuracy, reliability, and safety. This includes testing against diverse scenarios, edge cases, and compliance with industry standards.

TSC (Traffic Signal Control)

TSC refers to the use of automated systems and technologies to manage the operation of traffic signals in a coordinated manner to optimize traffic flow, reduce congestion, and enhance road safety (**Figure T.10**). These systems are increasingly integrated with ADAS and V2I communication technologies in CVs.

FIGURE T.10 An example of a traffic signal.

Traffic signal controllers are devices that manage the operation of traffic lights at intersections. Sensors and detectors are used to gather real-time traffic data that inform signal timing adjustments. Communication networks enable the exchange of traffic data and control commands between traffic signals and central TMSs.

Signal timing optimization adjusts the duration of red, yellow, and green lights based on real-time traffic conditions to minimize delays and avoid congestion. Adaptive traffic control dynamically changes traffic signals based on continuous traffic flow measurements [14]. Priority and preemption give priority to emergency vehicles, public transportation, or designated vehicles by manipulating signal phases.

TSP (Traffic Signal Preemption)

TSP refers to the process by which the normal operation of traffic lights is overridden, typically to allow vehicles with higher priority (such as emergency vehicles, public transit, or certain fleet vehicles) to pass through intersections more quickly and safely. This technology is part of advanced TMSs and integrates with V2I communication to enhance response times and minimize disruptions.

TSP (Traffic Signal Priority)

TSP is a type of system implemented in traffic signals to allow preferential right of way to specific vehicles, typically emergency vehicles, public transportation, or other designated vehicles. It adjusts the traffic signal timing to minimize stopping and delays for these vehicles. TSP systems are widely used to enhance the efficiency of transit operations by reducing delays at signalized intersections. This is particularly beneficial for buses and trams in urban areas where frequent stops can lead to significant travel time increases.

Emergency vehicle preemption, a specific application of TSP, allows vehicles like ambulances and fire engines to navigate through traffic more quickly during emergencies by altering signal timings in their favor as they approach intersections.

Traffic signal priority systems, facilitated by standards such as **SAE J2735** and **SAE J2945/B**, play a crucial role in modern traffic management by improving transit efficiency and response times for emergency services [8, 15]. These systems require careful integration and management to maximize benefits while mitigating impacts on regular traffic.

TSR (Traffic Sign Recognition)

TSR is an ADAS technology that uses sensors and cameras to detect and interpret traffic signs along the road [16]. This system informs drivers about current road rules, such as speed limits, no-entry signs, or warning signs, by displaying detected information on the vehicle's dashboard or HUD (**Figure T.11**).

High-resolution cameras mounted on the vehicle capture real-time images of traffic signs. Advanced image processing software algorithms analyze the captured images to identify and classify various traffic signs. Display systems such as HUDs or IC screens within the vehicle relay traffic sign information to the driver. Sign detection identifies traffic signs from real-time video feeds using image recognition technologies. Sign interpretation classifies the type of sign detected (e.g., stop, yield, and speed limit) and determines its relevance to the current driving context. Driver notification communicates the detected sign information to the driver, often integrating with other vehicle systems to suggest or enforce compliance with the sign's directives.

FIGURE T.11 Training set gallery and corresponding feature image library [16].

TSS (Traffic Signal Synchronization)

TSS involves coordinating the traffic lights across multiple intersections to create a smooth flow of traffic, minimizing stops, and reducing congestion (**Figure T.12**). This engineering strategy aims to optimize the timing of traffic signals so that groups of vehicles can move through intersections with minimal stopping, thus improving travel times and reducing emissions.

FIGURE T.12 Examples of the variety of traffic flow sensing methods to make decisions on traffic flow [14].

(a) Induction loop based traffic flow sensing

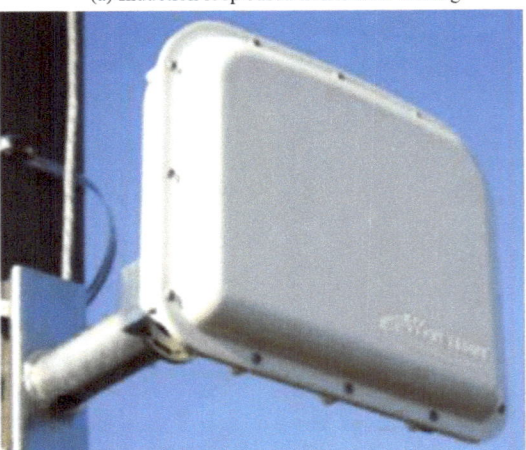
(b) Emitter receiver based traffic flow sensing

(c) Camera based traffic flow sensing

TSS typically involves the use of traffic controllers at intersections that are networked and programmed based on traffic flow data. Modern implementations often integrate adaptive signal control technology, which dynamically adjusts signal timings based on real-time traffic conditions detected via sensors or CV data.

Advanced systems may incorporate V2X communications to further enhance the responsiveness of traffic signals to the immediate environment, including the presence of emergency vehicles or heavy traffic volumes.

TTC/TTI (Time to Collision/Time to Impact)

TTC or TTI is often used interchangeably in the context of CAVs. CAVs must calculate the time it takes for an impact with another traffic participant, such as pedestrians, stationery, or moving obstacles. The separation distance and speed of the vehicle are used to calculate the TTI/TTC for each object in the driving path.

TC/TTI calculation is an essential part of determining which is the closest object in the driving path. A sensor fusion is performed on the data from radar, LiDAR, and camera. The distances and relative velocities are used to calculate the TTC/TTI. For AEB functions, based on TTC/TTI, escalation levels are implemented. When the TTC/TTI is less than a specific value, the driver is warned to take an action. When TTC/TTI reduces further, the vehicle applies partial braking or full braking to prevent a collision and bring the vehicle to a safe stop.

Type Approval

Type approval refers to the official certification process that ensures a CAV complies with relevant safety, performance, and regulatory standards established by organizations such as FMVSS, CMVSS, UNECE, and regulatory entities in Asia, Japan, and Korea as well. Standards published by SAE are used in developing evidence required by the regulations, which are often voluntary.

Type approval for CAVs often involves compliance with SAE standards related to vehicle communication protocols [8], autonomous driving functionalities [13], and cybersecurity [17]. The NHTSA in the US mandates compliance with FMVSS for vehicles sold in the US market. Relevant FMVSS standards for CAVs include FMVSS 126 (ESC systems), FMVSS 135 (light vehicle brake systems), and FMVSS 141 (minimum sound requirements for hybrid and EVs).

In Canada, Transport Canada enforces CMVSS standards for vehicle safety. CAVs seeking approval in Canada must adhere to standards such as CMVSS 500 (vehicle identification number), CMVSS 500.1 (ESC systems), and CMVSS 136 (hybrid and EV systems).

Certification Process:
CAVs undergo rigorous testing to evaluate their performance, safety features, cybersecurity measures, and compliance with specific standards outlined by SAE, FMVSS, and CMVSS. Manufacturers submit detailed documentation and technical reports demonstrating how their vehicles meet the requirements set forth by the identified standard organizations. This includes data from testing, simulations, and validation procedures. Regulatory authorities review the submitted documentation and conduct their assessments to ensure that the CAVs meet the prescribed safety and performance criteria. Upon successful review, the vehicles receive type approval certification.

REFERENCES

1. Kalra, V., Tulpule, P., and Giuliani, P., "Reducing Traffic Congestion with Reinforcement Learning-Driven Signal Control Systems," SAE Technical Paper 2025-01-0283 (2025), doi:https://doi.org/10.4271/2025-01-0283.

2. Department of Transportation, "FMVSS 126: Electronic Stability Control System," Washington, DC, 2007.

3. Fuchs, A., *Automotive Telematics: An Introduction into the Technical Aspects of Automotive Telematics with Reference to Business Model and User Needs* (Warrendale, PA: Society of Automotive Engineers, 2002).

4. Society of Automotive Engineers, "J2657 Tire Pressure Monitoring Systems for Light Duty Highway Vehicles," SAE Publishing, Warrendale, PA, 2019.

5. Society of Automotive Engineers, "J2848 Tire Pressure Monitoring Systems - for Medium and Heavy Duty Highway Vehicles," SAE Publishing, Warrendale, PA, 2023.

6. US Department of Transportation, "Tire Pressure Monitoring System FMVSS No. 138," Washington, DC, 2001.

7. Naidu, A., Mittal, A., Kreucher, R., Zhang, A. et al., "A Systematic Approach to Develop Metaheuristic Traffic Simulation Models from Big Data Analytics on Real-World Data," SAE Technical Paper 2021-01-0166 (2021), doi:https://doi.org/10.4271/2021-01-0166.

8. Society of Automotive Engineers, "J2735 V2X Communications Message Set Dictionary," SAE Publishing, Warrendale, PA, 2023.

9. Department of Transportation, "FMVSS 150 V2V Communications," Washington, DC, 2017.

10. United States Department of Transportation, "FMVSS 500: Low-Speed Vehicles," National Highway Traffic Safety Administration (NHTSA), Washington, DC, 2020.

11. Society of Automotive Engineers, "J2945/1 Dedicated Short Range Communication (DSRC) Systems Engineering Process Guidance for SAE J2945/X Documents and Common Design Concepts," SAE Publishing, Warrendale, PA, 2017.

12. Government of Canada, "Canada Motor Vehicle Safety Standard (CMVSS) No. 500: Low-Speed Vehicles. Motor Vehicle Safety Regulations (C.R.C., c. 1038), Schedule IV, Section 500," Transport Canada, Ottawa, Ontario, 2019.

13. Society of Automotive Engineers, "J3016 Taxonomy and Definitions for Terms Related to Driving Automation Systems for On-Road Motor Vehicles," SAE Publishing, Warrendale, PA, 2021.

14. Przybyla, J., Rush, T., Palframan, K., and Melcher, D., "Introduction to Traffic Signal Data Loggers and Their Application to Accident Reconstruction," SAE Technical Paper 2018-01-0527 (2018), doi:https://doi.org/10.4271/2018-01-0527.

15. Society of Automotive Engineers, "J2945 Performance Requirements for Cooperative Adaptive Cruise Control (CACC) and Platooning," SAE Publishing, Warrendale, PA, 2023.

16. Wang, L., "Road Sign Recognition System Based on Wavelet Transform and OPSA Point Set Distance," SAE Technical Paper 2018-01-1609 (2018), doi:https://doi.org/10.4271/2018-01-1609.

17. Society of Automotive Engineers, "J3061 Cybersecurity Guidebook for Cyber-Physical Vehicle Systems," SAE Publishing, Warrendale, PA, 2021.

U

Ultrasonic Sensors

Ultrasonic sensors use sound waves at frequencies higher than the human audible frequency range to detect objects and measure distances. These sensors are commonly used in automotive applications for various safety and assistance functions at low speeds.

Ultrasonic sensors emit high-frequency sound waves that reflect off objects and return to the receiver in the sensor (**Figure U.1**). The time it takes for the echo to return to the receiver is used to calculate the distance to the object [1]. This technology is crucial for parking assistance, low-speed collision avoidance, and other proximity-sensing applications. They have a smaller range compared to radar sensors and are not suitable for high-speed applications. They have a lower response time as the speed of sound waves is very small compared to radio waves.

FIGURE U.1 Block diagram of the ultrasonic sensor for automotive application [1].

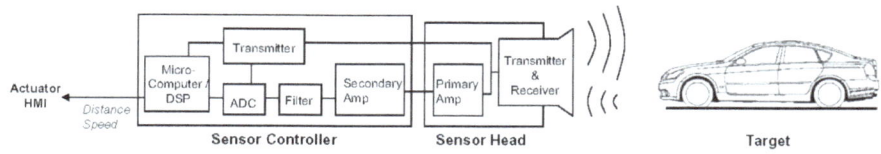

Ultrasonic sensors are an integral part of the sensor suite in modern vehicles, providing essential data that enhance vehicle safety systems and assist drivers in navigation and maneuvering tasks. This technology's continuous evolution, supported by ongoing research and standardization efforts by bodies like SAE International, ensures its relevance in developing safer, more efficient vehicles.

Unsupervised Learning

Unsupervised learning is an ML approach where algorithms are trained on unlabeled data to discover patterns without any explicit guidance from humans. In the context of CAVs, unsupervised learning plays a significant role in developing intelligent systems capable of making sense of complex and unstructured data from various sources.

Unsupervised learning algorithms, such as K-means clustering or hierarchical clustering, group data points based on similarities. In CAVs, inputs from sensors such as LiDAR are clustered to identify objects and their behavior patterns. Another application of unsupervised learning in CAVs is ground segmentation.

Unsupervised learning helps in detecting anomalies or unusual patterns in data. This is vital for CAVs to identify potential hazards or irregularities in sensor readings, contributing to safer autonomous driving. Unsupervised learning techniques like principal component analysis (PCA) or autoencoders extract meaningful features from high-dimensional data and are referred to as dimensionality reduction. This reduces data complexity and enhances the efficiency of decision-making algorithms in CAVs.

For more information, see **J3016** [2].

Urban Mobility

Urban mobility refers to the movement of people and goods within urban areas and cities, encompassing various modes of transportation and mobility solutions tailored to urban environments. In the context of connected cars, urban mobility involves integrating vehicles with smart infrastructure, digital platforms, and transportation networks to enhance the efficiency, accessibility, and sustainability of urban transportation systems.

Urban mobility solutions aim to seamlessly integrate different modes of transportation, including public transit, shared mobility services, walking, cycling, and private vehicles, to offer users flexible and efficient travel options. Connected cars in urban mobility systems leverage real-time data and communication technologies to provide drivers with up-to-date information on traffic conditions, parking availability, alternative routes, and public transit schedules, enabling informed decision-making and route optimization.

Connected cars interact with smart infrastructure elements such as traffic lights, road sensors, and TMS to improve traffic flow, reduce congestion, and enhance safety in urban environments. Urban mobility encompasses shared mobility services such as ride hailing, car sharing, and micro-mobility solutions (e.g., electric scooters and bikes), which offer convenient and cost-effective alternatives to traditional car ownership and promote sustainable urban transportation [3].

Usability

Usability in CAVs refers to the design and functionality of vehicle systems and interfaces to ensure intuitive and effective interactions between occupants and the vehicle's technology, contributing to safety, comfort, and user satisfaction.

The HMI encompasses the interfaces, displays, controls, and interaction mechanisms that allow occupants to interact with CAV systems, such as infotainment, navigation, vehicle status, and autonomous driving features [4]. Usability standards ensure that these interfaces are intuitive and easy to understand and promote safe interaction while minimizing driver distraction.

Usability standards also address the accessibility of CAV features to users with diverse needs, including individuals with disabilities. This includes control reachability, voice command capabilities, and interface adaptability to accommodate user preferences and requirements. Usability standards play a crucial role in ensuring the usability of safety-critical functions within CAVs, such as emergency braking, collision avoidance, and autonomous driving modes [5]. Clear and intuitive interfaces and effective user feedback mechanisms enhance user trust and confidence in these systems.

Usability standards often define standardized controls, symbols, and graphical representations used in CAV interfaces to convey information effectively to users. This includes compliance with relevant standards and graphical symbols.

UTC (Coordinated Universal Time)

UTC is the world's primary time standard regulating clocks and time. It is essential for synchronization across various technologies and systems involved in the operation of these vehicles.

- **Time Stamping Data:** UTC provides a consistent time stamp for data collected from various sensors and systems, ensuring that all data points are synchronized and accurate. This is vital for event logging, performance analysis, and diagnostic assessments.
- **Coordinating with Global Systems:** For vehicles that operate across different time zones, UTC serves as a stable time reference, helping in scheduling, fleet management, and logistics.
- **Integration with GPS:** AVs heavily rely on the GPS, which uses UTC to provide precise timing signals. Accurate GPS data are essential for navigation, mapping, and maintaining the integrity of communication between vehicles and TMSs.
- **Real-Time Communication:** In CV ecosystems, real-time communication between vehicles (V2V), infrastructure (V2I), and other elements relies on precise timing to avoid data conflicts and ensure the timely execution of commands and maneuvers. The V2X messages are synchronized based on UTC to ensure an appropriate real-time response.

UX (User Experience)

UX in CAVs refers to the overall quality of interaction and satisfaction that passengers and operators experience while utilizing the features and functionalities of ACVs. It encompasses various aspects such as interface design, human–machine interaction, comfort, safety, and accessibility.

The design of UI within CAVs is crucial in enhancing UX. This includes touchscreens, haptic feedback, audiovisual interfaces, voice-activated controls, physical buttons, and AR displays [4].

UX includes seating systems and their impact on occupant comfort and safety. Comfortable and ergonomic seating designs contribute to a positive UX during long journeys or autonomous driving scenarios [6].

Vehicle safety is significant in UX because it ensures a safe and secure driving environment. Safety features such as collision avoidance systems, AEB, and stability control enhance user confidence and satisfaction [5, 7].

CAVs should adhere to accessibility standards outlined in regulations such as the Americans with Disabilities Act (ADA) and the Canadian Accessibility Standards. This includes features like voice commands, adjustable controls, and interfaces optimized for users with diverse needs, ensuring an inclusive UX.

REFERENCES

1. Kamemura, T., Takagi, H., Pal, C., and Ohsumi, A., "Development of a Long-Range Ultrasonic Sensor for Automotive Application," *SAE Int. J. Passeng. Cars - Electron. Electr. Syst.* 1, no. 1 (2009): 301-306, doi:https://doi.org/10.4271/2008-01-0910.

2. Society of Automotive Engineers, "J3016 Taxonomy and Definitions for Terms Related to Driving Automation Systems for On-Road Motor Vehicles," SAE Publishing, Warrendale, PA, 2021.

3. Andreassi, L. and De Angelis, L., "The LCA Analysis Applied to Urban Mobility," SAE Technical Paper 2022-24-0022 (2022), doi:https://doi.org/10.4271/2022-24-0022.

4. Department of Transportation, "FMVSS 101 Controls and Displays," Washington, DC, 2000.

5. Department of Transportation, "FMVSS 126: Electronic Stability Control System," Washington, DC, 2007.
6. Department of Transportation, "FMVSS 202a Head Restraints for Passenger Vehicles," Washington, DC, 2006.
7. Department of Transportation, "FMVSS 208 Occupant Crash Protection," Washington, DC, 2019.

V2B (Vehicle-to-Business) Communication

V2B communication is a CV technology that enables data and service exchange between vehicles and business systems or enterprise back ends. It is a subset of the broader V2X. V2B supports a wide range of applications, from claim assistance and fleet management to UBI, predictive maintenance, and new business models leveraging real-time vehicle data.

Key Features and Architecture:

- **Integration Platforms:** The typical system employs a vehicle integration platform (VIP) on the business side and a back end integration manager (BIM) within the vehicle. The VIP acts as a web-service-based, event-driven message broker, enabling asynchronous, publish/subscribe communication between vehicles and enterprise systems.

- **Service-Oriented and Event-Driven:** The architecture is a SOA and event-driven architecture (EDA) to decouple vehicles and business applications, improving scalability and flexibility. This allows vehicles and businesses to subscribe to events of interest and retrieve or push information as needed.

- **Efficient and Secure Data Exchange:** The system supports secure, topic-based messaging, message prioritization, buffering (to handle intermittent connectivity), and identity management. It is designed for vehicle resource constraints and mobility, ensuring reliable delivery even when vehicles are temporarily offline.

- **Enterprise Application Integration:** V2B enables integration with various business functions, including claims processing (e.g., automatic accident notification for insurance), CRM, predictive maintenance, warranty management, and personalized service offerings.

- **Fleet and Office Communication:** In commercial and fleet contexts, V2B allows for the transfer of both business data (e.g., delivery status, compliance logs, and vehicle system state) and technical vehicle data (e.g., diagnostics and usage patterns) between vehicles and business offices, supporting cost-effective fleet operation and management.

V2C (Vehicle-to-Cloud) Communication

V2C communication is a technology that enables vehicles to exchange data with cloud-based applications and services in real time. This connectivity allows vehicles to send and receive information over broadband cellular networks, integrating onboard systems and sensors with cloud platforms to support ADAS, remote diagnostics, OTA updates, and a wide array of connected services [1].

Key Features:

- **OTA Updates:** Vehicles receive software and firmware updates remotely, ensuring systems remain secure, up to date, and feature-rich without requiring dealership visits.
- **Remote Diagnostics and Monitoring:** Real-time data from vehicle sensors are sent to the cloud for analysis, enabling proactive maintenance, fault detection, and performance monitoring.
- **Personalization and Shared Mobility:** User preferences (e.g., seat position, climate, and infotainment) can be stored in the cloud and automatically applied when a user accesses a shared or rental vehicle.
- **Integration with Smart Devices and IoT:** V2C allows vehicles to interact with other cloud-connected devices, such as home automation systems and digital assistants, for seamless UXs.
- **Enhanced Data Analytics:** The cloud enables powerful data processing, supporting predictive maintenance, intelligent energy management (especially for EVs), and improved autonomous driving capabilities.
- **Bidirectional Communication:** Not only do vehicles send data to the cloud, but cloud services can also send commands, updates, and information back to vehicles, enabling features like remote start, lock/unlock, and real-time navigation updates.

Onboard sensors and systems gather vehicle and environmental data. Data are transmitted to the cloud via cellular networks (V2Ns), Wi-Fi, or other broadband technologies. The cloud platform analyzes, stores, and acts on the data, providing insights, updates, and services. The vehicle receives updates, commands, or personalized settings from the cloud, enabling new features or responding to real-time events (**Figure V.1**).

FIGURE V.1 Zonal architecture for virtualized, CV applications [1].

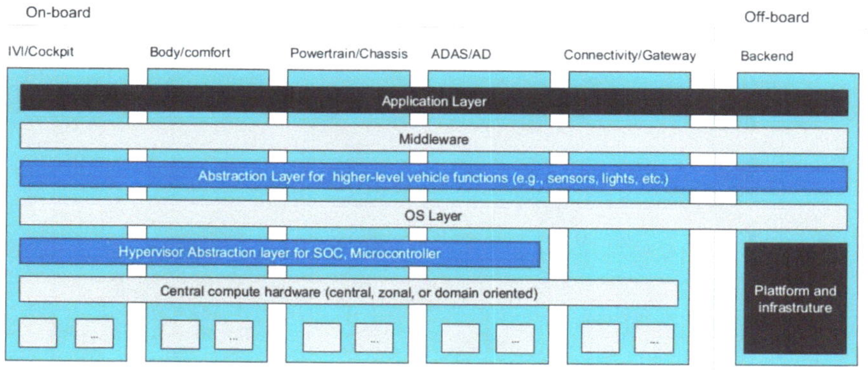

V2C (Vehicle-to-Cloud) Security

V2C security encompasses the protocols, technologies, and standards designed to protect data integrity, confidentiality, and availability in wireless communications between vehicles and cloud-based platforms. It addresses cybersecurity risks arising from the bidirectional exchange of telematics, OTA updates, infotainment, and autonomous driving data, ensuring safe and compliant integration of CVs into cloud ecosystems [2].

Key Security Challenges in V2C:
- **Expanded Attack Surface:** Cloud connectivity introduces vehicle APIs, OTA update mechanisms, and third-party service providers' vulnerabilities (**Figure V.2**). Threats include data interception, malware injection, and unauthorized access to vehicle controls [3].
- **Data Privacy Compliance:** Compliance with global regulations [e.g., GDPR, California Consumer Privacy Act (CCPA), and China's Personal Information Protection Law (PIPL)] is critical for handling sensitive user data, including location, biometrics, and driving behavior [3].
- **Supply Chain Risks:** Complex supply chains involving cloud providers, software vendors, and telecom networks increase exposure to compromised components or malicious firmware.
- **Real-Time Threat Detection:** Dynamic environments require continuous monitoring for anomalies in vehicle-cloud communications, such as unexpected API calls or data exfiltration [4].

FIGURE V.2 Attacker economics [2].

V2C (Vehicle-to-Cyclist) Communication
See **Cyclist Detection**.

V2E (Vehicle-to-Edge) Communication
V2E communication refers to the real-time data exchange between vehicles and edge computing nodes—servers or infrastructure close to the vehicles, such as cellular base stations or RSUs (Figure V.3). V2E is a key enabler for low-latency, high-reliability services within the broader V2X ecosystem, supporting ADAS, autonomous driving, and intelligent transportation applications by processing and analyzing data near its source.

FIGURE V.3 CV ecosystem [5].

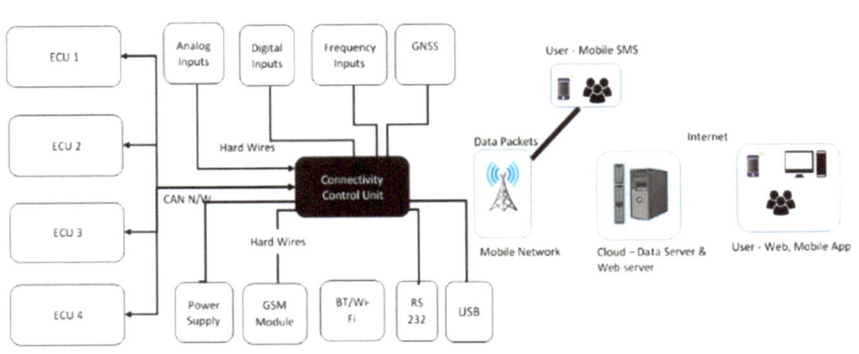

Key Features:
- **Low Latency and High Throughput:** By processing data at the edge (close to the vehicle), vehicle-to-edge minimizes communication delays, enabling near real-time alerts and decision-making for safety-critical applications such as collision avoidance, VRU detection, and traffic signal timing.
- **Bandwidth Optimization:** Edge computing reduces the volume of data sent to the central cloud by performing initial processing, filtering, and aggregation locally, conserving network resources and improving scalability.
- **Enhanced Security and Reliability:** Edge nodes can implement security measures and localized data storage, reducing exposure to network-wide cyberthreats and supporting resilient operation despite intermittent cloud connectivity.
- **Support for ITS:** Vehicle-to-edge enables CVs to interact with infrastructure, other vehicles, and pedestrians via edge platforms, supporting use cases like weather and roadway condition alerts, intersection management, and dynamic routing.
- **API-Driven Innovation:** Edge platforms (such as Verizon's Edge Transportation Exchange) provide APIs for automakers, technology developers, and governments to build, test, and scale new mobility applications efficiently.

Vehicles generate data from sensors, cameras, and onboard systems. These data are captured at the edge and processed to forward it to a regional ECU on the vehicle, which is analyzed in near real time. The analysis may result in some actions on the part of the vehicle or local vehicles, road users, regional infrastructure, and perhaps off-vehicle data into the cloud via the telemetry systems (**Figure V.4**).

FIGURE V.4 Connectivity control unit [5].

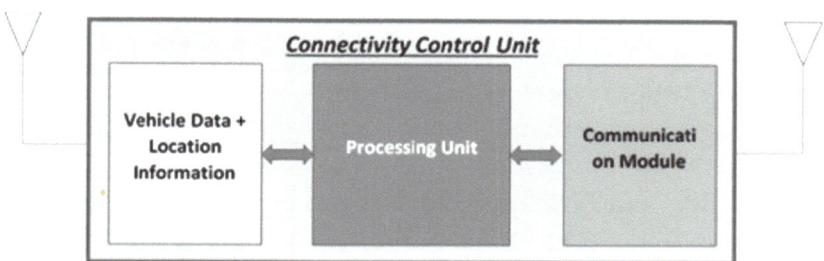

V2I (Vehicle-to-Infrastructure)

V2I is a wireless communication technology that enables vehicles to exchange data with infrastructure elements such as traffic lights, road signs, lane markers, RSUs, parking meters, and TMCs. The primary goal of V2I is to improve road safety, efficiency, and mobility by providing real-time, two-way data sharing between vehicles and the infrastructure that supports the transportation system.

RSUs, traffic signal controllers, management centers, and traffic/environmental sensors are installed along roadways to communicate with vehicles, which can provide warnings, such as CSW (**Figure V.5**). OBUs in vehicles interface with RSUs and process incoming/outgoing messages. Information is exchanged wirelessly, often using DSRC or C-V2X protocols, and processed to deliver coordinated messages and alerts to drivers and infrastructure managers [6, 7].

FIGURE V.5 CSW application information flow [6].

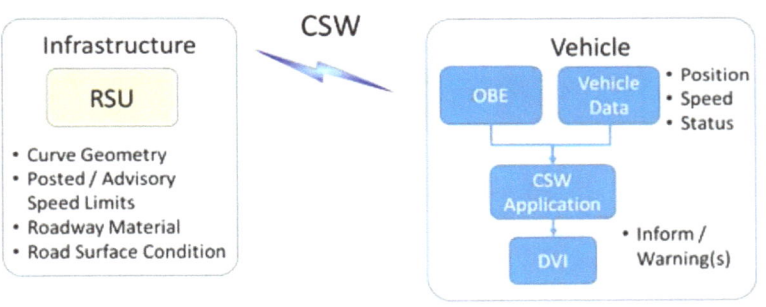

Key Features:

- **Bidirectional Wireless Communication:** V2I systems use DSRC (**Figure V.6**) and C-V2X to transmit and receive data between vehicles and infrastructure devices.
- **Real-Time Data Exchange:** Vehicles receive information about traffic signals, road hazards, speed limits, weather conditions, and construction zones, while infrastructure can collect traffic flow, vehicle speed, and other data for system optimization [9].
- **Safety Applications:** Includes red light violation warnings, CSWs, work zone alerts, pedestrian detection at crosswalks, and emergency vehicle preemption.
- **Mobility and Efficiency:** Enables eco-approach and departure at intersections, adaptive signal timing, dynamic routing, and public transport prioritization to reduce congestion and emissions.

- **Support for ACVs:** V2I is critical for higher-level automation, providing vehicles with reliable information about signal phases, hazards, and road status beyond the range of onboard sensors.

FIGURE V.6 V2I communication [8].

V2P (Vehicle-to-Pedestrian) Communication

V2P communication is a subset of V2X technologies focused on enhancing the safety of pedestrians and other VRUs by enabling real-time, wireless data exchange between vehicles and pedestrians (Table V.1). This is achieved through direct or infrastructure-mediated communication using smartphones, wearables, or dedicated devices carried by pedestrians, vehicle sensors, and connectivity.

TABLE V.1 V2P summary table: communication modes and technologies.

Mode	Technology	Use case	Strengths
Direct	DSRC, Bluetooth, BLE	Real-time safety alerts	Low latency, fast response
Indirect	Cellular, Wi-Fi	Broader coverage, data analytics	Scalable, more processing
Hybrid	Multi-protocol	Redundant, robust communication	Best of both

V2P systems use various wireless technologies, including Bluetooth, Wi-Fi, cellular networks, DSRC (IEEE 802.11p), long-range wide area network (LoRaWAN), Bluetooth Low Energy (BLE), and GPS IoT, to facilitate communication between vehicles and pedestrian devices.

Vehicles and VRU devices communicate directly, typically using DSRC or Bluetooth, for low-latency safety applications [9]. Indirect communication occurs via infrastructure (e.g., cellular networks and smart traffic lights), where data are processed and relayed between vehicles and pedestrians. A hybrid system combines direct and indirect modes for broader coverage and redundancy. Pedestrian devices transmit PSMs, including speed, location, and direction, which vehicles use to detect, track, and predict pedestrian movement. Vehicles may transmit up to 10 safety messages per second; VRUs adjust frequency based on movement and context [10].

V2P (Vehicle-to-Pedestrian) Safety
See **Pedestrian Detection** and **Pedestrian Collision Warning**.

V2R (Vehicle-to-Roadside)
See **Cooperative Intersection Management**.

V2R (Vehicle-to-Roadside) Communication
V2R communication refers to the wireless data exchange between vehicles and roadside infrastructure, such as RSUs, sensors, and TMSs. V2R is a core subset of the broader V2X ecosystem, often overlapping with V2I, and is foundational for advanced safety, mobility, and commercial vehicle operations (CVOs) (**Figure V.7**).

FIGURE V.7 Reference physical architecture [11].

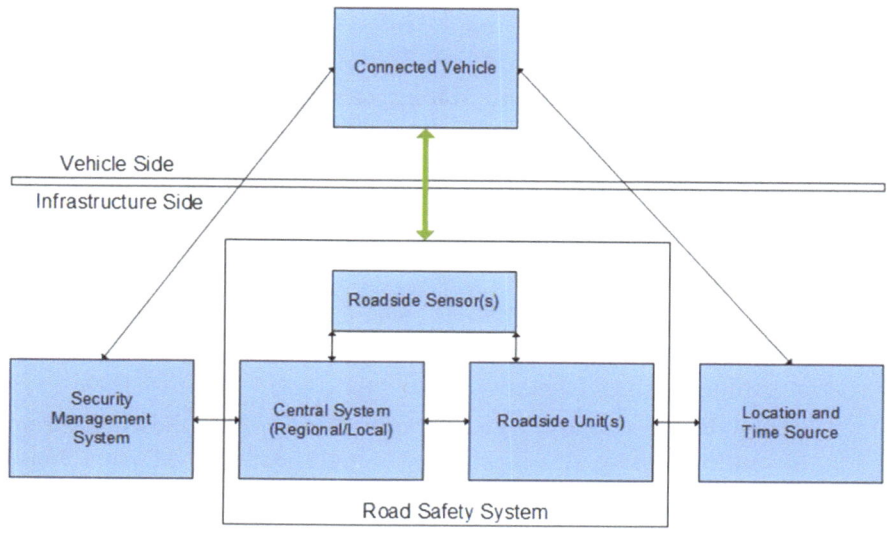

- **Wireless Technologies:** V2R typically uses DSRC (IEEE 802.11p/WAVE), cellular (C-V2X/5G), Bluetooth, or other wireless protocols to enable data exchange between vehicles and roadside equipment (Table V.2).
- **Data Exchanged:** Information includes vehicle speed, position, credentials, safety parameters, and sensor data. Roadside systems can also send alerts, signal timing, and hazard warnings to vehicles.
- **Infrastructure:** RSUs or wireless sensor networks are deployed along roadways to interact with passing vehicles, enabling applications such as traffic signal priority, digital alerts, and credential verification for commercial vehicles.

TABLE V.2 V2P summary table: V2R communication.

Application area	Example use case	Standard/protocol	Benefit
Road safety	Hazard alerts, accident prevention	SAE J2945/4, IEEE 802.11p [12]	Reduced crashes, faster response
Traffic management	Signal priority for buses/snowplows	SAE J2945/4	Fewer stops, improved flow
Work zone safety	Queue truck digital alerts	SAE J2945/4, IEEE 802.11p [12]	Fewer hard-braking events
CVOs	Credential checks, weigh in motion	NIST, SAE CVO standards	Faster, safer inspections
Pedestrian support	Mobile signal systems for crossings	SAE J2945/4	Safer, more accessible crossings

Key Applications and Use Cases

- **Road Safety and Hazard Alerts:** Roadside sensors detect hazards (e.g., accidents and slippery roads) and broadcast warnings to approaching vehicles to prevent accidents and investigate post accidents.
- **Traffic Signal Priority and Management:** School buses, snowplows, and emergency vehicles can request signal priority, reduce stops, and improve response times.
- **Work Zone and Queue Alerts:** V2R-equipped queue trucks and work zones issue digital alerts to reduce hard braking and improve safety for road workers and drivers.
- **CVOs:** V2R enables credential checks, weigh in motion, and real-time monitoring of safety parameters for trucks at highway speeds, streamlining inspections and reducing delays.
- **Pedestrian Assistance:** V2R supports systems that guide pedestrians (including those with disabilities) through intersections using mobile signal applications.
- **Hybrid Architectures:** Combining V2R with V2V and roadside sensor networks enhances system effectiveness and cost efficiency, especially in urban scenarios with limited infrastructure coverage.

V2V (Vehicle-to-Vehicle)

V2V communication is a wireless technology that enables vehicles to exchange real-time data, such as speed, location, direction, and braking status, with each other (**Figure V.8**). By allowing vehicles to anticipate and respond to the actions of vehicles, even beyond line of sight, the goal is to enhance road safety, optimize traffic flow, and support ADAS and autonomous driving.

FIGURE V.8 V2V communication [8].

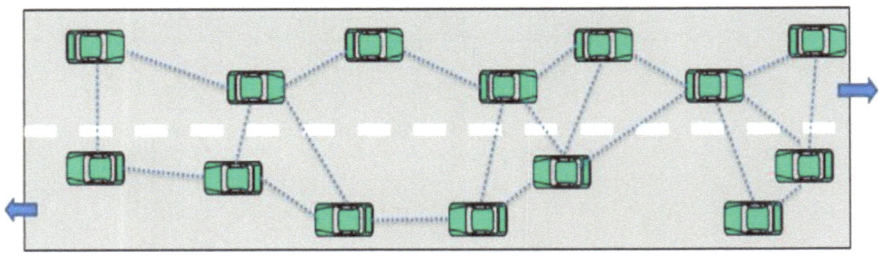

V2V uses DSRC (IEEE 802.11p) or C-V2X (5G) for low-latency, high-reliability data exchange [9]. Each vehicle processes received data to assess collision risks and may autonomously take corrective actions (e.g., braking and steering) (**Figure V.9**). Drivers receive visual, audible, or haptic warnings; sometimes, the vehicle may intervene automatically [6, 7].

FIGURE V.9 Vehicle classification using Federal Highway Administration (FHWA) 13—category scheme [13].

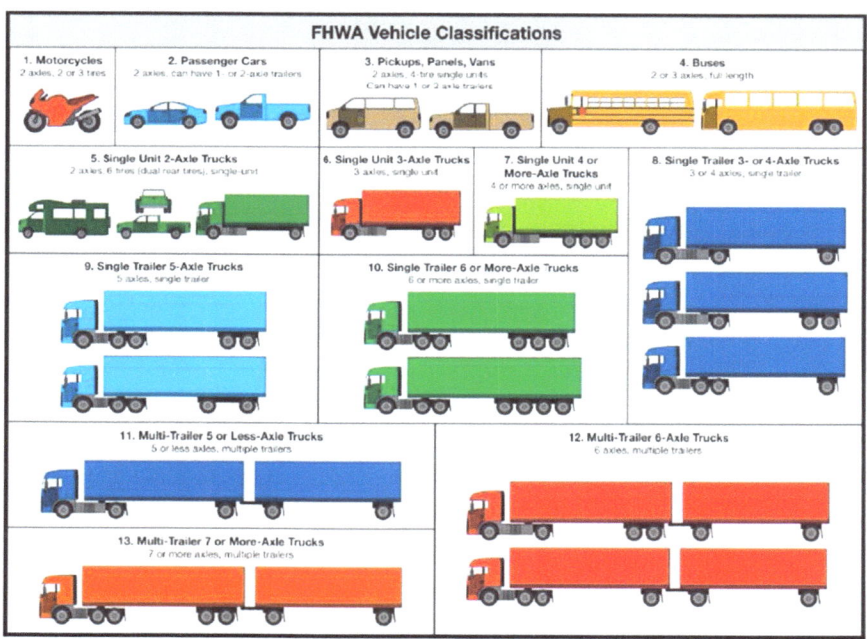

© SAE International.

Key Features:

- **Peer-to-Peer Mesh Network:** Vehicles form a decentralized, ad hoc network, continuously broadcasting and relaying messages to surrounding vehicles within a range of up to 300 m.
- **Real-Time Data Exchange:** Up to 10 times per second, information such as speed, vehicle position, acceleration, brake status, and intent (e.g., lane changes) is shared.
- **Collision Avoidance:** Vehicles receive instant alerts about sudden braking, intersection risks, blind spots, or vehicles in adjacent lanes, enabling drivers or systems to take preventive action.

- **Situational Awareness:** V2V extends a vehicle's awareness beyond the driver's FoV, detecting hazards around corners, in blind spots, or through adverse weather.
- **Traffic Optimization:** By coordinating speed and movement, V2V can reduce congestion, enable platooning (vehicles traveling closely together), and improve fuel efficiency

V2V (Vehicle-to-Vehicle) Communication
See **Cooperative Vehicle Communication**.

V2V (Vehicle-to-Vehicle) Communication Standards
See **Cooperative Vehicle Communication**.

V2X (Vehicle-to-Everything)
V2X is a transformative technology for ADAS. It enables vehicles to communicate not only with each other (V2V) but also with infrastructure (V2I), pedestrians (V2P), networks (V2N), homes (V2H), and even the cloud (V2C), much through 5G technology [15]. Unlike traditional ADAS sensors, such as radar, LiDAR, and cameras—limited by line-of-sight and environmental conditions—V2X significantly extends situational awareness by allowing vehicles to exchange real-time information about speed, position, and intent with their surroundings. This capability enables vehicles to detect and respond to potential hazards, such as an unseen pedestrian about to cross the street or a vehicle approaching an intersection, much earlier than conventional sensors can.

Integrating V2X with ADAS enhances safety by providing early warnings and enabling automated interventions, such as emergency braking, when the driver may not have enough time to react. Research indicates that automated braking triggered by V2X data is substantially more effective than driver-initiated responses, especially in scenarios involving obstructed or hidden hazards. Additionally, V2X enhances the precision of path prediction and route planning by providing absolute positioning and rich kinematic data, resulting in more reliable and accurate ADAS performance when combined with other sensor inputs.

As V2X matures, it is expected to become a safety-critical sensor within ADAS, requiring stringent functional safety standards and robust data integrity measures [16]. By allowing vehicles to "hear" and "speak" with their environment, V2X not only increases the digital horizon of ADAS but also paves the way for safer, more efficient, and more autonomous driving experiences.

V2X (Vehicle-to-Everything) Architecture

V2X architecture is a comprehensive, multilayered framework enabling real-time wireless communication between vehicles and their environment, including other vehicles (V2V), infrastructure (V2I), pedestrians (V2P), and networks/clouds (V2N/V2C). V2X is foundational for connected and AVs, enhancing safety, efficiency, and the driving experience by extending situational awareness beyond the line of sight and enabling coordinated, data-driven responses.

Core Architectural Elements

1. **Communication Technologies and Protocols**
   - **Legacy and Modern Standards:** Early V2X relied on IEEE 802.11p-based dedicated DSRC, which is suitable for V2V and V2I with limited range. Current architectures are shifting toward C-V2X, leveraging 4G LTE and 5G networks for short- and long-range, high-reliability connections, and broader interoperability [17].
   - **Layered System Architecture:**
     - **Physical Layer:** Handles the wireless transmission (modulation, encoding, and synchronization).
     - **Data Link Layer:** Manages reliable data frame transmission and medium access control (MAC).
     - **Network Layer:** Routes and forwards packets, manages addressing, and congestion.
     - **Transport Layer:** Ensures reliability and efficient packet delivery, error correction, and flow control.
     - **Application Layer:** Hosts V2X applications (collision avoidance, traffic signal coordination, infotainment).
   - **Direct and Network Modes:** C-V2X supports both direct (PC5/sidelink) and network (Uu/cellular) interfaces, enabling low-latency safety messaging and broader service integration [17].

2. **System Components**
   - **OBUs:** Embedded in vehicles, OBUs aggregate sensor data (from cameras, radar, LiDAR, etc.), process events, and handle V2X messaging.
   - **RSUs:** Infrastructure nodes that relay messages, extend coverage, and connect vehicles to broader networks and cloud services.
   - **Edge/Cloud Platforms:** Aggregate, analyze, and distribute data for traffic management, OTA updates, and advanced mobility services [17].

V2X (Vehicle-to-Everything) Communication

An LTE-V2X system uses the protocols specified in the set of ETSI standards based on 3GPP Release 14 and the IEEE Standard 1609 WAVE series of standards (**Figures V.10** and **V.11**). Applications that use the **SAE J2735** data dictionary (messages) may use LTE-V2X as illustrated in **Figure V.12**. In North America, 30-MHz bandwidth (corresponding to 5895 to 5925 MHz) has been allocated for deployment of ITS applications, which is divided into two channels: (1) LTE band 47 using E-UTRA absolute radio frequency channel number (EARFCN) 55140 with a 20-MHz channel width which corresponds to 5905 to 5925 MHz (also known as Channel 183 by IEEE), and (2) LTE band 47 using EARFCN 54990 with a 10-MHz channel width which corresponds to 5895 to 5905 MHz (also known as Channel 180 according to IEEE standards) [13, 14].

FIGURE V.10 LTE-V2X system architecture [13].

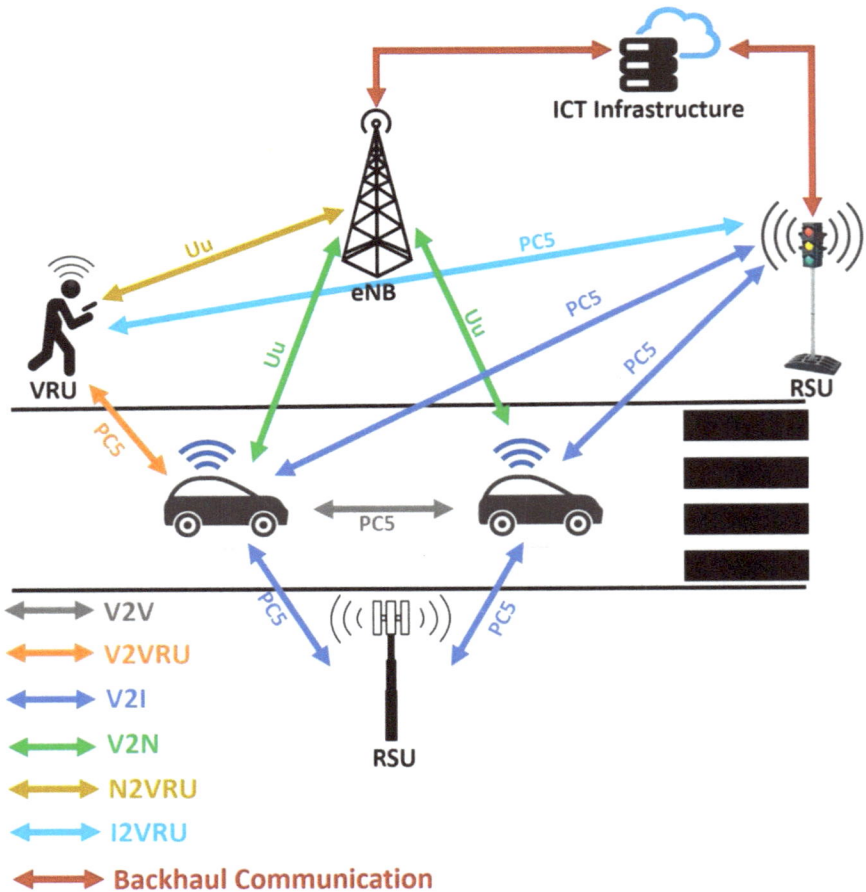

SAE International's Dictionary of ADAS and Connected Vehicles   311

FIGURE V.11 Relationship of system and SAE standards [13].

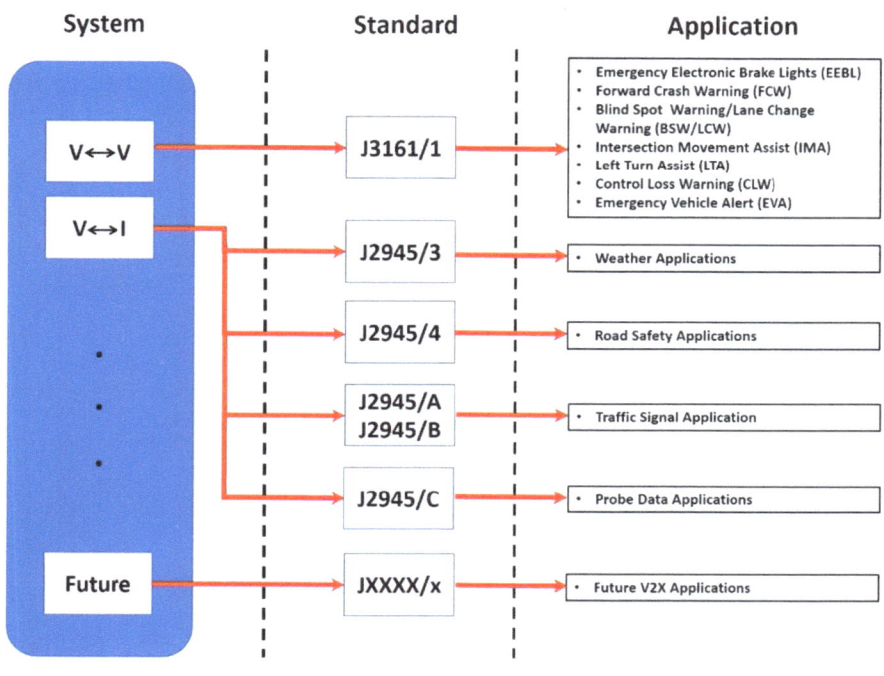

FIGURE V.12 Use of LTE-V2X for SAE J2735 message set [13].

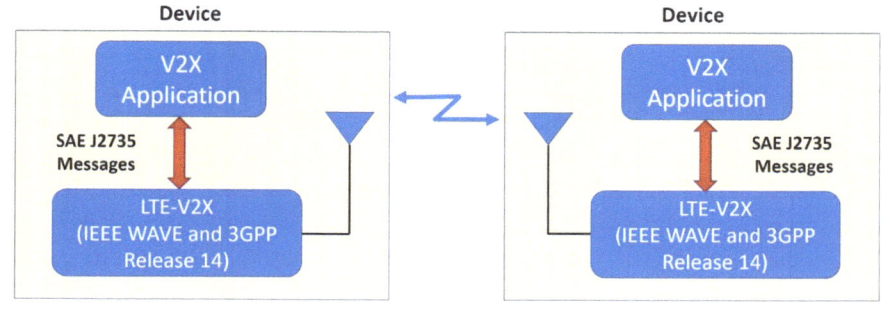

The ecosystem of ADAS communication is a complex, interconnected network that relies on a combination of sensors, real-time data exchange, and advanced processing to enhance vehicle safety and autonomy. At its core, ADAS integrates traditional line-of-sight sensors—such as cameras, radar, and LiDAR—with V2X communication protocols, which enable vehicles to interact not only with each other (V2V), but also with infrastructure (V2I), pedestrians (V2P), networks (V2N), homes (V2H), and the cloud (V2C), as illustrated in the provided diagram (**Figure V.13**). This broad connectivity allows vehicles to receive and transmit critical information about road conditions, traffic signals, nearby pedestrians, and hidden hazards, even those out of direct sensor range, significantly improving situational awareness and predictive safety.

FIGURE V.13 V2X communication protocol for self-driving cars [17].

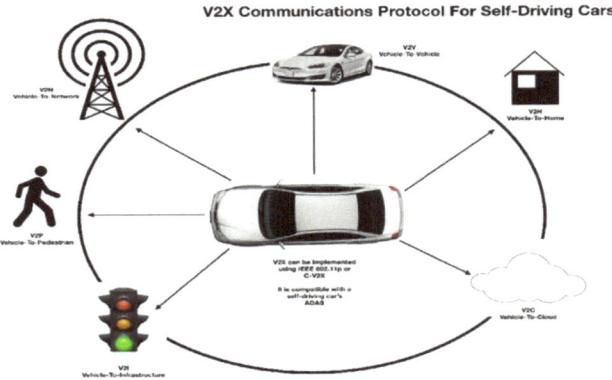

© SAE International.

Robust wireless, security, and processing subsystems support the ADAS communication ecosystem. These subsystems manage data integrity, message authenticity, and the timeliness of information, which are essential for real-time safety applications. Data from V2X are fused with inputs from other sensors in the vehicle, with sensor fusion algorithms weighing the confidence of each data source, typically prioritizing V2X for speed and acceleration, and traditional sensors for precise object localization. Integrating 5G and edge computing further enhances this ecosystem by enabling ultra-low latency, high bandwidth, and reliable OTA updates. This ensures vehicles can respond rapidly to dynamic driving environments and receive the latest software and map data without manual intervention [17].

V2X (Vehicle-to-Everything) Protocols

V2X protocols are standardized communication frameworks that enable vehicles to wirelessly exchange data with other vehicles (V2V), infrastructure (V2I), pedestrians (V2P), networks/clouds (V2N/V2C), devices (V2D), and more. These protocols are foundational for connected and AVs, supporting real-time safety, efficiency, and mobility applications.

Core V2X Communication Protocols

- WLAN-based (DSRC/802.11p) protocols
  - IEEE 802.11p (DSRC/WAVE/ITS-G5):
    Enables direct, low-latency communication between vehicles and infrastructure without relying on cellular networks.
    - Used for safety-critical messages such as
      - BSM/cooperative awareness message (CAM)
      - Decentralized environmental notification message (DENM)
      - SPaT
      - In-vehicle information (IVI)
      - Service request message (SRM)
    - Forms an ad hoc vehicular network, ideal for real-time hazard alerts and collision avoidance.
- Cellular-based (C-V2X) protocols
  - 3GPP C-V2X (LTE/5G NR):
    - Supports both direct (PC5/sidelink) and network-based (Uu/cellular) communication.
    - Advantages:
      - Greater range and coverage, including non-line-of-sight operation.
      - Scalability for massive numbers of CVs.
      - Forward compatibility with 5G, enabling ultra-reliable low-latency communication (URLLC) and new use cases12.
    - Used for both safety (direct) and extended services (network/cloud), such as dynamic traffic management and OTA updates.

VCU (Vehicle Control Unit)
See **ECU (Electronic Control Unit)**.

Vehicle Autonomy

Vehicle autonomy refers to the spectrum of automation in motor vehicles, ranging from no automation to full self-driving capability. The industry standard for defining and classifying vehicle autonomy is the SAE International **J3016** standard, which outlines six distinct levels (Level 0 to Level 5) based on the degree of driver involvement and system capability (**Figure V.14**).

FIGURE V.14 Examples of driving automation system features/types that could be available during a given trip [18].

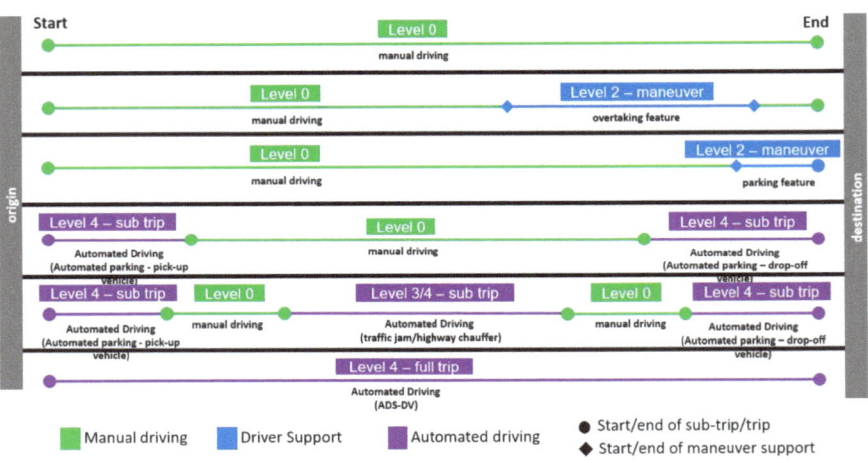

- **Levels 0–2:** The driver must always be engaged and supervise the vehicle. These are considered "driver support systems" and include ACC and lane-keeping features. The driver is always legally responsible.
- **Levels 3–5**: These are "ADS." At Level 3, the system can handle all driving tasks in certain conditions, but the driver must be ready to take over if the system requests. At Level 4, the system can operate without human attention within its ODD. At Level 5, the vehicle is fully autonomous in all conditions—no driver intervention or presence is needed.

As of 2025, no consumer vehicles offer Level 4 or Level 5 autonomy. The most advanced systems available to the public are Level 2 (partial automation) and a few rare cases of Level 3 (conditional automation) under specific conditions.

Vehicle Dynamics

Vehicle dynamics refers to the study and modeling of how a vehicle moves in response to driver inputs and external forces, encompassing aspects such as acceleration, braking, steering, and interaction with various road surfaces [19]. In the context of ADAS and automated driving, accurate vehicle dynamics modeling is crucial for simulating and validating the performance of these systems before deployment. This involves replicating complex physical interactions, such as roll, pitch, yaw, and jerk, to ensure that ADAS features, like LKA, ACC, and AEB, respond correctly under diverse driving conditions.

Modern vehicle dynamics simulations integrate with ADAS development tools, creating a holistic environment where sensor data, control algorithms, and actuator responses are tested together. This integration enables developers to accurately predict real-world vehicle behavior, reducing the need for expensive and time-consuming physical testing and thereby improving both safety and reliability. Without detailed vehicle dynamics, virtual tests may fail to capture critical behaviors, potentially compromising the effectiveness of ADAS features in real-world scenarios (**Figure V.15**).

FIGURE V.15 Light commercial vehicle with front longitudinal engine and rear wheel drive [19].

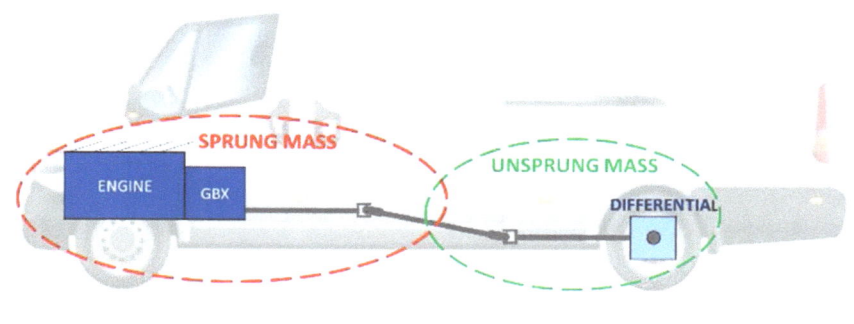

Unsprung mass refers to the portion of a vehicle's total mass not supported by the suspension system. This includes components such as the wheels, tires, wheel axles, wheel bearings, wheel hubs, a portion of the weight of driveshafts, springs, shock absorbers, and suspension links, essentially, all parts directly connected to the wheels and moving with them as they travel over bumps and road irregularities. Because unsprung mass moves independently of the vehicle body, it significantly impacts handling, traction, and ride quality. Higher unsprung mass can reduce the tires' ability to maintain consistent contact with the road, negatively affecting stability and control [19].

Sprung mass is the portion of a vehicle's total mass that is supported by the suspension system. This typically includes the body, frame, engine, transmission, internal components, passengers, cargo, and approximately half of the suspension's weight. The sprung mass is isolated from road imperfections by the suspension, which helps absorb shocks and vibrations, improving ride comfort for occupants and protecting sensitive vehicle components. The ratio of sprung to unsprung mass is crucial: A higher ratio generally results in smoother ride quality, as the vehicle body and passengers are less affected by bumps and dips in the road [19].

Vehicle Information System
See also **Infotainment**.

Vehicle Software Architecture
Vehicle software architecture refers to the layered structure and design principles that govern how software components, hardware, and communication interfaces are organized and interact within a modern vehicle. With the rise of the SDV, automotive software architecture is transitioning from distributed, function-specific ECUs to centralized, modular, and updatable platforms that support ADAS, connectivity, and continuous feature evolution [20].

Layered Functional Model [20]:
Vehicle software is typically organized in layers:
- **Hardware Layer:** Sensors, actuators, and computing platforms.
- **Middleware Layer:** Communication, security, and abstraction (e.g., AUTOSAR and advanced software framework [ASF] middleware) (**Figure V.16**).
- **Application Layer:** ADAS, infotainment, body control, and user-facing features.

FIGURE V.16 Simplified diagram of adaptive AUTOSAR framework [20].

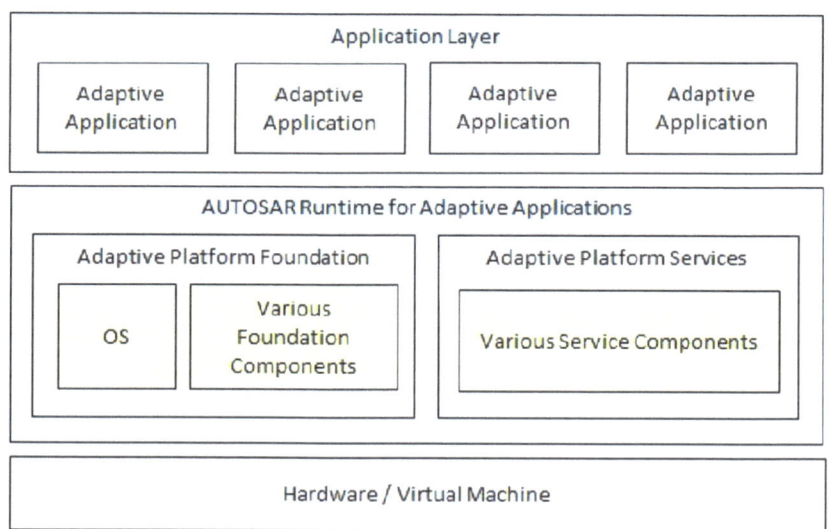

Traditional vehicles use dozens of distributed ECUs, each dedicated to a specific function. Modern architecture consolidates these into fewer, more powerful centralized ECUs or domain/zone controllers, each managing related functions (e.g., ADAS, infotainment, and chassis), reducing complexity, weight, and wiring (**Figure V.17**).

FIGURE V.17 Distributed E/E architecture [20].

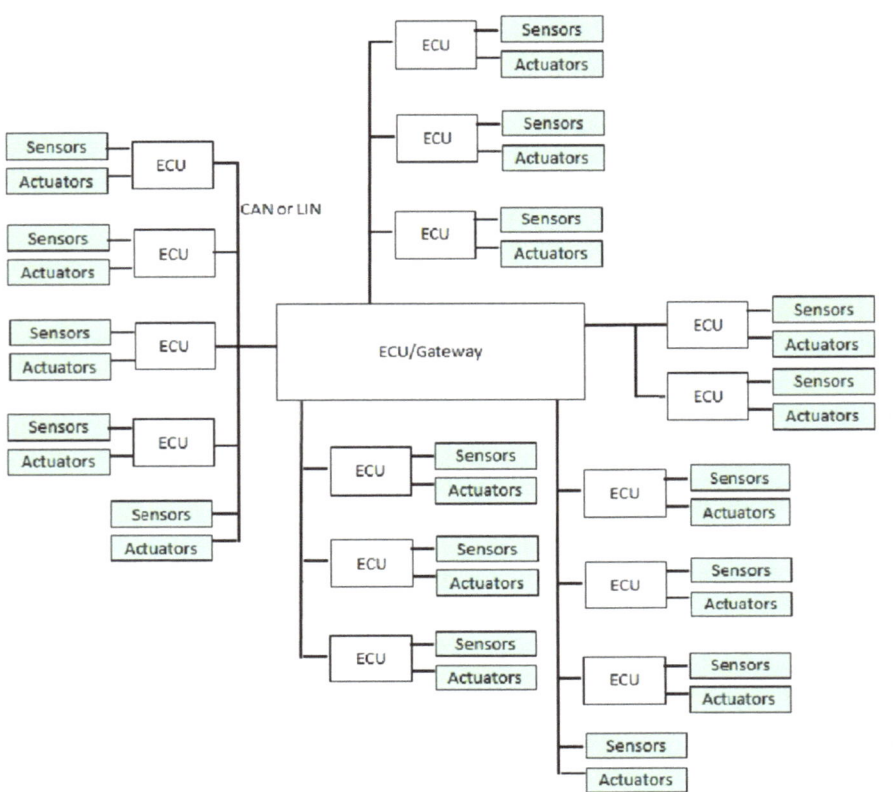

Voice Recognition

Voice recognition in vehicles refers to the technology that allows drivers and passengers to interact with in-car systems using spoken language. Modern automotive voice recognition systems leverage AI, ML, and advanced signal processing to interpret commands and queries. This enables hands-free control of navigation, entertainment, climate, communication, and more.

The system captures spoken input via microphones, processes the audio to filter noise, and converts speech to text using AI models. Advanced systems interpret the intent behind spoken commands, even if phrased conversationally or imprecisely, using NLU and context awareness. The recognized command is mapped to a vehicle function (e.g., adjusting temperature, playing music, and setting navigation) and executed by the appropriate subsystem.

VR (Virtual Reality)

VR technology is increasingly used as an immersive simulation platform to evaluate human interactions with AVs and to test autonomous driving algorithms safely and cost-effectively. Unlike traditional driving simulators, VR platforms can integrate real-time data from actual vehicles driving in real-world traffic, creating a highly realistic and interactive environment for testing both vehicle behavior and human factors [21].

FIGURE V.18 Vehicle-in-the-loop concept [21].

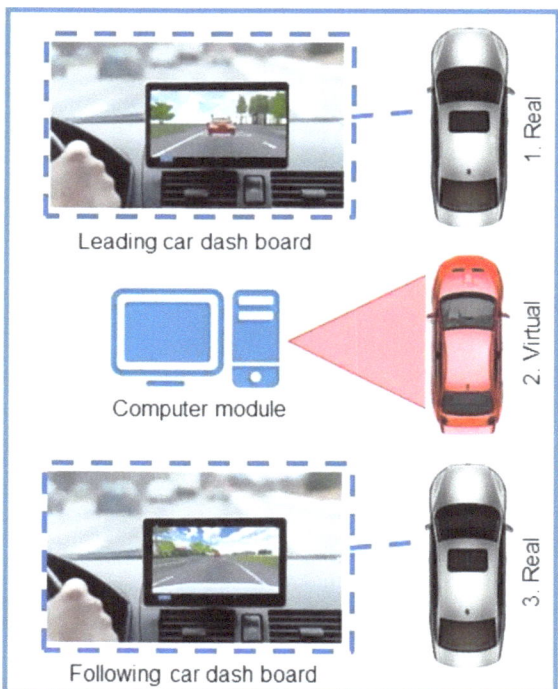

© SAE International.

Key Features and Benefits

- **Immersive Vehicle-in-the-Loop Simulation:** The VR platform combines a virtual AV (virtual vehicle) with a real human-driven vehicle (real vehicle) on real roads. The real vehicle driver can see the virtual vehicle's behavior in real time via a tablet interface, enabling dynamic interaction and feedback between real and virtual worlds (**Figure V.18**).

- **Real-Time Data Integration:** Real-world vehicle data (GPS, speed, acceleration, heading) are streamed live into the VR simulation, ensuring high fidelity and realism in the virtual environment.
- **High-Fidelity Visualization:** The virtual environment is built using game engines (e.g., Unity3D) with detailed 3D models of roads, buildings, traffic lights, and other urban elements, providing a realistic urban driving scenario.
- **Human Factor Evaluation:** VR enables safe testing of human responses to AV behaviors, including CACC scenarios, without the risks inherent in real-world testing.
- **Low-Latency Communication:** The system architecture employs DSRC-based communication infrastructure and MQTT (Message Queueing Telemetry Transport) messaging to maintain low latency (in milliseconds) between the real vehicle, control modules, and visualization modules, preserving synchronization and realism.
- **Flexible Visualization Options:** Users can experience the simulation via stereoscopic head-mounted displays (HMDs) or tablets mounted inside the vehicle, enhancing the sense of presence and immersion.

VR platforms allow developers to validate and refine AV control strategies under realistic traffic conditions, including interactions with human drivers. The system facilitates studies on passenger experience, driver takeover behavior, and trust in autonomous systems. VR simulation reduces the need for extensive on-road testing, minimizing participant risk and lowering development costs.

VRU (Vulnerable Road User)

VRUs refer to road users who are vulnerable to traffic accidents. This category of road users lack the safe guard of a vehicle body to protect them from harm due to vehicle accidents. VRUs include pedestrians, motorcyclists, skateboarders, people using scooters, hoverboards, wheelchair users or users of any kind of personal mobility. The term is used for motorized and non-motorized personal mobility users. Over the years, while the injury and fatality crashes are increasing in the US, there is a decreasing trend of such fatalities in other regions of the world such as Europe.

Key Features
- **AEB for VRU:** AEB-VRU system detects pedestrians and other vulnerable road users and applies brakes the brake autonomously. Cameras, radars, and LiDARs are mounted in the vehicles to detect VRUs. The VRU perception systems use machine learning models that are trained with the data of the VRU categories.

- The **Sensor Integration:** This technique uses data from cameras, steering angle sensors, speed sensors, and sometimes radar or LiDAR to inform lighting adjustments in real time [10].

FIGURE V.19 Top four precrash scenarios by functional years lost (FYL) grouped into four general scenarios [10].

Scenarios	Description	Cases	% Total FYL	Fatalities
S1 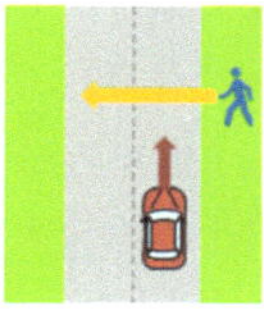	VRU crossing the road while vehicle is approaching	115000	84%	7000
S2 	VRU crossing while vehicle is turning right	2000	1%	16
S3 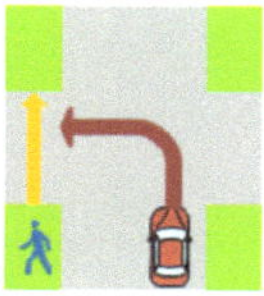	VRU crossing while vehicle is turning left	9000	1%	0
S4 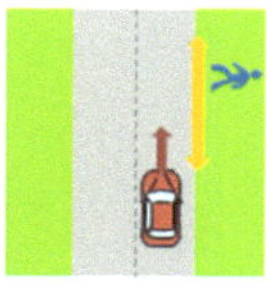	VRU traveling along the road	13000	10%	1000

REFERENCES

1. Weiß, M., Stümpfle, J., Dettinger, F., Jazdi, N. et al., "Simulating Cloud Environments of Connected Vehicles for Anomaly Detection," SAE Technical Paper 2024-01-2996 (2024), doi:https://doi.org/10.4271/2024-01-2996.

2. Ward, D. and Wooderson, P., *Automotive Cybersecurity: An Introduction to ISP/SAE 21434* (Warrendale, PA: SAE Publishing, 2022).

3. Society of Automotive Engineers, "J3061 Cybersecurity Guidebook for Cyber-Physical Vehicle Systems," SAE Publishing, Warrendale, PA, 2021.

4. Society of Automotive Engineers, "J3101 Hardware Protected Security Environment - Application Programming Interface Analysis - Information Report," SAE Publishing, Warrendale, PA, 2024.

5. Gobi, S., Lavakumar, S., and Raj, B., "Automated Test Setup for Edge Compute Connectivity Devices by Recreating Live Connected Ecosystem on the Bench," SAE Technical Paper 2021-26-0498 (2021), doi:https://doi.org/10.4271/2021-26-0498.

6. Parikh, J., Kailas, A., Adla, R., Rajab, S. et al., "Development of Wireless Message for Vehicle-to-Infrastructure Safety Applications," SAE Technical Paper 2018-01-0027 (2018), doi:https://doi.org/10.4271/2018-01-0027.

7. Society of Automotive Engineers, "J2945 Dedicated Short Range Communication (DSRC) Systems Engineering Process Guidance for SAE J2945/X Documents and Common Design Concepts," SAE Publishing, Warrendale, PA, 2017.

8. Ghosal, A., Debouk, R., Zeng, H., and Bai, F., "Reliability and Safety/Integrity Analysis for Vehicle-to-Vehicle Wireless Communication," *SAE Int. J. Passeng. Cars - Electron. Electr. Syst.* 4, no. 1 (2011): 156-165, doi:https://doi.org/10.4271/2011-01-1045.

9. Society of Automotive Engineers, "J2735 Dedicated Short Range Communications (DSRC) Message Set Dictionary," SAE Publishing, Warrendale, PA, 2006.

10. Society of Automotive Engineers, "J2945/9 Vulnerable Road User Safety Message Minimum Performance Requirements," SAE Publishing, Warrendale, PA, 2017.

11. Society of Automotive Engineers, "J2945/4 Road Safety Applications," SAE Publishing, Warrendale, PA, 2023.

12. Arena, F., Pau, G., and Severino, A., "A Review on IEEE 802.11p for Intelligent Transportation Systems," *Journal of Sensor and Actuator Networks* 9, no. 2 (2020): 22.

13. Society of Automotive Engineers, "J3161 (R) LTE Vehicle-to-Everything (LTE-V2X) Deployment Profiles and Radio Parameters for Single Radio Channel Multi-Service Coexistence," SAE Publishing, Warrendale, PA, 2024.

14. Society of Automotive Engineers, "J3161/1 Onboard System Requirements for LTE-V2X V2V Safety Communications," SAE Publishing, Warrendale, PA, 2024.

15. Jin, P., Guan, J., and Zhong, G., "Research on the Enhanced ADAS Functions with Modular Design Based on 5G-V2X," SAE Technical Paper 2020-01-5026 (2020), doi:https://doi.org/10.4271/2020-01-5026.

16. Pimentel, J., "S2V2X – A New V2X Framework to Support and Enable the Design of Automated Vehicles with a Sufficient Level of Safety," SAE Technical Paper 2023-01-1905 (2023), doi:https://doi.org/10.4271/2023-01-1905.

17. Sundar, S., Pundalik, K., and Unnikrishnan, U., "Contextual Study of Security and Privacy in V2X Communication for Architecture & Networking Products," SAE Technical Paper 2024-28-0038 (2024), doi:https://doi.org/10.4271/2024-28-0038.

18. Society of Automotive Engineers, "J3016 Taxonomy and Definitions for Terms Related to Driving Automation Systems for On-Road Motor Vehicles," SAE Publishing, Warrendale, PA, 2021.

19. Zerbato, L., Galvagno, E., Tota, A., Mancardi, L. et al., "Light Commercial Vehicle ADAS-Oriented Modelling: An Optimization-Based Conversion Tool from Multibody to Real-Time Vehicle Dynamics Model," SAE Technical Paper 2023-01-0908 (2023), doi:https://doi.org/10.4271/2023-01-0908.

20. Jiang, S., "Vehicle E/E Architecture and Key Technologies Enabling Software-Defined Vehicle," SAE Technical Paper 2024-01-2035 (2024), doi:https://doi.org/10.4271/2024-01-2035.

21. Rastogi, V., Merco, R., Kaur, M., Rayamajhi, A. et al., "An Immersive Vehicle-in-the-Loop VR Platform for Evaluating Human-to-Autonomous Vehicle Interactions," SAE Technical Paper 2019-01-0143 (2019), doi:https://doi.org/10.4271/2019-01-0143.

W

WAVE (Wireless Access in Vehicular Environment)

WAVE is a set of DSRC standards and technologies designed to enable high-speed, low-latency wireless communication between vehicles (V2V) and between vehicles and roadside infrastructure (V2I) [1]. This includes the CICAS for Violation (CICAS-V) (**Figure W.1**) [1].

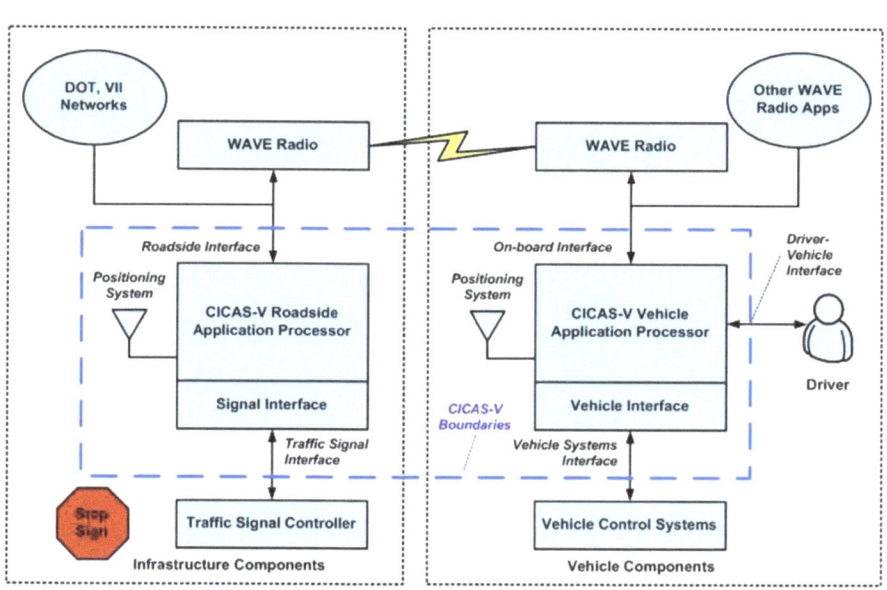

FIGURE W.1 CICAS-V system with interfaces [from CICAS-09-0118.PDF] [1].

WAVE is foundational for ADASs and CV applications, supporting safety, mobility, and infotainment services in ITSs (**Table W.1**) [2].

TABLE W.1 Normative references [1].

Document identifier	Document title
FMCSA MCS-150	Federal Motor Carrier Safety Administration (FMCSA), Form MCS-150, Motor Carrier Identification Report (Revision 06/12/07)
FIPS Publication 5-2	Federal Information Processing Standards Publication 5-2, May 29 1987
IEEE Std 1609.2-2006	IEEE Trial-Use Standard for WAVE - Security Services for Applications and Management Messages.
IEEE Std 1609.3-2010	IEEE Standard for WAVE - Networking Services
IEEE Std 1609.4-2010	IEEE Standard for WAVE - Multi-channel Operation
ISO 3166-2:2007	ISO 3166-2:2007, Codes for the Representation of Names of Countries and their Subdivisions - Part 2: Country Subdivision Code.
ISO 3780:2009	Road Vehicles - World Manufacturer Identifier (WMI) Code
NMEA 0183 Version 4.00	NMEA 0183 The Standard for Interface Marine Electronics
NTCIP 1202, v2.19f	National Transportation Communications for ITS Protocol - Object Definitions for Actuated Traffic Signal Controller Units, v02.19f
NTCIP 1204, v3.08r2	National Transportation Communications for ITS Protocol - Environmental Sensor Station Interface Protocol, Version v03
RTCM 10402.3	RTCM 10402.3 Recommendation Standards for Differential GNSS Service
RTCM 10403.1	Differential GNSS Services - Version 3
SAE J2354	Message Set for Advanced Traveler Information System (ATIS)
SAE J2540/2	ITIS Phrase List (International Traveler Information Systems)
SAE J2735	DSRC Message Set Dictionary
Title 49 CFR 395.16 - Regulation for On Board Recorders, Appendix A	Title 49 CFR 395.16 - Regulation for On Board Recorders, Appendix A, April 5, 2010

© SAE International.

Key Components:

- **IEEE 802.11p:**
  This is the physical and MAC layer amendment to the IEEE 802.11 standard, tailored explicitly for vehicular environments. It enables data exchange between fast-moving vehicles and RSUs in the 5.9-GHz ITS band, supporting communication ranges up to 1 km and relative speeds up to 200 km/h [1].

- **IEEE 1609 Family:**
  These standards define the architecture, communication model, management structure, security mechanisms, and interfaces for WAVE.

They include specifications for multi-channel operation (1609.4), resource management (1609.1), security (1609.2), and networking (1609.3) [2].

- **OBUs and RSUs:**
  OBUs are devices installed in vehicles, while RSUs are deployed along roadways. Both are essential for supporting V2V and V2I communications, exchanging data such as safety messages, traffic information, and service advertisements.

- **WAVE Short Message Protocol (WSMP):**
  A protocol within WAVE for rapid message exchange in highly dynamic and mobile environments, particularly for safety-critical applications requiring low latency (**Figure W.2**).

FIGURE W.2 Message dispatcher (MD) sender (upper figure) and receiver (lower figure) components—in this example, the data probe application is missing on the receiver side, and its data are simply discarded [2].

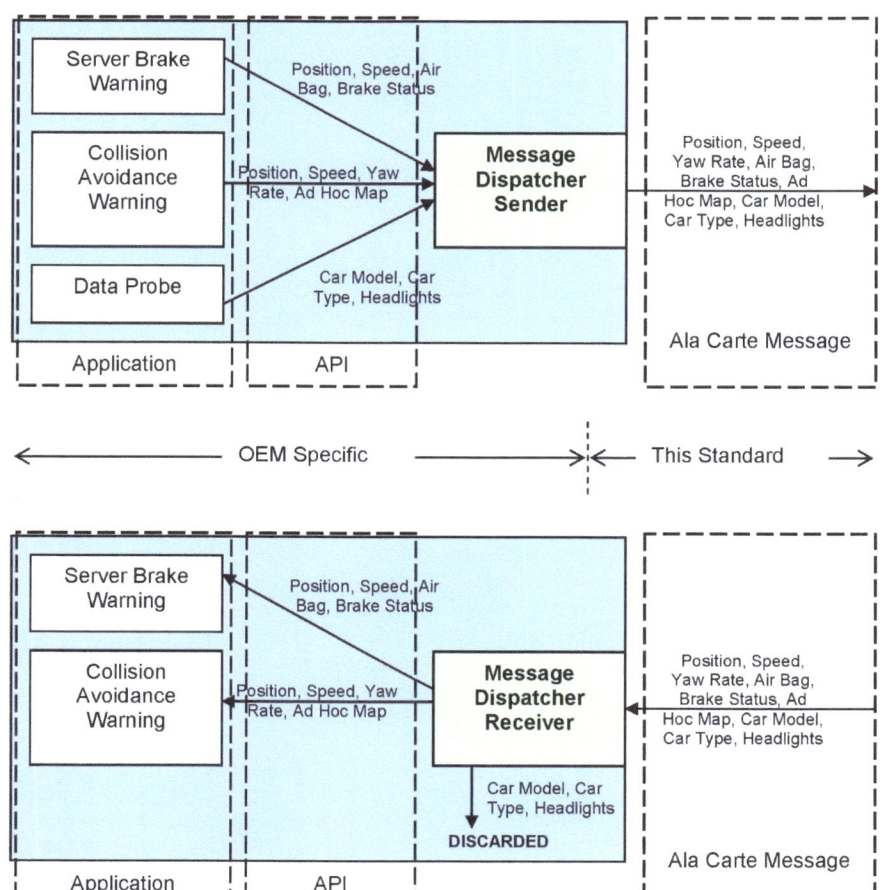

WAVE is a core technology that enables real-time, reliable, and secure wireless communication, essential for ADAS features such as CCA, lane change assistance, and traffic signal communication. It also underpins broader CV ecosystems, supporting both safety-critical and value-added services.

REFERENCES

1. SAE International, "J3067 Candidate Improvements to Dedicated Short Range Communications (DSRC) Message Set Dictionary [SAE J2735] Using Systems Engineering Methods," SAE Publishing, Warrendale, PA, 2020.

2. Society of Automotive Engineers, "J2735 Dedicated Short Range Communications (DSRC) Message Set Dictionary," SAE Publishing, Warrendale, PA, 2006.

Y

Yaw

Yaw refers to the rotation of a vehicle around its vertical axis, which passes through the center of gravity from the roof to the undercarriage. In simpler terms, yaw describes the left or right-turning motion of a vehicle, similar to how an airplane turns left or right while maintaining its level flight [1].

In ADAS and CV technologies, yaw is a critical parameter for understanding and controlling vehicle dynamics on the road or off-road [1]. Yaw rate sensors, often part of an IMU, measure the rate at which the vehicle is rotating around its vertical axis, typically expressed in degrees per second (°/s).

Applications:

- ESC: Uses yaw data to detect and correct skidding or loss of control.
- LKA: Relies on yaw measurements to maintain proper lane position.
- Autonomous Driving: Yaw information helps the vehicle follow planned trajectories accurately.
- V2X Communication: Sharing yaw and other motion data with nearby vehicles or infrastructure enhances safety and coordination.

FIGURE Y.1 Vehicle axes of motion.

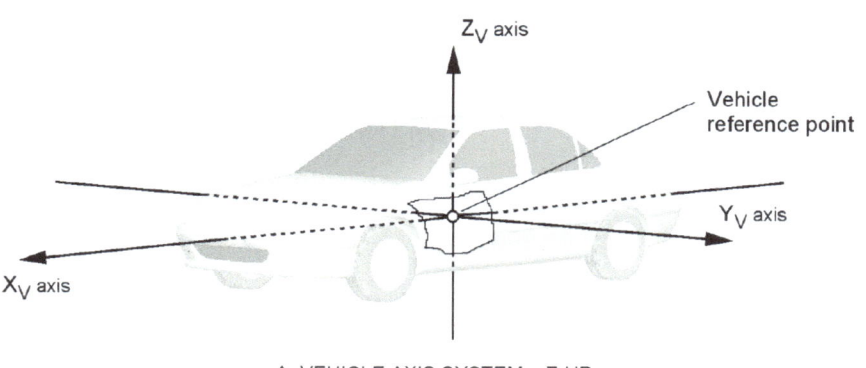

A. VEHICLE AXIS SYSTEM – Z-UP

REFERENCES

1. He, X., Yang, K., Liu, Y., and Ji, X., "A Novel Direct Yaw Moment Control System for Autonomous Vehicle," SAE Technical Paper 2018-01-1594, 2018, https://doi.org/10.4271/2018-01-1594.

Index

A

ABS (anti-lock braking system), 1
ACC (adaptive cruise control), 1
Access control, 187
Active safety, 233
Active speed limiting, 152–153
Adaptive cruise control (ACC), 1, 10, 71, 182
Adaptive driving beam (ADB), 2, 127
Adaptive front lighting system (AFS), 126
Adaptive headlights, 2
ADAS (advanced driver assistance systems), 3
ADS (automated driving system), 3
Advanced automatic collision notification (AACN), 43
Advanced driver assistance systems (ADAS), 3, 8
Advanced energy management, 114
Advanced features, 143
Advanced navigation systems, 4
Advanced parking assistance, 4–5
Advanced sensor data, 251
Advanced traffic management, 5–6
AEB (automatic emergency braking), 7, 71
AHB (automatic high beam), 7
AI (artificial intelligence), 8, 15, 125, 133
AI-driven adaptation, 137
Alert escalation, 129–130
Alert system, 70
ANNs (artificial neural networks), 8–9
Anomaly detection, 187
Antenna, 9
Anti-lock braking system (ABS), 1
API-driven innovation, 301
App access, 146
AR (augmented reality), 10
Artificial intelligence (AI), 8, 15, 125, 133, 218
Artificial neural networks (ANNs), 8–9
Assisted driving, 10–11
Attention monitoring, 70
Audio and video playback, 146
Augmented reality (AR), 10
Augmented reality (AR) navigation, 4
Authentication, 133, 187
Automated braking, 11
Automated driving, 11
Automated driving system (ADS), 3
Automated emergency braking systems (AEBSs), 43
Automated parking and retrieval, 149
Automated payment, 152
Automated valet parking (AVP), 15, 16
Automatic climate control, 142
Automatic crash notification (ACN), 223
Automatic emergency braking (AEB), 7, 320
Automatic high beam (AHB), 7, 127
Automatic recognition, 131
Automatic self-parking, 192
Automation support, 46
Automotive middleware, 12
Automotive operating systems, 12
Automotive vehicle legislation, 13
Autonomous and connected vehicles (ACVs), 303
Autonomous braking, 7
Autonomous driving, 8, 13, 216
Autonomous mobile robots (AMRs), 149
Autonomous parking, 14
Autonomous shuttles, 157
Autonomous software, 14
Autonomous vehicle (AV), 14, 198, 218
AUTOSAR, 317
AVP (automated valet parking), 15, 16
AV simulation, 14
AV technology, 15
Axes of motion vehicles, 16–17

B

Bandwidth optimization, 301
Basic safety message (BSM), 23
Battery management system (BMS), 22
Behavioral analysis, 8
Behavioral and usage data, 131
Behavioral biometrics, 128
Behavioral metrics, 72
Behavioral monitoring, 75
Behavioral sensing, 141
Behavior analysis, 70
Bidirectional communication, 298
Bidirectional wireless communication, 302
Big data, 19
Big data analytics, 26
Biometric authentication, 19

331

Index

Biometric vehicle access, 20
Biometric vehicle start, 20
Blind spot detection, 20–21
Blind spot intervention, 21
Blockchain, 21–22
BMS (battery management system), 22
Brake assist, 23
Brake-by-wire (BbW), 66
Breath alcohol concentration (BrAC), 128
Breath-based detection, 129
BSM (basic safety message), 23, 71

C

CACC (cooperative ACC), 26–27
CA (connected and autonomous) data analytics, 25–26
Camera integration, 146
Cameras, 15, 27–28
Canadian Motor Vehicle Safety Standards (CMVSSs), 94
Carbon footprint, 28–30
CarMaker line sensor, 165
Carpooling, 30
Car sharing, 128, 132
Car-to-X (C2X), 25
CCA (cooperative collision avoidance), 31
Cellular networks, 54, 212
Cellular vehicle-to-everything (C-V2X), 52
Centralized and modular architecture, 151
Centralized fusion structure, 245
Centralized interface, 143
Central management software, 152
Central processing unit (CPU), 49
Charging infrastructure, 31–32
Charging station, 32–33
CICAS (Cooperative Intersection Collision Avoidance System), 33–34
CICAS for Violation (CICAS-V), 325
CIM (Cooperative Intersection Management), 34
CIS (Cooperative Intersection Safety), 34
C-ITS (Cooperative Intelligent Transport System), 35–36
Climate and vehicle controls, 130
Climate control, 216
Cloud-based fleet management, 37
Cloud-based management, 149
Cloud-based vehicle management, 38
Cloud computing, 26, 36
Cloud platform, 36–37
Cloud storage, 37
Cognitive and emotional state detection, 134
Cognitive computing, 130
Collaborative driving, 38
Collaborative intelligence, 131
Collision avoidance, 54, 76, 182, 183, 219, 307
Collision avoidance systems, 11, 35, 39
Collision energy management, 39–40
Collision impact reduction, 40
Collision mitigation system, 41
Collision reconstruction, 41
Collision sensing system, 41–42
Collision severity prediction, 43
Collision severity reduction, 43
Collision warning, 7
Collision warning system, 43–44
Combined redundancy, 214
Commercial vehicle operations (CVOs), 306
Communication networks, 54
Communication technologies, 45
Compliance management, 37
Computer vision, 44
Computer vision algorithms, 59
Conditional automation, 162, 242
Conditional driving automation, 11
Connected alerts, 130
Connected and autonomous (CA) data analytics, 25–26
Connected intersection, 44–45
Connected payments, 140
Connected services, 136–138
Connected vehicle ecosystem, 45–46
Connected vehicle infrastructure (CVI), 53–54
Connected vehicles (CVs), 54, 204
Connectivity control unit, 301
Connectivity platforms, 4
Contextual awareness, 136
Contextual understanding, 130
Continuous learning, 131, 137
Continuous monitoring, 130
Cooperative ACC (CACC), 26–27
Cooperative collision avoidance (CCA), 31
Cooperative driving automation (CDA), 52
Cooperative Intelligent Transport System (C-ITS), 35–36
Cooperative Intersection Collision Avoidance System (CICAS), 33–34
Cooperative Intersection Management (CIM), 34
Cooperative Intersection Safety (CIS), 34
Cooperative V2C (vehicle-to-cloud), 46
Cooperative vehicle communication (CVC), 52–53
Cooperative vehicle safety, 47–48
Cooperative vehicle safety systems (CVSSs), 48, 55–56
Cooperative vehicle security (CVS), 55
Cooperative vehicle testing (CVT), 56
Cooperative V2G (vehicle-to-grid), 46–47
Cooperative V2X (vehicle-to-everything), 47
Coordinated Universal Time (UTC), 293
Cost-effective operation, 199
Cost efficiency, 169
CPS (cyber-physical system), 48
CPU (central processing unit), 49
Cross-platform integration, 49–50

Index

Cross-Vehicle Sync, 131
CSA (curve speed assistance), 50–51
CSW (curve speed warning), 51
CTA (cross-traffic alert)-rear, 51–52
Curve speed assistance (CSA), 50–51
Curve speed warning (CSW), 51
Customized UX, 9
CV (connected vehicle), 54
CVC (cooperative vehicle communication), 52–53
CVI (connected vehicle infrastructure), 53–54
CV (connected vehicle) platform, 53
CVS (cooperative vehicle security), 55
CVSSs (cooperative vehicle safety systems), 55–56
CVT (Cooperative vehicle testing), 56
C-V2X (cellular vehicle-to-everything), 52
C2X (Car-to-X), 25
Cyber-physical system (CPS), 48
Cybersecurity auditing, 56–57
Cybersecurity for vehicles, 57–58
Cyclist detection, 58–59

D

Data analytics, 63–64, 298
Data analytics techniques, 26
Database management, 170–171
Data collection, 161, 216
Data collection management, 138
Data distribution services (DDSs), 228, 229
Data fusion, 64
Data integrity, 187
Data labeling, 64–65
Data logging, 129
Data management, 37
Data minimization, 139
Data processing, 25, 26, 161
Data security, 65, 171
Data services, 45
Data sharing, 46, 134
Data storage, 26

DbW (drive-by-wire), 66
DDT (dynamic driving task), 66–67
Decision-making, 172
Dedicated short-range communication (DSRC), 54, 75–76
Deep learning (DL), 160
Delivery robots, 157
Digital infrastructure, 67–68
Digital maps, 175
Digital signage and guidance, 149
Digital twin, 68–69, 205
Digital wayfinding signage, 152
Direct current fast charging (DCFC), 96
Direct horizontal FoV, 110
Distributed E/E architecture, 318
DMS (driver monitoring system), 69–70
Drive-by-wire (DbW), 66
Driver alertness monitoring, 70–71
Driver assistance, 11, 162, 242
Driver assistance systems, 10
Driver assistance technologies, 71–72
Driver behavior analysis, 72
Driver fatigue monitoring, 73
Driver feedback and alerts, 152
Driverless car regulations, 73
Driverless transportation, 74
Driver monitoring and safety, 132
Driver monitoring systems (DMSs), 9, 69–70, 141
Driver performance monitoring, 37
Driving automation, 243
DRO (dynamic route optimization), 74
Drowsiness detection, 74–75
DSRC (dedicated short-range communication), 75–76
Dynamic driving task (DDT), 3, 66–67
Dynamic driving test (DDT), 87
Dynamic pricing and reservations, 149
Dynamic route optimization (DRO), 74

Dynamic routing, 205
Dynamic traffic management, 36

E

EBA (emergency brake assistance), 79–80
EBW (emergency brake warning), 81
ECN (electric charging network), 81–82
Eco-driving, 82
Eco-friendly routing, 114
ECU (electronic control unit), 83
Edge/cloud platforms, 309
Edge computing, 26, 84
Edge detection algorithms, 221
Edge devices, 84
Edge security, 84
Edge sensors, 85
EEBL (emergency electronic brake light), 85
E-Horizon (Electronic Horizon), 85
Electric charging network (ECN), 81–82
Electric vehicles (EVs), 31, 96
Electronic brakeforce distribution (EBD), 80
Electronic control unit (ECU), 12, 83
Electronic Horizon (E-Horizon), 85
Electronic stability control (ESC), 40, 91–93
ELK (emergency lane keeping), 86
Embedded software, 86
Emergency assist, 87
Emergency brake assistance (EBA), 79–80
Emergency brake lights, 88
Emergency brake warning (EBW), 81
Emergency braking systems (EBSs), 43, 80–81
Emergency electronic brake light (EEBL), 85
Emergency services integration, 88
Emergency stop signal (ESS), 94–95
Emergency vehicle communication (EVC), 95

Emergency vehicle priority (EVP), 96
Emergency vehicle traffic management (EVTM), 97
Emergency vehicle traffic signal priority (EVTSP), 97
Emission monitoring, 88–89
Emotion recognition system, 69
Encryption, 139, 186
Energy efficiency, 142
Energy-efficient driving, 90
Energy-efficient routing, 90–91
Energy management, 89
Energy storage, 90
Enhanced connectivity, 37
Enhanced safety, 46
Enhanced security, 132
Enhanced visibility, 179
Enrollment, 133
Enterprise application integration, 297
Environmental and energy benefits, 149
Environmental impact, 169
Environmental sensors, 91
Error correction, 198
ESC (electronic stability control), 91–93
e-Sim (Subscriber Identity Module), 93–94
ESS (emergency stop signal), 94–95
EVC (emergency vehicle communication), 95
Event-driven architecture (EDA), 297
EVP (emergency vehicle priority), 96
EV range optimization, 95
EVs (electric vehicles), 31, 96
EV supply equipment (EVSE), 32
EVTM (emergency vehicle traffic management), 97
EVTSP (emergency vehicle traffic signal priority), 97

F

Facial recognition systems, 75
Fail-operational systems, 101–102
Fail-safe systems, 102–103
FCW (forward collision warning), 103
Federal Highway Administration (FHWA), 307
Feedback loops, 161
Feedback mechanisms, 72
Field of view (FoV), 109–110
Firmware, 103–104
Firmware OTA (FOTA), 241
Fleet and expense management, 136
Fleet management, 104–105, 128
Fleet management software, 105
Fleet route optimization, 105–107
Fleet safety management, 107–108
Fleet telematics, 108–109
Fleet tracking, 109
Flexible visualization options, 320
Forward collision alert (FCA), 42
Forward collision warning (FCW), 7, 72, 103
Forward obstruction warning (FOW), 110
FoV (field of view), 109–110
FOW (forward obstruction warning), 110
Freight and delivery services, 74
Fuel management, 37
Full automation, 162, 242
Full driving automation, 11
Functional years lost (FYL), 321

G

General Data Protection Regulation (GDPR), 139
Gesture, 9
Gesture controls, 146
Global navigation satellite system (GNSS), 113
Global positioning system (GPS), 4, 9, 114
Global systems, 293
GNSS (global navigation satellite system), 113
GPS (Global Positioning System), 114
Green driving, 114
Green route planning, 115
Green transportation, 115

H

Hands-free communication, 146
Hardware redundancy, 214
HAVs (hybrid autonomous vehicles), 117
Hazard alerts, 139
Head-up display (HUD), 10, 121
High automation, 162, 242
High driving automation, 11
High-fidelity visualization, 320
High-occupancy vehicle (HOV) lanes, 30
High reliability, 75
High-resolution cameras, 59
Highway assist, 118
Highway pilot/highway chauffeur, 118
HMI (human–machine interface), 118–119
Home charging, 32
Homologation, 119–121
Homologation 4.0 Framework, 121
HUD (head-up display), 121
Human-centric lighting, 135
Human factor evaluation, 320
Human factors, 122
Human–machine interface (HMI), 118–119
Hybrid autonomous vehicles (HAVs), 117
Hyper-personalization, 137

I

IEEE 1609 family, 326–327
IEEE 802.11p, 326
Ignition, 132
Ignition interlock, 129
Immersive vehicle-in-the-loop simulation, 319
IMU (inertial measurement unit), 125
In-car adaptive lighting, 126–127
In-car biometric authentication, 128
In-car breathalyzer, 128–129
In-car child safety alerts, 129–130
In-car cognitive computing, 130–131
In-car driver profiles, 131
In-car entertainment, 132
In-car facial recognition, 132–133

Index   335

In-car gesture recognition, 133–134
In-car health analytics, 134
In-car mood lighting, 135
In-car payments, 128, 132, 135–136
In-car personal assistant, 136–137
In-car personalization, 137–138
In-car privacy controls, 138–139
In-car safety notifications, 139–140
In-car shopping, 140
In-car sleep monitoring, 141
In-car temperature control, 141–142
In-car touchscreen controls, 142–143
In-car voice assistant, 143
In-car voice biometrics, 143
In-car voice command, 143
In-car weather updates, 144
In-car Wi-Fi, 144
In-car wireless charging, 145
Inertial measurement unit (IMU), 125
Information redundancy, 214
Infotainment, 145–148
Infotainment controls, 138–139
Infotainment integration, 135
Infrastructural connectivity, 149
Infrastructure integration, 157
Injury severity score (ISS), 43
Integrated vehicle health management (IVHM), 49
Intelligent parking guidance system (IPGS), 151–152
Intelligent parking solutions, 149
Intelligent speed adaptation (ISA), 152–153
Intelligent transportation infrastructure. See V2I (vehicle-to-infrastructure)
Intelligent transportation systems (ITSs), 153, 205
Interconnectivity, 46
Internet, 146
Internet of Things (IoT), 54, 151, 218, 298

Intersection collision warning. See V2I (vehicle-to-infrastructure); V2V (vehicle-to-vehicle)
Intersection management. See V2I (vehicle-to-infrastructure); V2V (vehicle-to-vehicle)
Intersection movement assist (IMA). See V2I (vehicle-to-infrastructure); V2V (vehicle-to-vehicle)
Intuitive operation, 143
In-vehicle software platforms, 150–151
IoT (Internet of Things), 151
IoT-based sensors, 149
IPGS (intelligent parking guidance system), 151–152
ISA (intelligent speed adaptation), 152–153
ITS (intelligent transportation system), 153
I2V (infrastructure-to-vehicle) communication. See V2I (vehicle-to-infrastructure)
IVI (in-vehicle infotainment). See In-car entertainment

K
Keyless access, 132

L
Lane centering control (LCC), 11, 164–166
Lane departure warning (LDW), 71, 158–160
Lane-keeping assistance (LKA), 10, 71
Last-mile connectivity, 157
Last-mile delivery, 158
Layered functional model, 317
LCC (lane centering control), 164–166
LDW (lane departure warning), 158–160
Learning, 160–161
Left turn assist (LTA), 34
Levels of automation, 161–162
License plate recognition (LPR), 149, 152

LiDAR (light detection and ranging), 15, 162, 243, 247
LiDAR mapping, 163–164
Life cycle and data management, 151
Light detection and ranging (LiDAR), 162
LKA (lane-keeping assist). See LCC (Lane centering control)
Localization, 166–167
Lower carbon emissions, 201
Lower emissions, 199
Low latency, 75
Low-latency communication, 320
Low power, 125

M
MaaS (mobility as a service), 169
Machine learning (ML), 8, 26, 133, 160, 172–173
Maintenance alerts, 37
Maintenance scheduling, 38
MAP (map data), 169
Maximum horizontal swivel angle, 126
MEC (multi-access edge computing), 170
Message dispatcher (MD) sender, 327
MFM (mobile fleet management), 170–171
Micro-mobility, 171
Middleware, 172
ML (machine learning), 8, 26, 133, 160, 172–173
Mobile app, 138–139, 152
Mobile app integration, 173–174
Mobile fleet management (MFM), 170–171
Mobile wallet and card support, 136
Mobility, 302
Mobility as a service (MaaS), 169
Model training, 161
Multi-access edge computing (MEC), 170
Multimodal interaction, 133
Multimodal systems, 128
Multi-point preview model of automatic parking, 5
Multi-user management, 132

N

National Highway Traffic Safety Administration (NHTSA), 94
Natural language interaction, 130
Natural language processing (NLP), 4, 8, 179
Natural language understanding (NLU), 179
Navigation, 175
NCAP (New Car Assessment Program), 175–176
Neural networks, 176–177
New Car Assessment Program (NCAP), 175–176
Night vision, 178–179
Night vision system (NVS), 179
NLP (natural language processing), 179
NVS (night vision system), 179

O

Object and event detection and response (OEDR), 70, 184–185
Object detection, 179, 197
Object fusion and tracking, 181–182
Object recognition, 10, 182–183
Object tracking, 182, 197
Obstacle detection, 183–184
OBUs (onboard units), 53, 184, 309, 327
Occupant detection, 129
Occupant health and wellness, 134
ODD (operational design domain), 184
OEDR (object and event detection and response), 184–185
Onboard units (OBUs), 53, 184, 309
On-demand mobility, 185–186
Operating system and middleware, 151
Operational design domain (ODD), 184
Optimal speed management, 114
Opt-out and deletion requests, 138
Original equipment manufacturers (OEMs), 240
OTA (over-the-air) security, 186–187
OTA (over-the-air) updates, 151, 187–188
Overhead indicators, 152
Over-the-air (OTA), 36
Over-the-air (OTA) security, 186–187
Over-the-air (OTA) updates, 187–188, 298

P

Parallel park assist, 192
Parking assistance, 183
Parking assistance system (PAS), 72, 191–192
Parking reservation system, 191
Partial automation, 162, 242
Partial driving automation, 11
PAS (parking assistance system), 191–192
Passenger miles traveled (PMT), 36
Passenger participation, 140
Passive monitoring, 129
Passive safety, 233
Path planning and prediction, 198
Pattern recognition, 72
PCS (pre-collision system), 192–193
Pedestrian and cyclist detection, 7
Pedestrian collision avoidance system, 194
Pedestrian collision warning, 194
Pedestrian crosswalk monitoring, 195
Pedestrian detection, 182, 196
Pedestrian safety, 183
Peer-to-peer mesh network, 307
Perception algorithms, 196–198
Perception systems, 198
Performance analysis, 38
Performance monitoring, 72
Personalization, 128, 130, 132, 136, 140, 142, 143, 298
Personalized settings, 131
Personal mobility devices, 157
Personal safety message (PSM), 206
Personal transportation, 74
PHEV (plug-in hybrid electric vehicle), 198–199
Physical biometrics, 128
Physiological monitoring, 75, 134
Platooning, 76, 199–201
Plug-in hybrid electric vehicle (PHEV), 198–199
Power electronics, 201
Pre-collision systems (PCSs), 43, 192–193
Predictive analytics, 9, 26, 130, 201
Predictive collision alerts, 202
Predictive fuel management, 202
Predictive maintenance, 8, 203
Predictive maintenance alerts, 203
Predictive personalization, 137
Predictive road maintenance, 203
Predictive roadside assistance, 203
Predictive traffic alerts, 204
Predictive traffic congestion, 204–205
Predictive traffic modeling, 205
Predictive traffic patterns, 205
Predictive traffic signal timing, 206
Prioritization, 140
Privacy, 131, 171
Privacy protection, 206
Proactive assistance, 136
Product life cycle emissions, 30
PSM (personal safety message), 206
Public charging, 32
Public transportation, 74

R

Radar, 15, 209–210, 243, 247
Radar sensors, 59
Range anxiety, 210
RCTA (Rear Cross Traffic Alert), 210–211
Read-only memory (ROM), 103
Real-time alerts, 141
Real-time analysis, 216
Real-time communication, 293
Real-time data access, 174
Real-time data exchange, 302, 307
Real-time emotion detection, 70
Real-time environmental analysis, 221

Index

Real-time health assessment, 134
Real-time operating system (RTOS), 151, 227–228
Real-time threat detection, 299
Real-time tracking, 37
Real-time traffic management, 170
Real-time traffic updates, 4, 212
Rear collision warning, 212
Rear Cross Traffic Alert (RCTA), 71, 210–211
Rearview camera, 213
Recognition, 133
Reduced traffic congestion, 169, 200
Redundancy, 125, 213–214
Regenerative braking, 215
Regulatory compliance, 171
Regulatory frameworks, 215
Regulatory issues, 157
Reinforcement learning (RL), 161, 173, 218–219
Reliability, 125
Remote control, 173
Remote diagnostics and monitoring, 298
Remote monitoring, 216
Remote start, 216
Remote vehicle diagnostics, 217
Ride sharing, 218
Risk assessment, 72
Road departure mitigation. *See* Lane departure
Road edge detection, 219–221
Road infrastructure, 222
Road safety, 223
Roadside assistance, 223
Roadside assistance app, 223
Road side equipment. *See* RSU (roadside unit)
Roadside unit (RSU), 44, 53, 226–227, 309
Road sign assist (RSA), 224–226
Road Weather Information System (RWIS), 228–230
Rolling retests, 129
Route planning, 224
Routine automation, 138
Routing, 8
RSA (road sign assist), 224–226
RSU (roadside unit), 226–227, 327

RTOS (real-time operating systems), 227–228
Rural mobility, 228
RWIS (Road Weather Information System), 228–230

S

Safety alerts, 174
Safety applications, 170, 302
Safety of the intended functionality (SOTIF), 260
Safety standards, 233–234
Scenario-based testing, 234–235
SDC (software-defined chassis), 236
SDC (software-defined cockpit), 236
SDC (software-defined connectivity), 237
SDCAV (software-defined connected and autonomous vehicle), 237
SDS (software-defined suspension), 238
SDT (software-defined telemetry), 238
SDTM (software-defined traffic management), 238–239
SDVD (software-defined vehicle diagnostics), 240–241
SDVs (software-defined vehicles), 239–240
Secure access control, 128
Secure boot, 241–242
Security, 131
Self-driving car systems, 183
Self-driving vehicle, 242–243
Semantic segmentation, 197
Sensor, 15, 243–244
Sensor-based detection, 152
Sensor calibration, 198
Sensor data processing, 8
Sensor fusion, 7, 125, 182, 192, 197, 228, 244–246
Sensor integration, 127, 133
Sensor-level analysis, 173
Sensor range, 246
Sensor resolution, 247–248
Service-oriented architecture (SOA), 240
Shared mobility, 298

Side collision warning, 248
Side-impact airbags, 248–250
Signalized intersection, 250–251
Signal phase and timing (SPaT), 260–261
Simultaneous localization and mapping (SLAM), 252–253
Situational awareness, 251–252, 308
Six-degree-of-freedom sensing, 125
SLA (speed limit assist), 252
SLAM (simultaneous localization and mapping), 252–253
Smart city, 198, 253–254
Smart devices, 298
Smart grid, 254–255
Smart infrastructure interaction, 54
Smart intersections, 255
Smart mobility, 205
Smart navigation, 204
Smart parking solutions, 255–256
Smartphone integration, 146
Smart technology, 135
Software-defined battery management, 256
Software-defined braking system, 256–258
Software-defined chassis (SDC), 236
Software-defined climate control, 258
Software-defined cockpit (SDC), 236
Software-defined connected and autonomous vehicle (SDCAV), 237
Software-defined connectivity (SDC), 237
Software-defined IC (instrument cluster), 258–259
Software-defined sensors, 259
Software-defined suspension (SDS), 238
Software-defined telemetry (SDT), 238
Software-defined traffic management (SDTM), 238–239

Software-defined vehicle diagnostics (SDVD), 240–241
Software-defined vehicles (SDVs), 239–240
Software-defined V2X (vehicle-to-everything) architecture, 259–260
Software management, 37
Software redundancy, 214
SOTIF (safety of the intended functionality), 260
SPaT (signal phase and timing), 260–261
Special modes, 131
Speed limit assist (SLA), 252
Speed limit detection, 152
Sprung mass, 316
Standardized interfaces, 151
Steer-by-wire (SbW), 66
Steering assist, 261
Suburban mobility, 261–262
Supervised learning, 173, 262
Supply chain risks, 299
Surround view system (SVS), 263
Sustainability, 262–263
SVS (surround view system), 263
System status notifications, 139

T

TBP (traffic behavior prediction), 267
TCA (traffic congestion assistance), 268
TCS (traction control system), 268–269
Technological barriers, 157
Telematics, 25, 146, 269–271
Telematics insurance, 271–272
Teleoperated assistance, 272
Teleoperation, 272–273
Text-to-speech (TTS), 179
Thermal imaging, 179, 273–274
Threat detection, 274
Time redundancy, 214
Time stamping data, 293
Time to collision (TTC), 40, 41
Time to collision/time to impact (TTC/TTI), 287
Time to impact (TTI), 43
Tire pressure monitoring system (TPMS), 276–277
TJP/TJA (traffic jam pilot/traffic jam assist), 274

TMS (traffic management system), 275
Touchless control, 133
TPMS (tire pressure monitoring system), 276–277
Traction control system (TCS), 268–269
Traffic analysis, 277–278
Traffic analytics, 278
Traffic behavior analysis, 278–279
Traffic behavior prediction (TBP), 267
Traffic congestion, 279–280
Traffic congestion assistance (TCA), 268
Traffic efficiency, 46
Traffic flow analysis, 205
Traffic jam assist (TJA), 11
Traffic jam pilot/traffic jam assist (TJP/TJA), 274
Traffic management, 54, 183
Traffic management centers (TMCs), 54
Traffic management systems (TMSs), 6, 204, 275
Traffic optimization, 308
Traffic pattern recognition, 280
Traffic prediction, 8, 280
Traffic queue assistance, 281
Traffic signal control (TSC), 97, 283–284
Traffic signal preemption (TSP), 284
Traffic signal priority (TSP), 76, 284, 306
Traffic signal synchronization (TSS), 285–287
Traffic sign recognition (TSR), 72, 285
Traffic simulation, 205, 281
Traffic violation detection, 282
Training ML (machine learning) model, 282–283
TSC (traffic signal control), 283–284
TSP (traffic signal preemption), 284
TSP (traffic signal priority), 284
TSR (traffic sign recognition), 285
TSS (traffic signal synchronization), 285–287

TTC/TTI (time to collision/time to impact), 287
Type approval, 287–288

U

Ultrasonic sensors, 244, 247, 291
Unsprung mass, 316
Unsupervised learning, 173, 292
Urban mobility, 292
Usability, 293
User data, 25
User experience (UX), 8, 294
User interface (UI), 8, 37
User profiles, 136, 137
UTC (Coordinated Universal Time), 293
UX (user experience), 294

V

V2B (vehicle-to-business) communication, 297
V2C (vehicle-to-cloud) communication, 298–299
V2C (vehicle-to-cyclist) communication. See Cyclist detection
V2C (vehicle-to-cloud) security, 299–300
VCU (vehicle control unit). See ECU (electronic control unit)
V2E (vehicle-to-edge) communication, 300–301
Vehicle access and ignition, 128
Vehicle autonomy, 314
Vehicle axes of motion, 329
Vehicle control handover, 70
Vehicle controls, 141, 146
Vehicle diagnostics, 38
Vehicle dynamics, 315–316
Vehicle information system. See Infotainment
Vehicle integration platform (VIP), 297
Vehicle-in-the-loop, 319
Vehicle safety systems, 134
Vehicle software architecture, 316–318
Vehicle telematics, 37
Vehicle-to-business (V2B) communication, 297
Vehicle-to-cloud (V2C) communication, 298–299

Vehicle-to-cloud (V2C) security, 299–300
Vehicle-to-edge (V2E) communication, 300–301
Vehicle-to-everything (V2X), 237, 308
Vehicle-to-everything (V2X) architecture, 309
Vehicle-to-everything (V2X) communication, 310–312
Vehicle-to-everything (V2X) protocols, 313
Vehicle-to-infrastructure (V2I), 302–303
Vehicle-to-network (V2N) communication, 35
Vehicle-to-pedestrian (V2P) communication, 303–304
Vehicle-to-roadside (V2R) communication, 304–306
Vehicle-to-vehicle (V2V), 306–308
Vehicle-to-X (V2X) communication, 9
V2I (vehicle-to-infrastructure), 302–303
V2I communication, 114
Virtual displays and controls, 10
Virtual reality (VR), 319–320
Voice-activated control, 136
Voice-driven commerce, 140
Voice recognition, 4, 9, 146, 318
V2P (vehicle-to-pedestrian) communication, 303–304
V2P (vehicle-to-pedestrian) safety. *See* Pedestrian collision warning; Pedestrian detection
VR (virtual reality), 319–320
V2R (vehicle-to-roadside). *See* Cooperative intersection management
V2R (vehicle-to-roadside) communication, 304–306
VRU (vulnerable road user), 320–321
Vulnerable road users (VRUs), 7, 320–321
V2V (vehicle-to-vehicle), 306–308
V2V (vehicle-to-vehicle) communication. *See* Cooperative vehicle communication
V2V (vehicle-to-vehicle) communication standards. *See* Cooperative vehicle communication
V2X (vehicle-to-everything), 237, 308
V2X (vehicle-to-everything) architecture, 309
V2X (vehicle-to-everything) communication, 310–312
V2X integration, 140
V2X (vehicle-to-everything) protocols, 313

W

WAVE (wireless access in vehicular environment), 325–328
WAVE Short Message Protocol (WSMP), 327
Wireless access in vehicular environment (WAVE), 325–328
Wireless local area network (WLAN), 31
Wireless technologies, 305
Workplace charging, 32
Work zone warnings, 76

Y

Yaw, 329
You Only Look Once— Pedestrian Detection (YOLO-PD), 196

Z

Zoning control, 45

About the Authors

Jon M. Quigley PMP (204278) CTFL is a product development (from idea-intellectual property to product retirement) and cost improvement expert. Jon has an engineering degree from the University of North Carolina at Charlotte, master's degrees from the City University of Seattle, and two globally recognized certifications. In addition, Jon has more than 30 years of experience in product development, project management, and manufacturing.

Jon won the Volvo 3P Technical Award in 2005 and the 2006 Volvo Technology Award. He has secured seven U.S. and international patents covering multiplexing systems, human–machine interfaces, telemetry systems, and driver's aids.

Jon has been on the Western Carolina University Master of Project Management Advisory Board and Forsyth Technical Community College Advisory Board. He has also been a guest lecturer at Wake Forest University, Charlotte, NC, and Eindhoven Technical University (Holland). He has taught in technical schools, local and federal government entities, universities, SimpliLearn, B2B, and technical project management topics for Stafford Consulting, mostly federal government acquisitions efforts.

Jon has authored more than 20 product development and project management books. These books are used in bachelor's and master's level classes worldwide, including Manchester Metropolitan University, San Beda College Manila in the Philippines, and Tecnológico de Monterrey. He has guest lectured at Eindhoven Technical University on configuration and project management at the Charlotte campus of Wake Forest.

In addition to more than 70 magazines, e-zines, and other outlets, he writes four recurring columns and one recurring podcast with more than 10 million visitors:

1. PMTips Quigley and Lauck's Expert Column
2. Assembly Magazine, P's and Q's on project management and quality
3. Automotive Industries, Quigley's Corner on automotive product development
4. MPUG—Transformation Corner
5. Automotive Testing Technology International
6. IEEE Reliability Magazine
7. SPaMCAST (software process and measurement cast)—Alpha Omega Product Development

Jon has given numerous presentations at PMI and technical conferences on various domains of product development, including product testing, learning, agile, and project management. He has also frequently been interviewed by numerous business and project magazines, podcasts, and webinars.

Jon is the co-author or contributed to the books:

1. Coauthor of the Taylor & Francis book—*Project Management of Complex and Embedded Systems: Ensuring Product Integrity and Program Quality*, ISBN: 1420070256
2. Coauthor of the Taylor & Francis book—*Scrum Project Management*, ISBN: 1439825157
3. Coauthor of the Taylor & Francis book—*Testing of Complex and Embedded Systems*, ISBN: 1439821402
4. Coauthor of the Software Test Professionals/Redwood Collaborative Media Professional Development Series—*Saving Software with Six Sigma*, ISBN: 978-0-9831220-0-5
5. Coauthor of the Software Test Professionals/Redwood Collaborative Media Professional Development Series—*Aggressive Testing for Real Pros*, ISBN: 978-0-9831220-1-2
6. Coauthor of the Taylor & Francis book—*Total Quality Management for Project Managers*, ISBN: 978-1-4398-85055-5
7. Coauthor of the Taylor & Francis book—*Reducing Process Costs with Lean, Six Sigma, and Value Engineering Techniques*, ISBN: 978-1-4398-8725-7
8. Coauthor of the Society of Automotive Engineering book—*Project Management for Automotive Engineers: A Field Guide*, eISBN PDF: 978-0-7680-8315-6; eISBN prc: 978-0-7680-8316-3; eISBN epub: 978-0-7680-8317-0

9. Coauthor of the Taylor & Francis book—*Configuration Management: Theory, Practice, and Application*, ISBN: 978-1482229356 (second edition underway)
10. Contributor to the e-book—*Opening the Door: 10 Predictions on the Future of Project Management in the Professional Services Industry* with MavenLink and ProjectManagers.net
11. Coauthor of the Taylor & Francis book—*Configuration Management, Second Edition: Theory and Application for Engineers and Managers*, August 2019
12. Coauthor of the Taylor & Francis book—*Continuous and Embedded Learning for Organizations*, August 2020
13. Coauthor of the Taylor & Francis book under contract—*Risk Management Handbook: A Handbook for Managing Tomorrow's Threats to Project Success Today*, March 14, 2024.
14. Co-contributor to the "Scrum Project Management" topic in the *Encyclopedia of Software Engineering*, ISBN: 1-4200-5977-7 and e-ISBN: 1-4200-5978-5
15. Contributor to the e-book—*The Project Manager Who Smiled* by Peter Taylor. Catalog record for the book is available from the British Library, ISBN: 978-0-9576689-0-4 production, June 2013
16. Coauthor of the Society of Automotive Engineering book—*Modernizing Product Development Processes: Guide for Engineers*, production June 31, 2023, ISBN: 978-1-4686-0541-9
17. Coauthor of the Society of Automotive Engineering book—*Dictionary of Test and Verification*, ISBN: 978-1-4686-0590-7
18. Coauthor of the Society of Automotive Engineering book—*Dictionary of Commercial Vehicles*, production, ISBN: 978-1-4686-0788-8
19. Coauthor of the Society of Automotive Engineering book—*Dictionary of Electric Vehicles*, production Q4 2024, ISBN: 978-1468605907
20. Coauthor of the Society of Automotive Engineering book—*Dictionary of Automated and Connected Vehicles*, production under contract 2025
21. Coauthor of *Cost Efficiency Unleashed*, ISBN: 978-1-939641-00-7, October 2024

He has authored, co-authored, or interviewed in scores of magazine articles on various product development, teams, and manufacturing topics in more than seventy magazines, ezines, and other outlets (see below).

Embedded Systems Design
Software Development Times
Project Magazine
Advanced Trading
Tech Online India
Villanova University Computer Science
ECN
Automotive Design Line
Design Reuse
EDA Design Line
Tellurium Test Talk Blog
Software Test Professionals
The Art Career Project
EE Times
PANNAM Imaging
Embedded
MavinLink
Fit Small Business
Rasmussen College
Smartsheet
Business.com
Rescue A CEO
Zendesk
Codegiant
Automotive Industries
Pro-Sky
Freight Pop
iFour Technolab Pvt. Ltd.
Oro Inc
BCP Builder
Owl Guru
Food Manufacturing
Jalopnik
Verve Times
The Epoch Times
EMS Solutions
Schedule Reader
IEEE Reliability Magazine
Product Design and Development
All Business Online Magazine
DSP Design Line
IMPO Magazine
Software Magazine
The Standish Group P2Go
Software Test and Performance
Electronics Weekly
Quality Magazine
Taylor and Francis Newsletter
CM Trends
Manufacturing.net
Project Management Institute
Smartfile
DevOps.com
Society of Cost Management
Assembly Magazine
ProjectManagers.org
Business 2 Community
FileStageIO
Upjourney
Top Firms
PMTips
JotForm
6 River Systems
Reworked
Functionize
JW Surety Bonds
BuiltIn
Range
Gocious
Clickup
All In1 News
HealthLine
OrgCommerce Blog
Internet of Business
MPUG
Automotive Testing Technology International

    Jon enjoys the beauty of nature, hiking in the woods, and playing the bass. Jon lives in Lexington North Carolina, with his wife, Nancy.

Amol Gulve is a chartered engineer and a recognized eminent engineer, celebrated for his exceptional contributions to automotive engineering. With over 20+ years of experience in the automotive and commercial heavy-duty truck sectors, Amol has consistently demonstrated innovative thinking, technical excellence, and transformative leadership across product development, process optimization, manufacturing, embedded systems, software engineering, and quality improvement.

Courtesy of Amol Gulve.

He earned his bachelor's degree from the University of Mumbai and a master's degree from the University of Detroit Mercy, academic achievements that laid the foundation for his global impact on engineering. In 2022, Amol was honored with the prestigious SAE Forest R. McFarland Award for his pivotal role in advancing technical communication, editorial initiatives, and standards development within the Society of Automotive Engineers (SAE).

Beyond his technical leadership, Amol has coauthored and curated editorial content for leading international journals. His published work addresses critical challenges in mobility and innovation, including vehicle rollover safety, carbon dioxide (CO_2) emission reduction, and modularity strategies. His recent SAE publication, *Modernizing Product Development Processes: Guide for Engineers* (SAE #R-548), reflects his commitment to shaping the future of automotive product development.

Outside the engineering sphere, Amol is a passionate cyclist, pickleball enthusiast, avid reader, and dedicated volunteer. Based in Dallas, Texas, he leads cutting-edge initiatives in software-defined vehicle transformation using artificial intelligence (AI), driving the next era of mobility technology. Amol actively shares his expertise through international conferences and events, contributing thought leadership to the global engineering community.

For speaking engagements, mentorship opportunities, or professional collaboration, Amol can be reached at agulve@gmail.com.

Jayalekshmi (Jaya) Krishnamoorthy is a system safety expert with over 20 years of experience in autonomous and connected vehicle technologies. Her career spans the development, validation, and safety assurance of advanced driver assistance systems, intelligent mobility platforms, connected vehicles, and autonomy-enabling architectures.

With a Bachelor of Technology in Applied Electronics and Instrumentation Engineering from Kerala University, Jaya brings deep technical expertise to the design and deployment of safety-critical systems. She started her career as a software developer for automotive systems and later transitioned into system engineering and safety roles. She has interpreted and applied international standards such as ISO 26262, ISO 21448, UL 4600, and ISO/PAS 8800. Her work has guided the creation of safety concepts and system-level safety cases for a diverse range of applications—from passenger vehicles and commercial fleets to autonomous delivery bots and truck convoys.

Jaya has developed comprehensive training content for ISO 26262, ISO 21448, and ISO/PAS 8800 and has trained hundreds of engineers worldwide in safe system development. Her trainings are enriched with real-world anecdotes drawn from her consulting engagements and engineering roles. She has worked across the distributed automotive supply chain—including OEMs, Tier 1 suppliers, and global technology companies helping teams navigate complex safety standards and pass rigorous audits and assessments.

Her publications explore critical topics in safety, including fault tolerance, machine learning safety, and autonomy safety acceptance criteria. She is also a reviewer for SAE International. Fluent in multiple languages, Jaya brings a rare combination of technical depth, global perspective, and ethical insight to the field of intelligent transportation.

Her work continues to shape the standards, systems, and conversations that define the future of safe mobility. Through authorship and thought leadership, Jaya is committed to making complex concepts accessible to engineers, researchers, and innovators worldwide.

www.ingramcontent.com/pod-product-compliance
Lightning Source LLC
LaVergne TN
LVHW070045070526
838200LV00028B/399